Advances in measuring soil health

It is widely recognised that agriculture is a significant contributor to global warming and climate change. Agriculture needs to reduce its environmental impact and adapt to current climate change whilst still feeding a growing population, i.e. become more 'climate-smart'. Burleigh Dodds Science Publishing is playing its part in achieving this by bringing together key research on making the production of the world's most important crops and livestock products more sustainable.

Based on extensive research, our publications specifically target the challenge of climate-smart agriculture. In this way we are using 'smart publishing' to help achieve climate-smart agriculture.

Burleigh Dodds Science Publishing is an independent and innovative publisher delivering high quality customer-focused agricultural science content in both print and online formats for the academic and research communities. Our aim is to build a foundation of knowledge on which researchers can build to meet the challenge of climate-smart agriculture.

For more information about Burleigh Dodds Science Publishing simply call us on +44 (0) 1223 839365, email info@bdspublishing.com or alternatively please visit our website at www.bdspublishing.com.

Related titles:

Managing soil health for sustainable agriculture Volume 1: Fundamentals
Print (ISBN 978-1-78676-188-0); Online (ISBN 978-1-78676-191-0, 978-1-78676-190-3)

Managing soil health for sustainable agriculture Volume 2: Monitoring and management
Print (ISBN 978-1-78676-192-7); Online (ISBN 978-1-78676-195-8, 978-1-78676-196-1)

Improving soil health
Print (ISBN 978-1-78676-670-0); Online (ISBN 978-1-78676-672-4, 978-1-78676-673-1)

Achieving sustainable crop nutrition
Print (ISBN 978-1-78676-312-9); Online (ISBN 978-1-78676-314-3, 978-1-78676-315-0)

Chapters are available individually from our online bookshop:
https://shop.bdspublishing.com

climate•SMART•publishing
IN AGRICULTURAL SCIENCE

BURLEIGH DODDS SERIES IN AGRICULTURAL SCIENCE
NUMBER 92

Advances in measuring soil health

Edited by Professor Wilfred Otten, Cranfield University, UK

burleigh dodds
SCIENCE PUBLISHING

Published by Burleigh Dodds Science Publishing Limited
82 High Street, Sawston, Cambridge CB22 3HJ, UK
www.bdspublishing.com

Burleigh Dodds Science Publishing, 1518 Walnut Street, Suite 900, Philadelphia, PA 19102-3406, USA

First published 2021 by Burleigh Dodds Science Publishing Limited

Library of Congress Control Number: 2020945583

British Library Cataloguing in Publication Data
A catalogue record for this book is available from the British Library

ISBN 978-1-78676-426-3 (Print)
ISBN 978-1-78676-429-4 (PDF)
ISBN 978-1-78676-428-7 (ePub)
ISSN 2059-6936 (print)
ISSN 2059-6944 (online)

DOI 10.19103/AS.2020.0079

Typeset by Deanta Global Publishing Services, Dublin, Ireland

Contents

Series list

Title	Series number
Achieving sustainable cultivation of maize - Vol 1 From improved varieties to local applications Edited by: Dr Dave Watson, CGIAR Maize Research Program Manager, CIMMYT, Mexico	001
Achieving sustainable cultivation of maize - Vol 2 Cultivation techniques, pest and disease control Edited by: Dr Dave Watson, CGIAR Maize Research Program Manager, CIMMYT, Mexico	002
Achieving sustainable cultivation of rice - Vol 1 Breeding for higher yield and quality Edited by: Prof. Takuji Sasaki, Tokyo University of Agriculture, Japan	003
Achieving sustainable cultivation of rice - Vol 2 Cultivation, pest and disease management Edited by: Prof. Takuji Sasaki, Tokyo University of Agriculture, Japan	004
Achieving sustainable cultivation of wheat - Vol 1 Breeding, quality traits, pests and diseases Edited by: Prof. Peter Langridge, The University of Adelaide, Australia	005
Achieving sustainable cultivation of wheat - Vol 2 Cultivation techniques Edited by: Prof. Peter Langridge, The University of Adelaide, Australia	006
Achieving sustainable cultivation of tomatoes Edited by: Dr Autar Mattoo, USDA-ARS, USA; and Prof. Avtar Handa, Purdue University, USA	007
Achieving sustainable production of milk - Vol 1 Milk composition, genetics and breeding Edited by: Dr Nico van Belzen, International Dairy Federation (IDF), Belgium	008
Achieving sustainable production of milk - Vol 2 Safety, quality and sustainability Edited by: Dr Nico van Belzen, International Dairy Federation (IDF), Belgium	009
Achieving sustainable production of milk - Vol 3 Dairy herd management and welfare Edited by: Prof. John Webster, University of Bristol, UK	010
Ensuring safety and quality in the production of beef - Vol 1 Safety Edited by: Prof. Gary Acuff, Texas A&M University, USA; and Prof. James Dickson, Iowa State University, USA	011
Ensuring safety and quality in the production of beef - Vol 2 Quality Edited by: Prof. Michael Dikeman, Kansas State University, USA	012
Achieving sustainable production of poultry meat - Vol 1 Safety, quality and sustainability Edited by: Prof. Steven C. Ricke, University of Arkansas, USA	013
Achieving sustainable production of poultry meat - Vol 2 Breeding and nutrition Edited by: Prof. Todd Applegate, University of Georgia, USA	014

Acknowledgements

We wish to acknowledge the following for their help in reviewing particular chapters:

- Chapter 1: Dr Maria J. I. Briones, University of Vigo, Spain
- Chapter 2: Dr Lisa Lobry De Bruyn, University of New England, Australia
- Chapter 4: Dr Jeremy P. Emmet-Booth, University College Dublin, Ireland
- Chapter 7: Dr Dennis Flanagan, USDA-ARS, USA
- Chapter 8: Professor Rainer Horn, University of Kiel, Germany
- Chapter 9: Professor Alan Franzluebbers, North Carolina State University, USA
- Chapter 11: Emerita Professor Lynette Abbott, The University of Western Australia, Australia

Introduction

The UN Agency's Intergovernmental Technical Group on Soils (ITPS) has defined healthy soil as 'the ability to sustain productivity, diversity and environmental services of terrestrial ecosystems'. Understanding and measuring the different dimensions of soil health is key to sustain agricultural productivity and protect environmental resources. There has been a wealth of research on developing better analytical techniques to measure the biological, physical and chemical properties of soils. This volume reviews these developments and their implications for better management of soils. Chapters in Part 1 examine advances in measuring soil biological activity such as earthworms and fungi as indicators of soil health. Part 2 addresses developments in measuring soil physical properties through advances in visual, imaging and geophysical techniques, as well as the methods used to measure chemical properties such as soil organic carbon. Part 3 of the book looks at how measurement can be translated into farming practice through soil health indicators and decision support systems.

Part 1 Measuring soil biological activity

Part 1 opens with a chapter on assessing soil health by measuring fauna. Chapter 1 provides examples of the impact of soil fauna on soil health within different ecosystems and how the soil habitat changes in relation to this biodiversity. It focuses specifically on mesofauna in agriculture, grasslands, woodlands and as bioindicators, before concluding with an overview of how the development of mesofauna as bioindicators is important in establishing a healthy soil.

The next chapter reviews quantifying earthworm community structures as indicators of soil health. Chapter 2 begins by describing the three epigeic, endogeic and anecic groups of earthworms before going on to discuss the importance of these earthworms in soil health and management. The challenges in collecting data on earthworms are also discussed, specifically focusing on the resources required and the ability to accurately identify earthworm species. This is followed by a section on developing improved assessment methods for earthworms by farmers. The chapter concludes by highlighting how important it is to develop earthworm observation networks in the future.

Chapter 3 focuses on the characterisation of fungal communities and functions in agricultural soils. The chapter begins by highlighting the challenges in characterising fungal communities, such as investigating species-rich communities and our knowledge of fungal community structure, spatial distribution and sampling issues. The chapter then discusses molecular characterisation of fungal communities followed by a review of the range of proxies for fungal abundance. The chapter includes a case study on the practicalities of investigating soil fungal communities, emphasising the importance of fungal communities in soil health.

Part 2 Measuring soil physical and chemical properties

Soil structure is a complex and dynamic property that constitutes a key aspect of soil health. Soil structure assessment can be evaluated in the field by visual soil evaluation techniques. The first chapter of Part 2 discusses advances in these techniques which have been a focus of significant international research. Chapter 4 reviews the development and protocols of selected topsoil and soil profile methods to show differences in methodological approaches, and includes guidance on selection of the appropriate method for particular situations. Two case studies show how visual techniques can be used to improve management of soil, and to assess and monitor soil health in developing countries.

The use of non-invasive imaging techniques offers new approaches to characterize soil health, complementing information from traditional soil structure analyses. Chapter 5 discusses imaging soil structure using X-ray computed micro-tomography (X-ray μCT) to assess soil health and functioning. Imaging techniques are particularly suitable for characterising soil pore architecture which drives processes such as water and gas fluxes, chemical transport as well as soil biota activity. The chapter shows how the technique can be used to measure features such as macroporosity, pore connectivity, pore shape and solid-to-pore distance. It includes examples of the application of pore measurement for soil characterization and practical advice on methods to use.

Chapter 6 explores the use of geophysical methods to assess soil physical characteristics. The chapter begins by reviewing geophysical properties of soil such as clay and organic matter content. It then moves on to discuss methods such as electromagnetic induction (EMI), electrical resistivity and acoustic-to-seismic coupling and their applications in measuring soil properties such as water content and rates of soil drying. The chapter shows how these methods can be used to monitor soil health and identify within-field variations required for successful application of precision farming.

Soil erodibility is the susceptibility of soil to the erosive forces of rain splash, runoff and wind. Chapter 7 surveys advances in techniques to assess soil erodibility, beginning by examining the factors that affect soil erodibility, including soil properties, land use and management practices and the effect of soil amendments and conditioners. The chapter then summarises the various techniques that can be used to assess soil erodibility, including both static and dynamic laboratory and field tests. The chapter concludes with an overview of future research needs in developing improved methods to assess soil erodibility.

The next chapter focuses on advances in measuring mechanical properties of soil in relation to soil health. Chapter 8 begins by discussing soil rheology, showing how a rotational rheometer test can be used to measure the microstructural stability of soil. Cone penetration resistance tests and their importance in characterising the variability of soil strength within the soil are also discussed. A section on a uniaxial confined compression test is also included, followed by an analysis of miniature indentation tests and how they can be used to determine the mechanical properties of soil at the millimetre scale. The use of indirect tensile strength tests is also examined, before the chapter concludes by emphasising the importance of measuring both soil mechanical properties and soil matric potential.

Chapter 9 reviews advances in near-infrared (NIR) spectroscopy to assess soil health. Infrared absorbance frequencies of soil constituents such as organic matter and clay minerals form the basis for developing reliable calibrations for predicting soil health indicators (SHI). Diffuse reflectance spectroscopy in the near-infrared (NIR: 350-2500 nm) region offers a relatively rapid, non-destructive and high-throughput alternative to wet chemistry measurements of soil health. To demonstrate the potential for using NIR for soil health measurements, this chapter describes the use of a NIR spectral dataset of diverse United States soils (n=709) from the USDA NRCS National Soil Survey Center to develop chemometric prediction models of representative SHI: total organic C (TOC), aggregate stability, clay content, and β-glucosidase activity. Future directions for NIR prediction of SHI and infrared spectroscopy-based soil health assessment are also discussed.

The final chapter of Part 2 examines spectral mapping of soil organic carbon. Chapter 10 first reviews recent pilot studies testing the potential of this technique. The chapter then focuses on the challenges in large-scale application of spectral mapping when the soil and parent material are heterogeneous and surface conditions are unknown. To deal with these constraints, the chapter assesses the calibration of spectral models based on large spectral libraries, the surface conditions that disturb the soil signal and a time series of images in order to delimit cropland fields and increase the extent of bare soil that can be mapped. A case study developing a soil organic carbon prediction map derived from the spectra of a Sentinel-2 image and calibrated using the LUCAS spectral library is also included.

Part 3 From measurement to management

Chapter 11 concentrates on developing soil health indicators for improved soil management. It starts by discussing the use of current indicators for soil health, such as land use capability and suitability approaches, environmental quality monitoring and concepts derived from food quality management. The chapter includes a case study on developing a practical toolkit of soil health indicators for UK farmers to rapidly assess biological, chemical and physical measures of soil health. The chapter concludes by emphasising the importance of developing these soil health indicators for better soil quality management in the future.

The final chapter of the book examines developing decision support systems for farm soil management. Chapter 12 provides a review of the potential role of decision support systems (DSS) and current systems. Different aspects of agricultural DSS design, implementation and operation are covered. These aspects include spatial planning, the need for and use of sensor technology, modelling and software components, system-operator interfaces. The also discusses what is needed to make DSS more successful and widely used in agriculture.

Part 1

Measuring soil biological activity

Chapter 1

Assessing soil health by measuring fauna

Felicity Crotty, Royal Agricultural University, UK

1 Introduction

Soil biodiversity, including abundance, species diversity, genetic diversity and functional diversity of fauna living within the soils, is an important aspect of soil health. Soil biodiversity can also act as an indicator of soil health, with a functioning biodiverse soil food web exemplifying a healthy soil which will be able to deliver many ecosystem services (the benefits provided by the ecosystem). The biodiversity of soil animal communities may exceed above-ground biodiversity by several orders of magnitude in many habitats (Anderson, 2009). Conserving soil biodiversity is key to improving and sustaining soil health, through maintaining nutrient cycles, decomposition and plant growth within the soil habitat (Firbank et al., 2008; Handa et al., 2014; Crotty et al., 2015). Soil quality and soil health, have been used frequently and interchangeably within the scientific literature. Initially, soil health was considered in relation to organic agriculture (Howard, 1947), but since the early 1990s and the seminal work by Doran and Zeiss (2000), soil health has been thought of as 'the continued capacity of a soil to function as a vital living system'. Soil health is broader, covering multiple soil functions with greater emphasis on the whole soil ecosystem, as a changing interlinked network. The response of diversity, abundance and function of soil organisms to soil management constitutes an important aspect of soil quality (Mbuthia et al., 2015). A healthy soil provides

http://dx.doi.org/10.19103/AS.2020.0079.03

many ecosystem services; these include provisioning services (such as food, timber and fibre); regulating services (modifying water quality, flooding or climate); and supporting services (such as soil formation, photosynthesis and nutrient cycling) (Millennium Ecosystem Assessment, 2005). These ecosystem services are driven by the soil fauna, particularly the cycling of mineral nutrients and water regulation. These decomposers provide the basis for soil fertility through recycling plant material and mineralizing soil nutrients (Zhang et al., 2017). Tibbett et al. (2020), reviewed potential threats to soil biodiversity and found in 70% of the papers investigated, that a decline in soil health is directly related to soil biodiversity loss.

One of the most important features of the soil is that the fauna are immersed in the environment they live in, surrounded by their food, living space and excretion products (Crotty, 2011). This is why soil fauna have such a large impact on the ecosystem services provided by the soil. Soil biodiversity is also recognised as one of the cornerstones for soil security, as it is one of the seven main functions (McBratney et al., 2014), as well as the presence of life within the soil being necessary for the soil to be considered soil (Coleman, 2008). Soils are the most species-diverse habitat on the Earth (Bender et al., 2016), and there are more microorganisms residing in a teaspoon of soil than there are humans globally. To visualise this quantification, for example, microbial biomass in 1 ha of soil, 20 cm depth, weighs approximately 1 tonne (Kaczmarek, 1984), equivalent to the weight of 17 sheep (Schon et al., 2011b), while protozoa can be calculated to weigh approximately 400 kg per ha (Ekelund and Ronn, 1994) (or six sheep). Mesofauna (springtail and mites) vary greatly in numbers and biomass across habitats but have been found to equate to around 52 kg per ha (Crotty, 2011; Crotty et al., 2016) (almost one sheep), whilst nematode biomass equates to 317 kg per ha (Crotty, 2011; Crotty et al., 2016) or five sheep. However, earthworms account for the majority of soil fauna biomass, with weights calculated to reach 1.5 tonnes per ha (Crotty et al., 2016) and over 2.5 million individuals per ha, equivalent to 24 sheep. Therefore, there is a greater biomass of soil fauna residing below ground (equivalent to 53 sheep) than there would be sheep grazing above ground, with conventional stocking densities around 16 sheep per ha (Schon et al., 2011b). These calculations are for arable soils, and in a grassland soil the fauna biomass can be up to 10 times these amounts. This gives an indication of the importance of soil biodiversity that is often overlooked due to the focus of researchers on just one group of soil organisms rather than the whole soil food web. A decline in soil biodiversity has implications for soil health, as it impedes the soil's ability to perform ecosystem service functions and lowers its innate resistance, resilience to perturbations and the capacity to recover from these perturbations (Tibbett et al., 2020).

The structure of the soil profile has a large influence over the habitable space for soil fauna, particularly mesofauna and macrofauna. As large pores are less abundant than pores with smaller diameters, large animals have access to fewer crevices than smaller animals (Kampichler, 1999), leading to the abundance of a vast array of mesofauna of different diameters. Mesofauna are one of the most abundant and diverse groups of animal fauna within the soil, mainly consisting of springtails (Collembola), mites (Acari), Enchytraeidae and other primitive arthropods (Protura, Diplura, Symphyla) with less than 2 mm body width (Swift et al., 1979) (Fig. 1). Mesofauna are an important component of litter transformation, processing organic inputs through the detrital food web. Collembola are found within the fossil record as one of the earliest examples of terrestrial life, almost 400 million years ago (Brown, 2001), and although they are different species to those of modern soils, the functionality of the soil food web was likely to be similar to today. Springtails and mites can be frequently found at large abundances of 150 000 per m^2, independent of the above-ground biome (Coleman and Crossley, 2003). Diversity of species can also be high, with up to 100 species of springtails and mites found within the same site (or soil

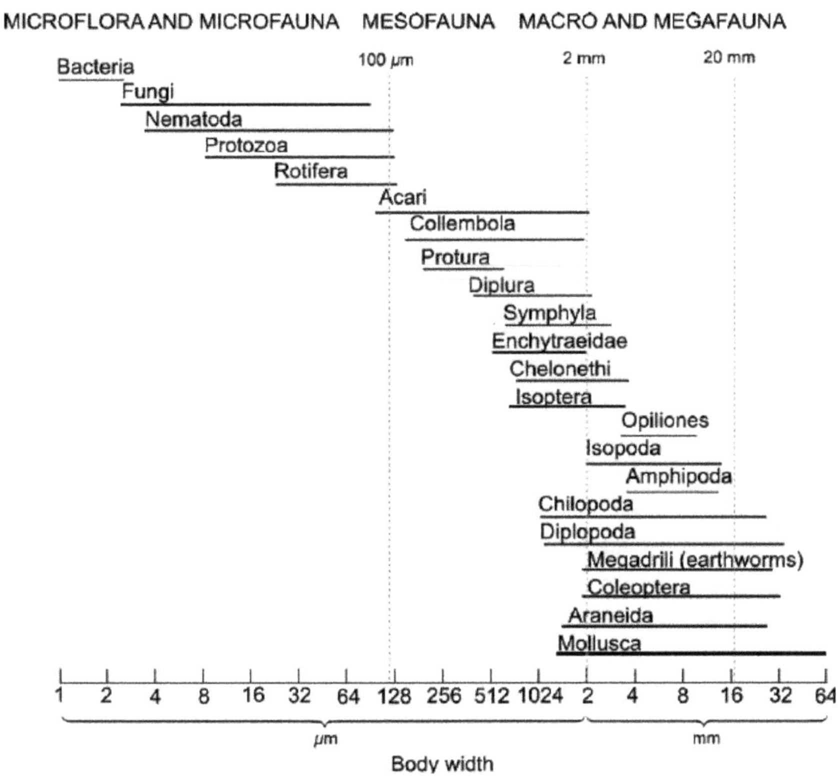

Figure 1 Size classification of soil organisms according to body width (from Swift et al., 1979).

sample). Springtails are mainly microbivorous and have an important role in controlling microbiological populations, selectively grazing, having an impact on plant litter decomposition processes and affecting the formation of soil microstructures (Rusek, 1998). Mites are more diverse functionally than springtails, with small microbivorous or saprotrophic species as well as large predatory species. Mites also have greater differences in life strategy, with Oribatid species more likely to be found in stable environments due to their slower (K-strategists) lifecycles compared to Prostigmata that are often found in disturbed environments and have rapid development, fast growth cycles and greater fecundity (R-strategist). Enchytraeidae worms are also part of the mesofauna and are found in large abundances in highly organic (peat) soils. These soils have limited nutrient cycling and decomposition due to cold temperatures and/or waterlogging and are fairly acidic, limiting the diversity of soil fauna living within them. Enchytraeidae worms exert a large influence on the soil structure and promote decomposition and nutrient accumulation in their casts, while also redistributing nutrients throughout the soil profile (van Capelle et al., 2012).

Awareness of soil biodiversity and its importance in ecosystem function has been increasing exponentially over the last few decades. Recently, the European Soil Biodiversity Atlas (Jeffery et al., 2010) and the Global Soil Biodiversity Atlas (Orgiazzi et al., 2016) have been published, as well as the

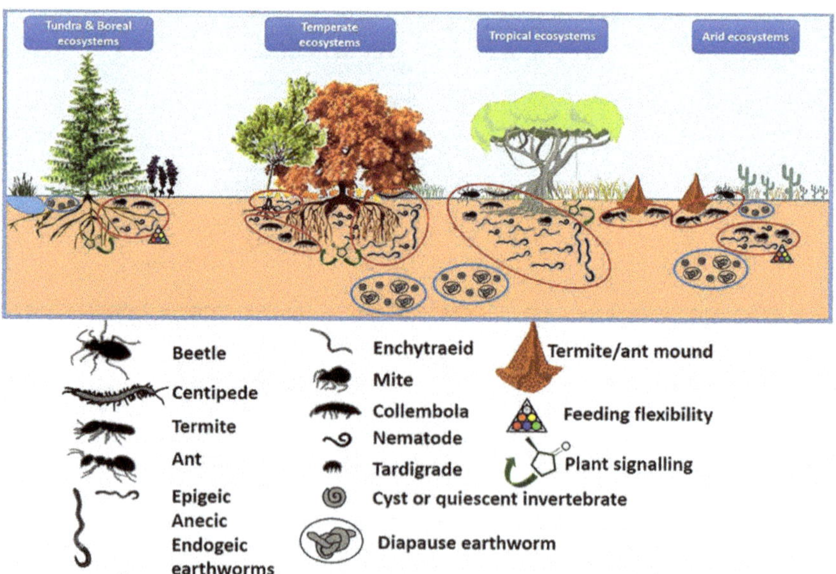

Figure 2 Illustration of how soil biodiversity and activity differ across biomes and within the soil profile; red ellipses indicate activity hotspot within the soil food web; blue ellipses indicate inactivity (adapted from Briones, 2018).

creation of the Global Soil Biodiversity Initiative (launched in 2011) with tri-annual conferences discussing the topic. However, the interactions occurring within the soil between soil fauna and how they exemplify soil health remain unclear; making the soil more biodiverse does not directly equate to making a soil healthier. Throughout this chapter, examples will be provided of the impact of soil fauna on soil health within different ecosystems and how the soil habitat changes in relation to this biodiversity. Soil mesofauna can be found globally from the equator to the poles, although the diversity and activity of the mesofauna change in relation to the above-ground habitat; this is due to the amount of food available to the fauna and the stability of the ecosystem. Soil fauna are most active in temperate ecosystems, where there is a large input of plant litter and organic matter throughout the year; however, a large diversity of organism can still be found in tundra, boreal and arid ecosystems, although more fauna may be inactive (Fig. 2).

2 The impact of mesofauna on the soil habitat

Soil fauna, as well as being surrounded by the food that they eat, are also surrounded by the habitat they live within and have a far greater impact on this habitat in comparison to other organisms in other ecosystems. Soil mesofauna will comminute and fragment plant litter (Fig. 3), processing plant debris

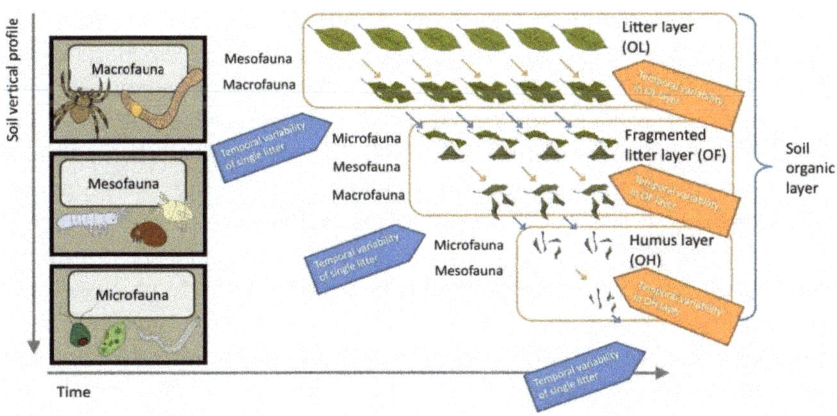

Figure 3 Change in plant litter after decomposition enhanced by the action of soil fauna, the main actors highlighted at each level of fragmentation. Adapted from Fujii et al. (2020), due to the nature of the soil habitat, soil fauna effectively live in 'a house made out of food'; therefore, consumption of food changes not just the availability of food resources but the composition of resources, of which each is affected by the temporal variability of single litter species and the whole layer. For example, a decreasing litter volume during decomposition will negatively affect mesofauna and macrofauna, but the increase in surface area during fragmentation can positively affect microfauna (Fujii et al., 2020).

and depositing faecal pellets for further decomposition by microfauna. The comminuted plant material and faeces increases the surface area available to be processed by fungi and bacteria; these microorganisms, through the process of mineralization, convert the plant nutrients into simpler compounds available for plant uptake (Culliney, 2013). Macrofauna are considered to be the main drivers of organic matter fragmentation and burial and mixing of plant residues; however, mesofauna play a key role. Burrowing activities are also not solely due to macrofauna (earthworms). Enchytraeidae have been found to increase the porosity and aeration of the soil as well as the pore size distribution in relation to their body width (Porre et al., 2016), through burrowing. Enchytraeidae often become the dominant influence on the soil habitat, in the absence of earthworms, or when earthworm numbers are reduced due to increased acidity of soils or soil disturbance. Collembola are not actively burrowing soil animals, but some species make active microtunnels within the soil matrix (Rusek, 1998). However, Phthiracaridae Oribatid mites have been found to actively tunnel within the litter debris layer and are considered an important group involved in the mechanical reduction of plant litter (Ponge, 1991, and the references therein). Faunal voids are formed by the consumption and removal of soil material or the burrowing activity of soil fauna, and these voids can be distinguished into three types: (1) channels, (2) chambers and (3) modified voids (Kooistra and Pulleman, 2010), commonly referred to as soil biopores. Soil macrofauna (earthworms, beetles, termites and ants) are the main organisms creating channels and chambers; however, the mesofauna (Collembola, mites and Enchytraeidae) usually modify voids, creating irregular shapes. These faunal voids and burrowing activities increase the aeration within the soil profile, as well as improve the bulk density and water infiltration rates, making the soil more hospitable to the other animals and plants living within it.

Mesofauna occur in high numbers, therefore producing a large number of faecal pellets. It has been calculated that some soils contain millions of Collembola faecal pellets per square metre (Hopkin, 2007). Mesofauna faecal pellets help improve soil structure as the pellets form water-stable macroaggregates (King and Hutchinson, 2007) and are a significant factor in the formation of humus (Culliney, 2013). Enchytraeidae faecal pellets contribute to soil structure as they transport, ingest and mix mineral and organic soil particles, adding to the stability of aggregates within the soil profile (Pelosi and Römbke, 2016). Coprophagy (consumption of faecal pellets) is a common phenomenon by the mesofauna. This is because the mesofauna, as well as consuming the faecal pellets within the soil, are also actively grazing the microorganisms growing on the faecal pellets within the soil, working like an 'external rumen' (Swift et al., 1979). Coprophagy has also been shown to increase when resources are scarce (Crotty and Adl, 2019). Consumption of partially decomposed faecal pellets allows the soil mesofauna (particularly

Enchytraeidae, Oribatid mites and Collembola) to obtain extra nutrients from what would normally be considered suboptimal food sources (Briones, 2018).

Around 90% of net primary production, as well as the bodies and excretion products of animals, is returned to the soil as dead organic matter (Coleman and Crossley, 2003). The primary agents of decomposition are bacteria and fungi (microbial biomass); however, the growth and proliferation of these microorganisms is selectively controlled by other organisms within the soil food web, namely protozoa, nematodes and the mesofauna. The consumption of microbial biomass leads to a release of nutrients that are more available for plant uptake, referred to as the 'microbial loop' effect (Bonkowski, 2004). Soil mesofauna are responsible for the fragmentation of plant litter, breaking it down, conditioning it and increasing its surface area, aiding microbial action (Fig. 3). Litter quantity and quality are considered to be the main drivers of decomposition and nutrient cycling (e.g. Jacob et al., 2009; Handa et al., 2014; Gergócs and Hufnagel, 2016). However, this relates closely to the abundance and diversity of soil mesofauna within that particular soil habitat – creating a positive feedback loop. The better the quality of litter and greater abundance of this high-quality litter, the faster and more effectively it will be broken down by microorganisms and the mesofauna, leading to higher quality and quantity of soil organic matter and plant-available nutrients and thus a healthier soil. Acidification is sometimes performed to change the soil habitat and 'restore' it to natural conditions, for example, the reversion of improved agricultural land to acidic grassland and heathland. However, this changes the soil fauna community, with a reduction in the mesotrophic components, for example, reductions in nematodes, rotifer and earthworm biomass (Tibbett et al., 2019). Changes in the acidity of the soil habitat have been shown to affect populations of mesofauna (King and Hutchinson, 2007). For example, Collembola were monitored within experimental liming of an acidic forest and showed changes in abundance and dominance of certain species, suggesting they could have value as a bioindicator of soil acidity (Chagnon et al., 2001).

3 Mesofauna in agriculture

Soil is at the base of all terrestrial biomes globally – whether it is tundra, savannah, forest, grassland or rainforest (Fig. 2), but it is of particular importance when considered within agriculture. It has been calculated that 95% of all food production comes directly (or indirectly) from the soil and around 11% of the globe's land surface (13.4 billion ha) is used in crop production (Bruinsma, 2003). Therefore, without the soil there will be limited agricultural production; however, agriculture is also the main driver of change within the soil habitat. Crop establishment methods lead to a disturbance within the soil habitat, which has the largest impact on soil organisms. For example, tillage disrupts the soil pore channels within the soil profile, affecting both mesofauna and macrofauna that live within the air-filled pore spaces. Tillage also increases nutrient mobility within the soil, providing a food source for bacteria, but disrupting the fungal hyphal network within the soil food web. This can promote microbial feeders like nematodes; however, mites and Collembola are known to consume nematodes and bacteria, and therefore this is likely to have neutral consequences in relation to food availability. Disturbance selects for shorter food chains; in farmed soils, disturbance can be chemical (pesticides, fertilisers) or physical (cultivation, organic matter incorporation, removal of the surface organic layer) (Crotty, 2020) as well as reduced arable plant diversity. This often results in a reduction in the complexity of the soil food web, therefore, reducing the functional redundancy and resilience within the system. The overall abundance of mesofauna may remain relatively static, but variation in abundance of the main groups will occur. The weed seed bank plays a key role in the complex interactions with mesofauna and macrofauna, such as a food source for earthworms and nematodes (Fagundez, 2014), and changes the soil habitat through differences in the root system and canopy cover. Weeds also provide a soil cover when it has been left bare, thus providing a potential food source for soil organisms. Overall, agricultural intensification (increased tillage, low diversity within the rotation, the dominance of winter wheat) has been found to reduce species richness in mesofauna (Collembola and oribatid mites), making soil food webs less diverse and composed of smaller-bodied organisms across Europe (Tsiafouli et al., 2015).

Enchytraeidae have been shown to be sensitive to agricultural management practices (tillage, fertiliser use), in relation to both abundance and species composition (Pelosi and Römbke, 2016). This is due to their close contact with the soil pore water, a high ingestion rate and a thin cuticle; they often react very sensitively to a broad range of pesticides (Römbke et al., 2017). Effects of pesticides on non-target mesofauna are variable and can be directly deleterious or indirectly beneficial. For example, Collembola populations have been shown to be reduced by neonicotinoid seed dressings (Pisa et al., 2017); however, herbicide use, leading to increases in litter, provides a large food supply for decomposer organisms like Collembola and mites (King and Hutchinson, 2007), leading to potential increases in numbers. Tillage affects different lineages of the Acari to a varying extent; for example, Oribatids can become locally extinct in regularly tilled arable soils, whereas the Heterostigmata (Prostigmata) and Astigmata can become extremely abundant (Crotty, 2020). Agricultural intensification has also led to a decline in arable plant species, with 582 species considered to have 'threatened status' across the EU (Richner et al., 2015). This loss might be seen by some as a measure of successful weed control (Storkey and Neve, 2018), but in reality it signifies the breaking down of the whole ecosystem and loss of rare plants rather than pernicious weeds.

This loss in arable plant species is likely to be a small indication of large-scale biodiversity loss as soil biodiversity remains unmonitored and could lead to reductions in crop yields in the long-term. In the EU, cereal monocultures are estimated to lose 0.43% of their crop productivity annually through soil erosion and biodiversity loss (Panagos et al., 2018). Therefore, it is critical to consider soil biodiversity in relation to food security, to maintain yields and be able to feed a growing global population.

Conventional agricultural practices act as a selection pressure for smaller organisms within the soil food web, and as agriculture intensifies, there is a tendency to remove most of the soil fauna. Agriculture in an ecological sense is an early-successional system, which is continually out of balance, therefore making stability hard to achieve (Erisman et al., 2016). Conservation agriculture's main aim is to increase stability within the environment through minimal soil movement (reduced tillage), continuous soil coverage (through residues, cover crops or both) and enhanced crop rotation (increased plant biodiversity). Conservation agriculture considers soil biodiversity as vital to the sustainable functioning of the whole ecosystem, and it aims to improve production, while also maintaining water quality and carbon sequestration, reducing soil erosion and enhancing biodiversity. Conservation agriculture has been shown to improve habitat and soil fauna biodiversity in comparison to conventional agriculture across Europe (Ghaley et al., 2018).

4 Mesofauna in grasslands

Permanent pasture has a continuous plant cover, providing a habitat for an abundant and diverse soil fauna that contribute to effective soil functioning, including the maintenance of sustainable agricultural fertility (Kibblewhite et al., 2008). Herbivory is an important and common feature of grasslands and has been shown to exert a strong but variable influence on species richness of soil fauna, and this effect is more variable than herbivory within woodlands (discussed later in this chapter). For example, a 3-year exclusion of all herbivory in sub-alpine grassland leads to a decrease in total springtail richness (Vandegehuchte et al., 2015), while comparisons of grazed versus ungrazed pasture has also been found to have higher diversity in the ungrazed pasture, with more endemic species (including large Oribatids) (Schon et al., 2011b). A meta-analysis of below-ground mesofauna response to above-ground herbivory found responses of soil fauna related to herbivory shifted from neutral to negative with increasing herbivore body size; however, mesofauna decomposers also tended to switch from negative to positive responses with increasing precipitation (Andriuzzi and Wall, 2017). Therefore, the negative effect of increasing grazing intensity was compounded by drought - a key consideration with the increasing risk of climate change. Differences have also been found dependent on grassland vegetation type, for example, tall-grass vegetation (developed where grazing did occur but herbivores did not rest) compared with short-grass vegetation (grazed by livestock and received high nutrient inputs). Tall-grass vegetation was found to have higher total springtail species richness and mesostigmatid mite abundance compared to short-grass vegetation (Vandegehuchte et al., 2015). Increased nutrient availability in fertilised grassland is compounded by other negative factors (e.g. increased stocking density of livestock), which tends to cause a decline in Oribatid mites as physical disturbance increases (Schon et al., 2010). Generally, nitrogen concentration within the soil has been found to be negatively related to the abundances of Collembola and mites across large spatial scales (Birkhofer et al., 2012).

Introduction of legumes within agricultural systems has been shown to consistently improve the abundance of earthworm populations (Schmidt et al., 2003; Crotty et al., 2015); however, it has a less obvious effect over mesofauna. D'Annibale et al. (2017) found changes in the community analysis of Collembola, but this varied more with sampling date and was thought to be associated with plant quality, rather than a preference for legumes in comparison to ryegrass. While Birkhofer et al. (2011) found Collembola abundance was influenced by species-rich legumes and grasses; however, Enchytraeidae abundance was influenced only by species-rich grasses (and was negatively affected by species-rich legumes). Studies have also shown that plant biomass is improved by soil fauna presence; for example, Kreuzer et al. (2004) found the inclusion of Collembola and earthworms within microcosms increased shoot biomass of both white clover and perennial ryegrass. The inclusion of legumes within agricultural grassland may have a greater effect on mesofauna depending on their functional group. For example, fungal-feeding nematodes and Poduromorpha Collembola were found in greater numbers in legume and chicory forage plots in comparison to ryegrass (Crotty et al., 2015). However, there have been limited studies investigating the effect of different grassland plant species on the Acari. Wissuwa et al. (2012, 2013) found Oribatid mites were more affected by soil properties than arable fallow plant species, while predatory mites (Mesostigmata) were more affected by the density of their prey (Collembola).

5 Mesofauna in woodlands

Woodland type and tree species diversity will have a major impact on the soil mesofauna community, through differences in root network, root exudates, litterfall and litter quality. Increased tree species richness and the introduction of broad-leaved species (within coniferous stands) have been found to have a positive effect on Collembola and Oribatid mite diversity and abundance

(Korboulewsky et al., 2016). Oribatid mites have been shown to differ in abundance and community structure depending on forest type (e.g. coniferous vs. deciduous woodlands) (Siira-Pietikainen et al., 2008), whereas Collembola have been found not to differ as greatly and forest age is more significant (Salamon et al., 2008). Litter quality and quantity change the humus characteristics - which will affect the soil food web. The decomposition rate of plant litter is determined by a number of factors - climate and precipitation are the main external factors, but also of equal importance is the chemical composition of the litter; plants with more recalcitrant compounds (e.g. lignin) will have slower decomposition rates. Different tree species will decompose at different rates, releasing nutrients in differing amounts (Jacob et al., 2009). Litter quality also affects the density of mesofauna, leading to population changes - mesofauna have been shown to enhance litter decomposition in more recalcitrant litter (Gergócs and Hufnagel, 2016) compared to higher-quality litter (which breaks down more easily naturally). Over a medium-term experiment (7 years), litter fall was found to significantly influence both Collembola and Oribatid species diversity, although Collembola were negatively correlated with total litter fall, while Oribatid mite diversity was positively correlated (Irmler, 2006).

Coniferous woodlands usually have thicker litter layers that are slower to decompose than deciduous woodlands. In the litter layer of Norwegian Spruce forests, about 40% of the needles were found to be excavated by Oribatid mites; these needles were filled with mite faecal pellets, which seemed to increase the resistance of these needles to decomposition (Hågvar, 2016). Soil communities have been referred to as the 'poor man's tropical rainforest' (Giller, 1996), due to the abundance and diversity of soil fauna located there. Paradoxically, where global distribution studies have been completed, there is usually less soil fauna found in tropical soil compared to temperate or arctic regions (e.g. microbes (Crowther et al., 2019), nematodes (van den Hoogen et al., 2019) and earthworms (Phillips et al., 2019)). Due to the nature of tropical rainforests, with a large amount of plant biomass, shallow nutrient-poor soils and increased leaching of nutrients due to the high amounts of precipitation occurring, it is likely that tropical soils are less favourable for an abundant soil fauna community compared to a deciduous or coniferous woodland. When considering mesofauna abundance as an indication of soil quality, it is important to take habitat type and climate into account.

Suspended soils are also a feature of many woodlands, and these accumulations of humus and organic detritus in trees, elevated above the soil level, often have similar mesofauna community assemblages when compared to the forest floor. For example, in a predominantly red cedar coniferous forest, Oribatid and Prostigmatid mite abundance was not significantly different between the forest floor and suspended soil habitat, with over 35% of species found in both habitats (Lindo and Winchester, 2006). Suspended soils have been found to contain higher nitrogen, phosphorus and potassium contents than the forest soils but lower moisture content (Abakumov et al., 2018), impacting the mesofauna community structure. Herbivory within woodlands can also have a strong influence over mesofauna populations. For example, the exclusion of deer herbivory for 10 years showed a significant increase in soil fauna species richness (Lessard et al., 2012). Deer browsing reduced the volume of understory plants, and this increased not only soil bulk density but also the relative humidity at the forest floor, leading to significant declines in soil mesofauna (including detritivores, fungivores and predators) through desiccation within the litter layer (Katagiri and Hijii, 2017). Fragmentation of woodlands may also lead to similar problems, and soil fauna activity has been found to vary due to edge effects, correlating with a decreasing soil moisture gradient, with the highest activity within the core of the woodland and lowest at the edges (Simpson et al., 2012), which had drier soils. Reductions in soil mesofauna abundance can lead to drastic reductions in litter comminution and decomposition (Fig. 3) within forest ecosystems.

6 Mesofauna as bioindicators

Soil mesofauna are considered a good biological indicator of the soil conservation status (Socarrás, 2013), but could soil mesofauna be used as a soil quality indicator? These small but diverse fauna are often overlooked within ecosystem management; however, this diversity and abundance is impacted by the different soil and ecosystem management that occurs -depending on whether the habitat is agricultural, grassland or woodland. Mesofauna are sensitive to the organic matter content, humidity percentage, pH, agricultural practices and insecticide use (Socarrás, 2013); therefore, the diversity and abundance will vary and could be monitored. Mesofauna (springtails and mites) were monitored as part of a national soil survey (George et al., 2017), which revealed a close relationship between mesofauna groups and soil organic matter class; however, limited differences were detected between similar habitats at a national scale. However, the identification of mesofauna within this national monitoring scheme was at very broad taxonomic levels (Acari sorted to Order; Collembola according to Superfamily). This is the inherent problem with using mesofauna as a bioindicator – to identify a single group of species takes too much time for it to be feasible across a large number of sampling points. For example, Wu et al. (2009) calculated that it took six person-years to identify individuals from one soil mesofauna group (mites) to species within a single experiment conducted at one grassland site (approximately 140 soil cores) (St John et al., 2006). Measurement of soil mesofauna usually consists of collecting 10 cm deep soil cores from the sampling site, extracting the fauna using a heat extraction method (e.g. Berlese tullgren funnels), before sorting through the samples using a dissecting microscope. The resolution of a dissecting microscope allows the identification of soil mesofauna to the main taxonomic groups or superfamilies, but further identification has to be performed using a compound microscope. Under compound microscopy each individual specimen is mounted on a slide (and sometimes cleared) before identification of genus or species can be achieved. High-throughput molecular characterisation has been utilised on soil nematodes, allowing assessments of community structure (Griffiths et al., 2012); however, this has not been achieved for mesofauna. For the mesofauna to be used as a bioindicator, both abundance and diversity currently have to be considered, as to date no 'key indicator species' have been identified; this is a clear research gap that warrants further investigation. Monitoring of mesofauna as a bioindicator and assessing the resilience of the soil system involve comparisons between reference sites (Orwin and Wardle, 2004), creating the need to take more samples.

The response of the soil ecosystem to disturbance is influenced by the resistance and resilience of the soil food web (Griffiths and Philippot, 2013). However, without an understanding of the potential biodiversity within the habitat, if species are lost but the functionality remains, this loss in biodiversity will be missed. A study defining a reference system for biological indicators of agricultural soil quality in Belgium (Krüger et al., 2018) overlooked the use of mesofauna (focusing on microbes or macrofauna instead). However, soil mite communities have been identified as potential bioindicators in a number of studies, usually focussing on Mesostigmatid and Oribatid mites (e.g. Manu et al., 2019; Meehan et al., 2019), although usually where remediation or large-scale disturbance has occurred. However, differences between mite assemblages (specifically ratios between Mesostigmata and Oribatids compared to Prostigmata and Astigmata) may be found to be a better indicator across less-drastic disturbance (e.g. tillage vs. no-till systems) (Bedano et al., 2011). Collembola have also been identified as a potential mesofauna indicator that varies with 'soil quality' (Vincent et al., 2018; Joimel et al., 2017). The use of other mesofauna groups as indicators was reviewed by Menta and Remelli (2020), including those rare groups that are often overlooked, for example, Protura, Diplura, Pseudoscorpionida, Pauropoda and Symphyla. Due to their variable abundance and sensitivity these organisms warrant further investigation as mesofauna bioindicators. To date, differences have been found between sites when comparing mesofauna abundance and species assemblage; the changes in species abundance and diversity have been shown to correlate to stability within the ecosystem (Cassani et al., 2020; Wahl et al., 2012) and changes in soil chemistry (Wierzbicka et al., 2019). Within the soil habitat there is a large amount of functional redundancy and resilience; when diversity decreases due to perturbations, understanding whether the niche spaces remain occupied will be a crucial indication of soil health.

7 Conclusion

Mesofauna are the most abundant and diverse group of animal fauna within the soil, the most abundant of these being springtails (Collembola), mites (Acari) and Enchytraeidae (Fig. 1). Soil mesofauna's importance is because of their abundance – Collembola, mites and Enchytraeidae are found globally from the equator to the poles, although their diversity and activity changes in relation to the above-ground habitat (Fig. 2). Reductions in soil mesofauna abundance can lead to drastic reductions in litter comminution (Fig. 3), reducing decomposition rates and affecting nutrient cycling – leading to reductions in plant growth. This chapter has discussed a number of different soil habitats, agricultural soil, grassland and woodland; however, soil type should also be a consideration. It is difficult to directly compare the effects of land use on soil fauna across different soil types, as management practices are often adjusted to compensate for soil differences (Schon et al., 2011a). Soil fauna also change in abundance and diversity dependent on the soil type, particularly as the soil type affects the water-holding capacity, with sandy soils inherently drier with less organic matter, thus having a smaller less diverse soil fauna community in comparison to heavy clay soil. Determination of a healthy soil therefore relies on using a benchmark for comparison. Development of an understanding of the 'soil health gap' (Maharjan et al., 2020) between a pristine undisturbed native soil and the soil that is being investigated, as well as 'key indicator species', will allow elucidation of the main differences in soil mesofauna community between these two ecosystems and the development of soil mesofauna as bioindicators of a healthy soil.

8 Where to look for further information

The Food and Agricultural Organisation (FAO) of the United Nations has recently published a report assessing the state of knowledge of soil biodiversity (FAO et al., 2020), the threats to this soil biodiversity, and the solutions that soil biodiversity can provide to problems in different fields. This report involved the work of over 300 scientists from around the world, the FAO, the Intergovernmental Technical Panel on Soils, the Convention on Biological Diversity, the Global Soil Biodiversity Initiative, and the European Commission. For further information there is also the newly launched soil information hub https://uksoils.org/ providing guidance and links to inspire, learn, and connect with soil health issues. Meanwhile the fascinating website https://www.chaosofdelight.org/ provides the reader with both information on soil mesofauna and also illustrates the beauty of the diversity that exists below-ground.

9 References

Abakumov, E. V., Rodina, O. A. and Eskov, A. K. 2018. Humification and humic acid composition of suspended soil in Oligotrophous environments in South Vietnam. *Applied and Environmental Soil Science* 2018, 1-8, 1026237.

Anderson, J. M. 2009. Why should we care about soil fauna? *Pesquisa Agropecuaria Brasileira* 44(8), 835-842.

Andriuzzi, W. S. and Wall, D. H. 2017. Responses of belowground communities to large aboveground herbivores: meta-analysis reveals biome-dependent patterns and critical research gaps. *Global Change Biology* 23(9), 3857-3868.

Bedano, J. C., Domínguez, A. and Arolfo, R. 2011. Assessment of soil biological degradation using mesofauna. *Soil and Tillage Research* 117, 55-60.

Bender, S. F., Wagg, C. and van der Heijden, M. G. A. 2016. An underground revolution: biodiversity and soil ecological engineering for agricultural sustainability. *Trends in Ecology and Evolution* 31(6), 440-452.

Birkhofer, K., Diekotter, T., Boch, S., Fischer, M., Muller, J., Socher, S. and Wolters, V. 2011. Soil fauna feeding activity in temperate grassland soils increases with legume and grass species richness. *Soil Biology and Biochemistry* 43(10), 2200-2207.

Birkhofer, K., Schoning, I., Alt, F., Herold, N., Klarner, B., Maraun, M., Marhan, S., Oelmann, Y., Wubet, T., Yurkov, A., Begerow, D., Berner, D., Buscot, F., Daniel, R., Diekotter, T., Ehnes, R. B., Erdmann, G., Fischer, C., Foesel, B., Groh, J., Gutknecht, J., Kandeler, E., Lang, C., Lohaus, G., Meyer, A., Nacke, H., Nather, A., Overmann, J., Polle, A., Pollierer, M. M., Scheu, S., Schloter, M., Schulze, E. D., Schulze, W., Weinert, J., Weisser, W. W., Wolters, V. and Schrumpf, M. 2012. General relationships between abiotic soil properties and soil biota across spatial scales and different land-use types. *PLoS ONE* 7(8), e43292.

Bonkowski, M. 2004. Protozoa and plant growth: the microbial loop in soil revisited. *New Phytologist* 162(3), 617-631.

Briones, M. J. I. 2018. The serendipitous value of soil fauna in ecosystem functioning: the unexplained explained. *Frontiers in Environmental Science* 6, 1-11.

Brown, B. V. 2001. Insects, overview. In: Levin, S. A. (Ed), *Encyclopedia of Biodiversity* (2nd edn.). Academic Press, Waltham, pp. 308-312.

Bruinsma, J. 2003. *World Agriculture: Towards 2015/2030 An FAO Perspective*. Earthscan Publications Ltd, London, UK, p. 444.

Cassani, M. T., Sabatté, M. L., González Arzac, A. and Massobrio, M. J. 2020. Mesofauna as an indicator of agroecosystem stability: degree of artificialization effect on land uses in Azul district, Argentina. *SN Applied Sciences* 2(3), 324.

Chagnon, M., Pare, D., Hebert, C. and Camire, C. 2001. Effects of experimental liming on collembolan communities and soil microbial biomass in a southern Quebec sugar maple (*Acer saccharum* Marsh.) stand. *Applied Soil Ecology* 17(1), 81-90.

Coleman, D. C. 2008. From peds to paradoxes: linkages between soil biota and their influences on ecological processes. *Soil Biology and Biochemistry* 40(2), 271-289.

Coleman, D. C. and Crossley, D. A. J. 2003. *Fundamentals of Soil Ecology* (2nd edn.). Academic Press, New York.

Crotty, F. V. 2011. *Elucidating the Relative Importance of the Bacterial and Fungal Feeding Channels within the Soil Food Web under Differing Land Managements*. Centre for Agricultural and Rural Sustainability, Plymouth University, Plymouth.

Crotty, F. V. 2020. Soil organisms within arable habitats. In: Wilson, P. (Ed), *Changing Status of Arable Habitats*. Springer, Berlin, Germany, pp. 123-138.

Crotty, F. V. and Adl, S.M. 2019. Competition and predation in soil fungivorous microarthropods using stable isotope ratio mass spectrometry. *Frontiers in Microbiology* 10, 1-10.

Crotty, F. V., Fychan, R., Sanderson, R., Rhymes, J. R., Bourdin, F., Scullion, J. and Marley, C. L. 2016. Understanding the legacy effect of previous forage crop and tillage management on soil biology, after conversion to an arable crop rotation. *Soil Biology and Biochemistry* 103, 241-252.

Crotty, F. V., Fychan, R., Scullion, J., Sanderson, R. and Marley, C. L. 2015. Assessing the impact of agricultural forage crops on soil biodiversity and abundance. *Soil Biology and Biochemistry* 91, 119-126.

Crowther, T. W., van den Hoogen, J., Wan, J., Mayes, M. A., Keiser, A. D., Mo, L., Averill, C. and Maynard, D. S. 2019. The global soil community and its influence on biogeochemistry. *Science* 365(6455), eaav0550.

Culliney, T. W. 2013. Role of arthropods in maintaining soil fertility. *Agriculture* 3(4), 629-659.

D'Annibale, A., Sechi, V., Larsen, T., Christensen, S., Krogh, P. H. and Eriksen, J. 2017. Does introduction of clover in an agricultural grassland affect the food base and functional diversity of Collembola? *Soil Biology and Biochemistry* 112, 165-176.

Doran, J. W. and Zeiss, M. R. 2000. Soil health and sustainability: managing the biotic component of soil quality. *Applied Soil Ecology* 15(1), 3-11.

Ekelund, F. and Ronn, R. 1994. Notes on protozoa in agricultural soil with emphasis on heterotrophic flagellates and naked amoebas and their ecology. *FEMS Microbiology Reviews* 15(4), 321-353.

Erisman, J. W., van Eekeren, N., de Wit, J., Koopmans, C., Cuijpers, W., Oerlemans, N. and Koks, B. J. 2016. Agriculture and biodiversity: a better balance benefits both. *AIMS Agriculture and Food* 1(2), 157-174.

Fagundez, J. 2014. The paradox of arable weeds: diversity, conservation, and ecosystem services of the unwanted. In: Benkelblia, N. (Ed), *Agroecology, Ecosystems and Sustainability*. CRC Press, Boco Raton, USA, pp. 139-150.

FAO, ITPS, GSBI, SCBD, and EC. 2020. *State of Knowledge of Soil Biodiversity - Status, Challenges and Potentialities, Report 2020*. FAO, Rome. https://doi.org/10.4060/cb1928en.

Firbank, L. G., Petit, S., Smart, S., Blain, A. and Fuller, R. J. 2008. Assessing the impacts of agricultural intensification on biodiversity: a British perspective. *Philosophical Transactions of the Royal Society of London. Series B, Biological Sciences* 363(1492), 777-787.

Fujii, S., Berg, M. P. and Cornelissen, J. H. C. 2020. Living litter: dynamic trait spectra predict fauna composition. *Trends in Ecology and Evolution* 35(10), 886-896.

George, P. B. L., Keith, A. M., Creer, S., Barrett, G. L., Lebron, I., Emmett, B. A., Robinson, D. A. and Jones, D. L. 2017. Evaluation of mesofauna communities as soil quality indicators in a national-level monitoring programme. *Soil Biology and Biochemistry* 115, 537-546.

Gergócs, V. and Hufnagel, L. 2016. The effect of microarthropods on litter decomposition depends on litter quality. *European Journal of Soil Biology* 75, 24-30.

Ghaley, B., Rusu, T., Sandén, T., Spiegel, H., Menta, C., Visioli, G., O'Sullivan, L., Gattin, I., Delgado, A., Liebig, M., Vrebos, D., Szegi, T., Michéli, E., Cacovean, H. and Henriksen,

C. 2018. Assessment of benefits of conservation agriculture on soil functions in arable production systems in Europe. *Sustainability* 10(3), 1-17.

Giller, P. S. 1996. The diversity of soil communities, the 'poor man's tropical rainforest'. *Biodiversity and Conservation* 5(2), 135-168.

Griffiths, B. S., Daniell, T. J., Donn, S. and Neilson, R. 2012. Bioindication potential of using molecular characterisation of the nematode community: response to soil tillage. *European Journal of Soil Biology* 49, 92-97.

Griffiths, B. S. and Philippot, L. 2013. Insights into the resistance and resilience of the soil microbial community. *FEMS Microbiology Reviews* 37(2), 112-129.

Hågvar, S. 2016. From litter to humus in a Norwegian spruce forest: long-term studies on the decomposition of needles and cones. *Forests* 7(12), 186.

Handa, I. T., Aerts, R., Berendse, F., Berg, M. P., Bruder, A., Butenschoen, O., Chauvet, E., Gessner, M. O., Jabiol, J., Makkonen, M., McKie, B. G., Malmqvist, B., Peeters, E. T. H. M., Scheu, S., Schmid, B., van Ruijven, J., Vos, V. C. A. and Hattenschwiler, S. 2014. Consequences of biodiversity loss for litter decomposition across biomes. *Nature* 509(7499), 218-221.

Hopkin, S. P. 2007. *A Key to the Collembola (Springtails) of Britain and Ireland*. Field Studies Council Publications, Shrewsbury.

Howard, A. 1947. *The Soil and Health: A Study of Organic Agriculture. Culture of the Land: a Series in the New Agrarianism*. The Devin-Adair Company, New York.

Irmler, U. 2006. Climatic and litter fall effects on collembolan and oribatid mite species and communities in a beech wood based on a 7 years investigation. *European Journal of Soil Biology* 42(1), 51-62.

Jacob, M., Weland, N., Platner, C., Schaefer, M., Leuschner, C. and Thomas, F. M. 2009. Nutrient release from decomposing leaf litter of temperate deciduous forest trees along a gradient of increasing tree species diversity. *Soil Biology and Biochemistry* 41(10), 2122-2130.

Jeffery, S., Gardi, C., Jones, A., Montanarella, L., Marmo, L., Miko, L., Ritz, K., Peres, G., Rombke, J. and van der Putten, W. H. 2010. European atlas of soil biodiversity. In: Commision, E. (Ed), *European Atlas of Soil Biodiversity*. Publications Office of the European Union, Luxembourg.

Joimel, S., Schwartz, C., Hedde, M., Kiyota, S., Krogh, P. H., Nahmani, J., Pérès, G., Vergnes, A. and Cortet, J. 2017. Urban and industrial land uses have a higher soil biological quality than expected from physicochemical quality. *Science of the Total Environment* 584-585, 614-621.

Kaczmarek, W. 1984. A comparison of bacterial and fungal biomass in several cultivated soils. *Acta Microbiologica Polonica* 33(3-4), 239-247.

Kampichler, C. 1999. Fractal concepts in studies of soil fauna. *Geoderma* 88(3-4), 283-300.

Katagiri, N. and Hijii, N. 2017. Effects of sika deer browsing on soil mesofauna in a thinned Japanese cypress plantation. *Journal of Forest Research* 22(3), 169-176.

Kibblewhite, M. G., Ritz, K. and Swift, M. J. 2008. Soil health in agricultural systems. *Philosophical Transactions of the Royal Society of London. Series B, Biological Sciences* 363(1492), 685-701.

King, K. L. and Hutchinson, K. J. 2007. Pasture and grazing land: assessment of sustainability using invertebrate bioindicators. *Australian Journal of Experimental Agriculture* 47(4), 392-403.

Kooistra, M. J. and Pulleman, M. M. 2010. Chapter 18. Features related to faunal activity. In: Stoops, G., Marcelino, V. and Mees, F. (Eds), *Interpretation of Micromorphological Features of Soils and Regoliths*. Elsevier, Amsterdam, pp. 397–418.

Korboulewsky, N., Perez, G. and Chauvat, M. 2016. How tree diversity affects soil fauna diversity: a review. *Soil Biology and Biochemistry* 94, 94–106.

Kreuzer, K., Bonkowski, M., Langel, R. and Scheu, S. 2004. Decomposer animals (Lumbricidae, Collembola) and organic matter distribution affect the performance of *Lolium perenne* (Poaceae) and *Trifolium repens* (Fabaceae). *Soil Biology and Biochemistry* 36(12), 2005–2011.

Krüger, I., Chartin, C., van Wesemael, B. and Carnol, M. 2018. Defining a reference system for biological indicators of agricultural soil quality in Wallonia, Belgium. *Ecological Indicators* 95, 568–578.

Lessard, J.-P., Reynolds, W. N., Bunn, W. A., Genung, M. A., Cregger, M. A., Felker-Quinn, E., Barrios-Garcia, M. N., Stevenson, M. L., Lawton, R. M., Brown, C. B., Patrick, M., Rock, J. H., Jenkins, M. A., Bailey, J. K. and Schweitzer, J. A. 2012. Equivalence in the strength of deer herbivory on above and below ground communities. *Basic and Applied Ecology* 13(1), 59–66.

Lindo, Z. and Winchester, N. N. 2006. A comparison of microarthropod assemblages with emphasis on oribatid mites in canopy suspended soils and forest floors associated with ancient western red cedar trees. *Pedobiologia* 50(1), 31–41.

Maharjan, B., Das, S. and Acharya, B. S. 2020. Soil Health Gap: a concept to establish a benchmark for soil health management. *Global Ecology and Conservation* 23, e01116.

Manu, M., Honciuc, V., Neagoe, A., Băncilă, R. I., Iordache, V. and Onete, M. 2019. Soil mite communities (Acari: Mesostigmata, Oribatida) as bioindicators for environmental conditions from polluted soils. *Scientific Reports* 9(1), 20250.

Mbuthia, L. W., Acosta-Martínez, V., DeBruyn, J., Schaeffer, S., Tyler, D., Odoi, E., Mpheshea, M., Walker, F. and Eash, N. 2015. Long term tillage, cover crop, and fertilization effects on microbial community structure, activity: implications for soil quality. *Soil Biology and Biochemistry* 89, 24–34.

McBratney, A., Field, D. J. and Koch, A. 2014. The dimensions of soil security. *Geoderma* 213, 203–213.

Meehan, M. L., Song, Z., Lumley, L. M., Cobb, T. P. and Proctor, H. 2019. Soil mites as bioindicators of disturbance in the boreal forest in northern Alberta, Canada: testing taxonomic sufficiency at multiple taxonomic levels. *Ecological Indicators* 102, 349–365.

Menta, C. and Remelli, S. 2020. Soil health and arthropods: From complex system to worthwhile investigation. *Insects* 11(1), 54. doi:10.3390/insects11010054.

Millennium Ecosystem Assessment 2005. *Ecosystems and Human Well-being: Synthesis*. Washington, DC.

Orgiazzi, A., Bardgett, R. D., Barrios, E., Behan-Pelletier, V., Briones, M. J. I., Chotte, J.-L., De Deyn, G. B., Eggleton, P., Fierer, N., Fraser, T., Hedlund, K., Jeffery, S., Johnson, N. C., Jones, A., Kandeler, E., Kaneko, N., Lavelle, P., Lemanceau, P., Miko, L., Montanarella, L., Moreira, F. M. S., Ramirez, K. S., Scheu, S., Singh, B. K., Six, J., van der Putten, W. H. and Wall, D. H. 2016. *Global Soil Biodiversity Atlas*. Publications Office of the European Union, Luxembourg.

Orwin, K. H. and Wardle, D. A. 2004. New indices for quantifying the resistance and resilience of soil biota to exogenous disturbances. *Soil Biology and Biochemistry* 36(11), 1907–1912.

Panagos, P., Standardi, G., Borrelli, P., Lugato, E., Montanarella, L. and Bosello, F. 2018. Cost of agricultural productivity loss due to soil erosion in the European Union: from direct cost evaluation approaches to the use of macroeconomic models. *Land Degradation and Development* 29(3), 471–484.

Pelosi, C. and Römbke, J. 2016. Are Enchytraeidae (Oligochaeta, Annelida) good indicators of agricultural management practices? *Soil Biology and Biochemistry* 100, 255–263.

Phillips, H. R. P., Guerra, C. A., Bartz, M. L. C., Briones, M. J. I., Brown, G., Crowther, T. W., Ferlian, O., Gongalsky, K. B., van den Hoogen, J., Krebs, J., Orgiazzi, A., Routh, D., Schwarz, B., Bach, E. M., Bennett, J., Brose, U., Decaëns, T., König-Ries, B., Loreau, M., Mathieu, J., Mulder, C., van der Putten, W. H., Ramirez, K. S., Rillig, M. C., Russell, D., Rutgers, M., Thakur, M. P., de Vries, F. T., Wall, D. H., Wardle, D. A., Arai, M., Ayuke, F. O., Baker, G. H., Beauséjour, R., Bedano, J. C., Birkhofer, K., Blanchart, E., Blossey, B., Bolger, T., Bradley, R. L., Callaham, M. A., Capowiez, Y., Caulfield, M. E., Choi, A., Crotty, F. V., Dávalos, A., Cosin, D. J. D., Dominguez, A., Duhour, A. E., van Eekeren, N., Emmerling, C., Falco, L. B., Fernández, R., Fonte, S. J., Fragoso, C., Franco, A. L. C., Fugère, M., Fusilero, A. T., Gholami, S., Gundale, M. J., López, M. G., Hackenberger, D. K., Hernández, L. M., Hishi, T., Holdsworth, A. R., Holmstrup, M., Hopfensperger, K. N., Lwanga, E. H., Huhta, V., Hurisso, T. T., Iannone, B. V., Iordache, M., Joschko, M., Kaneko, N., Kanianska, R., Keith, A. M., Kelly, C. A., Kernecker, M. L., Klaminder, J., Koné, A. W., Kooch, Y., Kukkonen, S. T., Lalthanzara, H., Lammel, D. R., Lebedev, I. M., Li, Y., Lidon, J. B. J., Lincoln, N. K., Loss, S. R., Marichal, R., Matula, R., Moos, J. H., Moreno, G., Morón-Ríos, A., Muys, B., Neirynck, J., Norgrove, L., Novo, M., Nuutinen, V., Nuzzo, V., Rahman P, M., Pansu, J., Paudel, S., Pérès, G., Pérez-Camacho, L., Piñeiro, R., Ponge, J. F., Rashid, M. I., Rebollo, S., Rodeiro-Iglesias, J., Rodríguez, M. Á, Roth, A. M., Rousseau, G. X., Rozen, A., Sayad, E., van Schaik, L., Scharenbroch, B. C., Schirrmann, M., Schmidt, O., Schröder, B., Seeber, J., Shashkov, M. P., Singh, J., Smith, S. M., Steinwandter, M., Talavera, J. A., Trigo, D., Tsukamoto, J., de Valença, A. W., Vanek, S. J., Virto, I., Wackett, A. A., Warren, M. W., Wehr, N. H., Whalen, J. K., Wironen, M. B., Wolters, V., Zenkova, I. V., Zhang, W., Cameron, E. K. and Eisenhauer, N. 2019. Global distribution of earthworm diversity. *Science* 366(6464), 480–485.

Pisa, L., Goulson, D., Yang, E. C., Gibbons, D., Sánchez-Bayo, F., Mitchell, E., Aebi, A., van der Sluijs, J., MacQuarrie, C. J. K., Giorio, C., Long, E. Y., McField, M., Bijleveld van Lexmond, M. and Bonmatin, J. M. 2017. An update of the Worldwide Integrated Assessment (WIA) on systemic insecticides. Part 2: impacts on organisms and ecosystems. *Environmental Science and Pollution Research International*, 1–49. https://link.springer.com/article/10.1007/s11356-017-0341-3.

Ponge, J. F. 1991. Food Resources and diets of soil animals in a small area of Scots pine litter. *Geoderma* 49(1–2), 33–62.

Porre, R. J., van Groenigen, J. W., De Deyn, G. B., de Goede, R. G. M. and Lubbers, I. M. 2016. Exploring the relationship between soil mesofauna, soil structure and N_2O emissions. *Soil Biology and Biochemistry* 96, 55–64.

Richner, N., Holderegger, R., Linder, H. P. and Walter, T. 2015. Reviewing change in the arable flora of Europe: a meta-analysis. *Weed Research* 55(1), 1–13.

Römbke, J., Schmelz, R. M. and Pélosi, C. 2017. Effects of organic pesticides on enchytraeids (Oligochaeta) in agroecosystems: laboratory and higher-tier tests. *Frontiers in Environmental Science* 5, 1–23.

Rusek, J. 1998. Biodiversity of Collembola and their functional role in the ecosystem. *Biodiversity and Conservation* 7(9), 1207–1219.

Salamon, J. A., Scheu, S. and Schaefer, M. 2008. The Collembola community of pure and mixed stands of beech (Fagus sylvatica) and spruce (Picea abies) of different age. *Pedobiologia* 51(5–6), 385–396.

Schmidt, O., Clements, R. O. and Donaldson, G. 2003. Why do cereal-legume intercrops support large earthworm populations? *Applied Soil Ecology* 22(2), 181–190.

Schon, N. L., Mackay, A. D. and Minor, M. A. 2011a. Effects of dairy cow treading pressures and food resources on invertebrates in two contrasting and co-occurring soils. *Soil Research* 49(8), 703–714.

Schon, N. L., Mackay, A. D. and Minor, M. A. 2011b. Soil fauna in sheep-grazed hill pastures under organic and conventional livestock management and in an adjacent ungrazed pasture. *Pedobiologia* 54(3), 161–168.

Schon, N. L., Mackay, A. D., Yeates, G. W. and Minor, M. A. 2010. Separating the effects of defoliation and dairy cow treading pressure on the abundance and diversity of soil invertebrates in pastures. *Applied Soil Ecology* 46(2), 209–221.

Siira-Pietikainen, A., Penttinen, R. and Huhta, V. 2008. Oribatid mites (Acari: Oribatida) in boreal forest floor and decaying wood. *Pedobiologia* 52(2), 111–118.

Simpson, J. E., Slade, E., Riutta, T. and Taylor, M. E. 2012. Factors affecting soil fauna feeding activity in a fragmented lowland temperate deciduous woodland. *PLoS ONE* 7(1), e29616.

Socarrás, A. 2013. Soil mesofauna: biological indicator of soil quality. *Pastos y Forrajes* 36, 14–21.

St John, M. G., Wall, D. H. and Hunt, H. W. 2006. Are soil mite assemblages structured by the identity of native and invasive alien grasses? *Ecology* 87(5), 1314–1324.

Storkey, J. and Neve, P. 2018. What good is weed diversity? *Weed Research* 58(4), 239–243.

Swift, M. J., Heal, O. W. and Anderson, J. M. 1979. *Decomposition in Terrestrial Ecosystems*. Blackwell Scientific Publications, Oxford.

Tibbett, M., Fraser, T. D. and Duddigan, S. 2020. Identifying potential threats to soil biodiversity. *PeerJ* 8, e9271.

Tibbett, M., Gil-Martinez, M., Fraser, T., Green, I. D., Duddigan, S., De Oliveira, V. H., Raulund-Rasmussen, K., Sizmur, T. and Diaz, A. 2019. Long-term acidification of pH neutral grasslands affects soil biodiversity, fertility and function in a heathland restoration. *CATENA* 180, 401–415.

Tsiafouli, M. A., Thébault, E., Sgardelis, S. P., de Ruiter, P. C., van der Putten, W. H., Birkhofer, K., Hemerik, L., de Vries, F. T., Bardgett, R. D., Brady, M. V., Bjornlund, L., Jørgensen, H. B., Christensen, S., Hertefeldt, T. D., Hotes, S., Gera Hol, W. H., Frouz, J., Liiri, M., Mortimer, S. R., Setälä, H., Tzanopoulos, J., Uteseny, K., Pižl, V., Stary, J., Wolters, V. and Hedlund, K. 2015. Intensive agriculture reduces soil biodiversity across Europe. *Global Change Biology* 21(2), 973-985.

van Capelle, C., Schrader, S. and Brunotte, J. 2012. Tillage-induced changes in the functional diversity of soil biota–a review with a focus on German data. *European Journal of Soil Biology* 50, 165–181.

van den Hoogen, J., Geisen, S., Routh, D., Ferris, H., Traunspurger, W., Wardle, D. A., de Goede, R. G. M., Adams, B. J., Ahmad, W., Andriuzzi, W. S., Bardgett, R. D., Bonkowski, M., Campos-Herrera, R., Cares, J. E., Caruso, T., de Brito Caixeta, L., Chen, X., Costa, S. R., Creamer, R., Mauro da Cunha Castro, J., Dam, M., Djigal, D., Escuer, M., Griffiths, B. S., Gutiérrez, C., Hohberg, K., Kalinkina, D., Kardol, P., Kergunteuil, A., Korthals,

G., Krashevska, V., Kudrin, A. A., Li, Q., Liang, W., Magilton, M., Marais, M., Martín, J. A. R., Matveeva, E., Mayad, E. H., Mulder, C., Mullin, P., Neilson, R., Nguyen, T. A. D., Nielsen, U. N., Okada, H., Rius, J. E. P., Pan, K., Peneva, V., Pellissier, L., Carlos Pereira da Silva, J., Pitteloud, C., Powers, T. O., Powers, K., Quist, C. W., Rasmann, S., Moreno, S. S., Scheu, S., Setälä, H., Sushchuk, A., Tiunov, A. V., Trap, J., van der Putten, W., Vestergård, M., Villenave, C., Waeyenberge, L., Wall, D. H., Wilschut, R., Wright, D. G., Yang, J. I. and Crowther, T. W. 2019. Soil nematode abundance and functional group composition at a global scale. *Nature* 572(7768), 194–198.

Vandegehuchte, M. L., Raschein, U., Schütz, M., Gwiazdowicz, D. J. and Risch, A. C. 2015. Indirect short- and long-term effects of aboveground invertebrate and vertebrate herbivores on soil microarthropod communities. *PLoS ONE* 10(3), e0118679.

Vincent, Q., Auclerc, A., Beguiristain, T. and Leyval, C. 2018. Assessment of derelict soil quality: abiotic, biotic and functional approaches. *Science of the Total Environment* 613-614, 990–1002.

Wahl, J. J., Theron, P. D. and Maboeta, M. S. 2012. Soil mesofauna as bioindicators to assess environmental disturbance at a platinum mine in South Africa. *Ecotoxicology and Environmental Safety* 86, 250–260.

Wierzbicka, A., Dyderski, M. K., Kamczyc, J., Rączka, G. and Jagodziński, A. M. 2019. Responses of soil mite communities (Acari: Oribatida, Mesostigmata) to elemental composition of mosses and pine needles and long-term air pollution in Scots pine (*Pinus sylvestris* L.) stands. *Science of the Total Environment* 691, 284–295.

Wissuwa, J., Salamon, J. A. and Frank, T. 2012. Effects of habitat age and plant species on predatory mites (Acari, Mesostigmata) in grassy arable fallows in Eastern Austria. *Soil Biology and Biochemistry* 50(6), 96–107.

Wissuwa, J., Salamon, J. A. and Frank, T. 2013. Oribatida (Acari) in grassy arable fallows are more affected by soil properties than habitat age and plant species. *European Journal of Soil Biology* 59, 8–14.

Wu, T. H., Ayres, E., Li, G., Bardgett, R. D., Wall, D. H. and Garey, J. R. 2009. Molecular profiling of soil animal diversity in natural ecosystems: incongruence of molecular and morphological results. *Soil Biology and Biochemistry* 41(4), 849–857.

Zhang, X., Ferris, H., Mitchell, J. and Liang, W. 2017. Ecosystem services of the soil food web after long-term application of agricultural management practices. *Soil Biology and Biochemistry* 111, 36–43.

Chapter 2

Quantifying earthworm community structures as indicators of soil health

Jacqueline L. Stroud, formerly Rothamsted Research, UK

1 Introduction

Earthworms are probably the most well-known soil organisms and there are a number of standard references providing further information on their importance and ecology (Stephenson, 1930; Lee, 1985; Edwards and Bohlen, 1996). A major advance in earthworm science was made in the 1970s with the classification of European earthworms into three ecological groups according to feeding and burrowing behaviours (Bouché, 1977):

- Epigeic;
- Endogeic; and
- Anecic.

Epigeic earthworms are surface dwelling, dentritivores (plant-residue feeding), for example, *Lumbricus rubellus*. Endogeic earthworms are topsoil-dwelling worms that make temporary horizontal burrows and are usually geophagous (soil feeding worms), for example, *Allolobophora chlorotica*. Anecic earthworms make permanent, deep vertical burrows, and can form middens by dragging plant leaves and residues into piles on the soil surface above their burrow, for example, *Lumbricus terrestris*. The purpose of a midden is unknown but is thought to be a source of food and influences the microclimate of the burrow (Darwin, 1882).

http://dx.doi.org/10.19103/AS.2020.0079.04

2 Earthworms, soil health and management

A range of studies have suggested positive effects of soil fauna such as earthworms on crop production (Scheu, 2003; Brown et al., 2004; van Groenigen et al., 2014; Bertrand et al., 2015). Earthworms are also an important source of food for many bird and animal species which may themselves benefit agriculture, for example, as pest predators or pollinators (Macdonald, 1983). Earthworms contribute indirectly to soil ecosystem services in various ways (Blouin et al., 2013). Earthworms mix soil and organic matter in their casts and create extensive networks of tunnels in the soil. By influencing soil structure and hydraulic properties, they support soil formation and the transport of water and nutrients which benefit crop growth (Brown et al., 2018). By processing soil and organic residues, they also support nutrient cycling and organic matter decomposition. There are also compost worms which are particularly effective in composting fresh organic residues to produce 'vermicompost' (Lazcano and Dominguez, 2011). Soil-dwelling earthworms can accelerate the aerobic digestion of soil contaminants such as polycyclic aromatic hydrocarbons (PCBs) and some pesticides (Hickman and Reid, 2008; Sanchez-Hernandez et al., 2014). Since earthworms play a central role in the trophic transfer of pollutants, they can be used as bioindicators of pollutants (Kuperman et al., 2009)

It is also important to be aware that earthworms can be hosts for organisms that cause parasitic diseases in wildlife, pets and commercially significant farm animals (Stroud, 2021). Examples include the gapeworm (*Syngamus trachealis*) which causes tracheo-bronchitis infections in fowl (Stephenson, 1930) and lung worms (*Metastrongylus* spp.) that infect swine and cause parasitic bronchitis (López and Martinson, 2017).

Research suggests that epigeic and anecic earthworms are depleted by intensive tillage in farmland soils (Edwards and Bohlen, 1996; Briones and Schmidt, 2017). In contrast, no-till techniques and Conservation Agriculture can promote earthworm numbers (Lahmar, 2010; Castellanos-Navarette et al., 2012; Baldivieso-Freitas et al., 2018). One of the first countries to mass adopt no-tillage was Brazil, with no-tillage farmers noticing an increase in earthworms and forming the 'worm club' perceiving these changes in soil biology as beneficial (Brown et al., 2003).

Conservation Agriculture has three principles: no-tillage, diverse rotation and keeping the soil permanently covered, for example, leaving plant residues on soil surfaces between cropping. A layer of this material can protect bare soil surfaces from erosion, but undecomposed plant residues on the soil surface can lead to poor crop emergence and build-up of pests and diseases such as *Fusarium* spp. Recent research has identified that better control of *Fusarium* spp. is likely to come from managing soil organisms and plant-residue consumption patterns (Hofgaard et al., 2016; Goncharov et al., 2020). Environmental economists have calculated that the effective biocontrol of *Fusarium* spp. and detoxification of its mycotoxins by anecic earthworms has an economic value of €75 ha^{-1} and would halve fungicide use (Plaas et al., 2019).

The perception of earthworms by farmers around the world ranges from indifference to seeing them as an important symbol of soil fertility, with views affected by gender (Zúñiga et al., 2013; Pauli et al., 2016). In Australia informal appraisals of soil quality used by farmers include organic matter (soil quality), crop appearance (productivity) and earthworms (soil life), plus soil chemical parameters (Bruyn and Abbey, 2003). This chapter focusses on the co-development of a quantitative earthworm ecological group assessment with UK farmers.

3 Challenges in collecting data on earthworms

There are two key problems associated with collecting data on earthworms:

- the labour required, and hence, the cost of collection; and
- accurate species identification.

For example, one of the largest biodiversity-monitoring efforts in Europe was the BioBio project (Biodiversity indicators for organic and low-input farming systems, KBBE-227161). Visiting the monitoring sites and sampling took 12–70 h per farm and cost between £180 (Bulgaria) and £2300 (Netherlands), not including the specialised taxonomy (microscope-based analysis) costs for species-level identifications (BioBio, 2012). Similarly, standard earthworm sampling in South Africa took 3–5 people, 45 min–1 h and the cost of laboratory taxonomic analysis per soil pit was Rl 4800 (£250) (Nxele et al., 2015).

In response to these challenges, scientists from the BioBio project suggested that there was a need to mobilise more farmers to assess their farmland soils and develop protocols that require lower technical expertise (Targetti et al., 2014). Similarly, in South Africa, scientists recommended developments including easier methods and earthworm keys so more people can participate to fill knowledge gaps (Nxele et al., 2015). Developments for the citizen science OPAL earthworm survey included a novel photographic species identification key for England (Archer et al., 2009). However, it is noticeable that just 16 arable fields were surveyed for the OPAL project (out of 2856 sites), indicating the potential challenges in engaging farmers with scientific species mapping projects (Archer et al., 2009).

In addition to cost, accurate identification is also a challenge, particularly given pressure to develop simpler identification methods to reduce cost and widen participation. The accuracy for 12 common earthworm species in the OPAL survey ranged from 0% to 98% by adults and from 0% to 64% by children (Archer et al., 2009). The OPAL project in the UK revealed that most people (88%) could accurately identify earthworms to the ecological group level (Archer et al., 2009). The data quality of species identification is further confounded by cryptic diversity (King et al., 2008). For example, the common worm, *Allolobophora chlorotica*, can be identified by 78% of people using the OPAL key (Archer et al., 2009). However, there is an 11.5% within-species divergence measured using molecular assessments meaning *Allolobophora chlorotica* is, in fact, many different species. A small study to check the identification of *Lumbricus terrestris* using molecular approaches (Stroud et al., 2016) cost £105 + tax per specimen, indicative of the analytical costs for accurate species identification.

4 Developing improved assessment of earthworms

This section focusses on the co-development of a quantitative earthworm ecological group assessment with UK farmers (Stroud, 2019). This soil health approach defined earthworms using a presence-absence occupancy measure. The concept was to dig multiple soil pits within a field to check that earthworms were present across the whole cultivation area, and check that the earthworms found included epigeic, endogeic and anecic earthworms. There were three stages of development:

a Pilot study (#60min worms): participatory science to create partnerships with farmers and inform co-development (Stroud, 2019);
b Co-development phase (#30min worms): using farmer feedback to improve the method and create the digital resources to support quality assessment, for example, an earthworm identification tutorial; and
c Application (#WorldWormWeek): the final protocol is detailed below.

The process of digging soil pits involved digging five soil pits per field using standard W shape field sampling. The following procedure applied:

1 At each soil pit spot, check the soil surface for the presence of middens (key shown) and tick/cross on the results sheet.
2 Dig out a 20 cm × 20 cm × 20 cm soil pit and place soil on mat (30 s). (20 cm = 8 inches).
3 Hand-sort soil (5 min), placing each whole earthworm into the pot. Note if pencil size vertical burrows are present and tick/cross on the results sheet. Optional: Please take a photo of the earthworms using the image analysis sheet (scale bar and colour correction). Please upload your photos at: www.wormscience.org for analysis.

4 Count the total number (adults and juveniles) of earthworms and write down.
5 For UK participants: select the adult earthworms (usually only a few) and return juveniles to soil pit. Only adults have a saddle - the reproductive ring near the head. Top tip: a saddle can be more obvious if the earthworm is turned upside-down.
6 Count the numbers of each type of adult earthworm (key shown) and write down.
7 Return worms to soil pit and back fill with soil.
8 Repeat steps 1–7, until five soil pits per field have been assessed.
9 Please input your data at: www.wormscience.org for results analysis.

The results analysis was via an online entry portal and was produced by a software company (Agrantec.com) to provide participants with an instant results dashboard with each individual results summary, and graphs of all the real-time earthworm data including numbers of earthworms and results entries submitted to facilitate peer-to-peer information exchange.

An interpretation of the result was instantly calculated. For a field to 'pass' the test, earthworms were present in at least four of the five soil pits (widespread) and at least one adult earthworm from each ecological group (or presence of middens/large burrows, for anecic earthworms) to indicate earthworm diversity (all three ecological types of earthworms) present in the field. If these conditions were not satisfied, the results were reported as a 'fail' - as in failing to find evidence, and a link to a downloadable interpretation booklet was provided. Results entry included basic field management information and optional postcode for regional and soil analyses, and by using the results entry form, informed consent was provided. The photographs generated were sent to researchers specialising in machine-learning technologies to try to develop digital earthworm ecological group identification tools.

4.1 #60minworms pilot study

The #60minworms pilot study has been published (Stroud, 2019) and fully describes the feedback method compliance and co-development. The pilot method included ten soil pits per field and a traffic light interpretation based on the presence of earthworms within the field. Briefly, this was a paper-based study with booklets physically posted to participants to avoid an untested and non-peer-reviewed method becoming widely available, and the role of Twitter was revealed as key to participation. Hence, this project evolved into an open digital science project, using a Twitter hub @wormscience, with # navigation to different project aspects.

4.2 #30minworms co-development

Co-development resulted in reducing the number of sampling pits to five per field and the abandonment of the traffic light interpretation because it was misunderstood. A website was created with free access to learning resources including an earthworm identification tutorial, YouTube demonstrations, the downloadable method booklet including a photographic identification guide and a link to the online form for results entry and survey feedback. Terms describing the earthworm ecological groups, for example, epigeic were disliked and requested to be changed to intuitive labels resulting in: surface = epigeic, topsoil = endogeic, deep burrowing = anecic to describe the different earthworm types. Through uploading their results, participants provided informed consent, and no identifying information was requested or recorded (i.e. the survey was anonymous). There were optional questions to provide feedback on the co-designed method and online training tools. The traffic light results' interpretation was replaced by a simple pass/fail interpretation. To 'pass' the test, earthworms were present in at least four of the five soil pits (widespread) and at least one adult earthworm from each ecological group (or presence of middens for anecic earthworms) to indicate earthworm diversity (all three ecological types of earthworms) is present in the field.

4.3 #WorldWormWeek (23–31 March 2019)

The feedback from participants of #30minworms indicated that no significant changes were needed. The final survey was advertised using Twitter (@ wormscience) and the independent dissemination of the #WorldWormWeek hashtag and 'how to sample' video views noted. A total of 30 scientists with publications in soils, agriculture or ecology were asked individually to anonymously generate a scientific consensus on the co-designed method and its applications. All invitees were emailed with details of the research project website and a summary of the results (including images), and the link to the Google form which had the 20 voluntary questions and space for specific comments. Again, through participation, informed consent was provided. There were three compulsory questions to estimate self-selection bias linked to experience, interests and familiarity with the survey. The survey was open for 1 week and the data populated a spreadsheet.

4.4 #WorldWormWeek at Rothamsted Research

To connect participatory data sets with scientific field trials, Rothamsted Research was surveyed. The 330 ha Rothamsted Research Farm (51.82 N and 0.37 W) in Harpenden, UK, has a temperate climate. The soil is characterised as a flinty silty clay loam of the Batcombe series and includes the longest running agricultural field experiment in the world, with full and open experimental details available from the electronic Rothamsted Archive (www.era.rothamsted.ac.uk). Every accessible field was surveyed by the author (scientist) using the #WorldWormWeek co-designed method. Modifications were made for the Fosters field experiment, a conventionally cultivated organic matter experiment with four replicate plots, and so one plot was randomly sampled twice to generate five data points for each treatment, and the data averaged for 'Fosters Field'. Individual plots (n = 9 plots, sections 0, 1 and 6, plots 2.1, 3 and 18) were sampled on the long-term experiment on Broadbalk field with five replicate pits per plot, with the data averaged for 'Broadbalk'.

5 Results and discussion

Invitations to advertise the co-developed #30minworms project were accepted, resulting in its promotion on BBC Farming Today (listening audience, 1 million), Farmers Weekly magazine (circulation, 53 000) and Twitter (@wormscience 3000 followers). Participants included new co-producers (43%) and pilot study participants (55%), who shared the belief that earthworm monitoring is likely to improve soil health (95%). The participants used the online learning resources provided and the feedback was that this was 'used and useful' (91%). They invested time (92% completing the field survey in under an hour, aligned with co-development improvement method (Stroud, 2019)) and physical labour to systematically measure their earthworm populations at field scales, resulting in 4647 earthworms assessed in 1628 ha. A diverse group using different tillage practices shared their results regardless of the explicit 'pass or fail' soil health interpretation.

In contrast, the application of #WorldWormWeek was solely advertised on Twitter. The 'How to sample' video was viewed 6779 times. The use of hashtag #WorldWormWeek' was independently disseminated by national organisations including BBC Spring Watch, Woodland Trust, Natural England, Soil Association, Buglife, AHDB and DefraSoils to a total social media audience of 1 million people. The *European Journal of Soil Science* directly contributed to information circulation by launching an online special issue of open access earthworm papers. A total of 216 surveys were received, which is a growth rate of 1 250% over the co-development phase.

The result of #WorldWormWeek was a globally coordinated community of people using the same co-developed method at the same time (23–31 March 2019) to sample farmland for earthworms (Fig. 1a, b). There were 18 entries from around the world including Australia, New Zealand, China, Bangladesh, India, Nigeria, Uruguay and the United States, that is, every continent except Antarctica. International management practices ranged from pasture, fruit and vegetables to arable production, ploughed to no-tillage soil management,

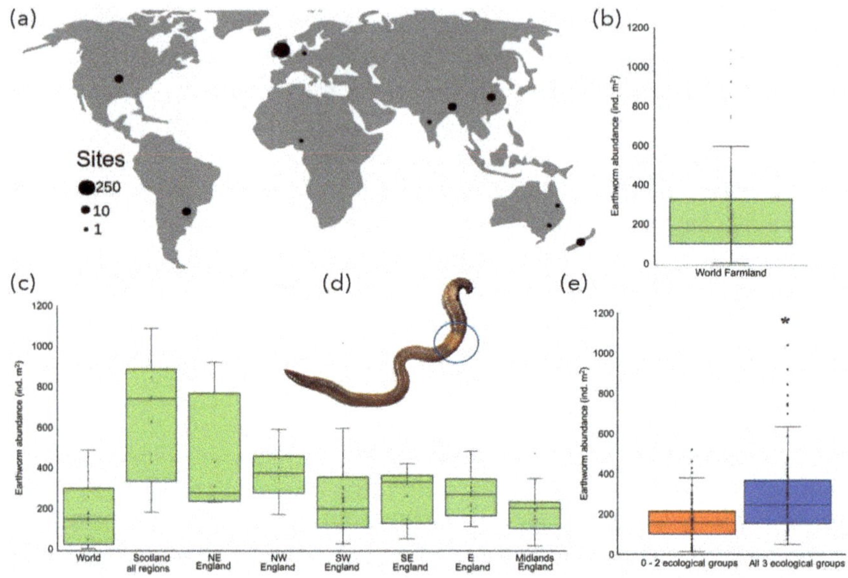

Figure 1 Farmland earthworm populations during #WorldWormWeek (23-31st March 2019). (a) Global participation. (b) World farmland earthworm abundance; typical field abundances <150 earthworms m^2. (c) UK farmland earthworm abundances. (d) Ecological group analysis performed on adult earthworms. (e) UK earthworm ecological groups within fields, *significantly different.

organic and conventional pesticide regimes. Earthworm numbers ranged from 0 earthworms m^2 (Nigeria), 15 earthworms m^2 (banana field in India), 120 earthworms m^2 (rice field in Bangladesh) to 377 earthworms m^2 (pasture field in New Zealand).

In the UK, both organic and conventional farmers shared information about 9755 earthworms living in 2000 has of their land (Fig. 1c, d). Highest global abundances were recorded in Scotland peaking at 1 085 earthworms m^2 (an average of 684 ± 100 earthworms m^2), while there was a UK average of 262 earthworms m^2. The geographical range was from Truro area in the South West of England up to Northern Scotland in the Orkneys, thus covering all UK regional climate areas. All UK regions included fields with the presence of adult epigeic, endogeic and anecic earthworms. However, in 5% fields earthworms were not widespread (only present in three or fewer of the five soil pits), 28% fields no epigeic earthworms were recorded, 8% fields no endogeic worms were recorded and 18% no anecic earthworms (or signs of them) recorded. There was a significant difference ($p < 0.001$) between the numbers of earthworms in fields and the presence of all three ecological groups. Fields containing all three ecological groups had an average of 310 ± 20 earthworms m^2. In contrast, fields where earthworms were not widespread or an ecological group was not recorded contained an average 188 ± 18 earthworms m^2 (Fig. 1e).

Overall, 61% UK fields recorded the presence of adult epigeic, endogeic and anecic earthworms (Fig. 2a), which was linked to the tillage system (Fig. 2b). Fields managed under no-till may have a problem from the ineffective decomposition of straw because 19% of these fields did not detect litter-feeding epigeic earthworms (Fig. 2c). For example, a no-till co-producer reported a crop failure caused by surface straw inhibiting seedling emergence, their results showed depleted litter-feeding earthworms and informed a change in management tactics (Fig. 2c). Many tilled fields did not detect litter-feeding earthworms (epigeic or anecic earthworms) (Fig. 2d, e). It should be noted an 18% absence rate of anecic earthworms was recorded, despite the method including indicators of their activities (middens and/or distinctive burrows) which are semi-permanent features on the soil surface and in the soil profile.

Earthworms were ubiquitous on the grass and arable fields at Rothamsted. The woodland areas had fewest earthworms (0–60 earthworms m^2), grass fields had 90–335 earthworms m^2, and arable fields had 15–485 earthworms m^2. There was good agreement between the Rothamsted survey and the Midlands regional data generated by the farmers. Fields that recorded the presence of the three ecological groups had average earthworm populations of 223 ± 39 (Rothamsted) and 225 ± 27 (Midlands) earthworms m^2. In contrast, fields without widespread earthworms and/or missing at least one ecological group had average earthworm populations of 100 ± 15 (Rothamsted) and 97 ± 39 (Midlands), respectively. Overall, 37% Rothamsted sampling locations recorded

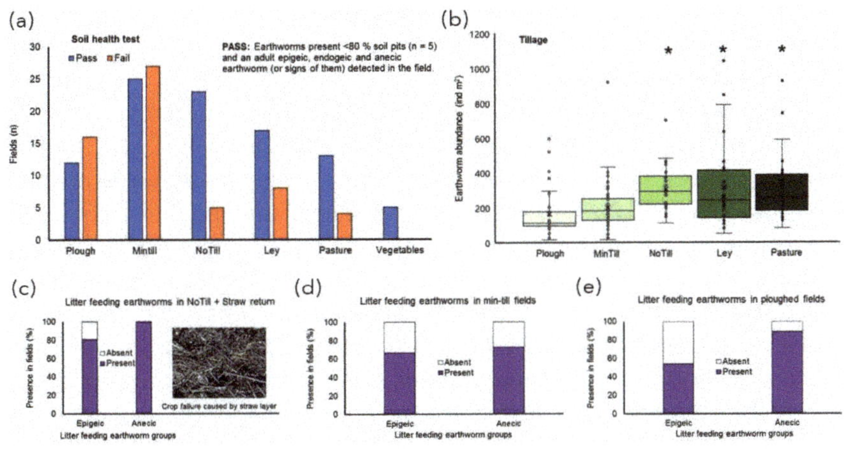

Figure 2 UK farmland earthworm populations and tillage. (a) The soil health test results. (b) Tillage influences on earthworm abundance, * indicates significantly different. (c–d): litter feeding earthworms in: (c) No-tillage with straw (photograph from a co-producer showing the straw problem); (d) MinTill; € Ploughed fields.

the presence of epigeic, endogeic and anecic earthworms. In terms of arable fields, Hoosfield spring barley (long-term experiment of continuous spring barley, with the soil ploughed and fallowed over winter every year) had 15 ± 10 earthworms m^2, and no epigeic or anecic earthworms (or signs of them) were detected. In contrast, West Barnfield (a minimum tillage rotation experiment, 400 m away from Hoosfield) had 465 ± 50 earthworms m^2 and every soil pit contained epigeic, endogeic and anecic earthworms.

The scientific consensus resulted in a 40% engagement rate (of 30 people). These experts could anonymously choose: 'The survey should be discontinued due to major flaws', but none selected this option. The expert consensus was that the results were likely to be caused by management practices (81%) and that they would recommend this activity to farmers (83%) and policymakers (58%); 82% thought the activity was likely to help to realise soil health in practice. However, 42% perceived the earthworm survey as 'not at all' or 'slightly' useful to scientists.

Social scientists have suggested that Twitter is used by farmers to inform sustainable soils management (Mills et al., 2019), in agreement with the findings reported here, that is, Twitter was used to advertise #WorldWormWeek resulting in 2000 ha surveyed by farmers. Data scientists have identified that the Twitter network topography shifts from a centralised network that is efficient at spreading agricultural information to a decentralised diffusion system supporting learning and community cohesion (Bastos et al., 2018). This means that the structure of agricultural communities on Twitter is aligned to traditional extension services, with technical experts as 'centralised' hubs (here established as @wormscience), which would explain the concentration of 3000 followers for this project. The number of followers suggests that earthworms are significant to soil management decisions in the UK, which is consistent with observations in Europe (Lahmar, 2010).

Social scientists have indicated that farmers use a range of information sources (Cristofari et al., 2017), scientific principles (Cristofari et al., 2018) and are open to co-creating knowledge with scientists to develop their farming practices (Schneider et al., 2012). The interaction and participation by UK farmers in this laborious activity over 18 months suggest that the co-creation of knowledge is viable. The scientific consensus indicated that the technical quality of the co-developed method was sufficient but there are tensions in the valuations of co-produced knowledge, that is, this initiative was 'not at all' or 'slightly' useful to scientists. In contrast, 87% participants were interested in comparing their data to the scientific field trials, for example, the long-term experiments at Rothamsted Farm (Stroud, 2019). This open survey included participants from around the world (Figs 1 and 2) and a broad range of soil management and pesticide regimes. Only earthworm numbers were recorded for the world sites because earthworm ecological group identification is beyond the scope of this study – some species behave as endogeic earthworms in the UK, but behave as anecic earthworms in New Zealand (Edwards and Bohlen, 1996).

The data co-produced by UK farmers revealed unexplained absences of surface-dwelling (epigeic) and surface-feeding (anecic) plant-residue feeding earthworms in 28% and 18% fields, respectively (Fig. 2c, d). Similarly, epigeic and/or anecic earthworms were not recorded for the majority Rothamsted farm sites surveyed by the author (Fig. 3). Potential causes include insufficient sampling of a patchy population, intensive tillage (Briones and Schmidt, 2017) and annelid diseases (Stroud, 2021). It is unknown whether patchy earthworm populations influence field-scale agricultural production. The legacy of this co-designed survey is the development of farmers' ecological skills and awareness of these unexplained absences. There is now a routine process for monitoring and recording their soil biological observations, building capacity to solve soil issues, as previously suggested by others (Bruyn and Abbey, 2003). Tillage (ploughing) has been linked with local extinctions of anecic earthworms (Kladivko et al., 1997) and air pollution and herbicides increase their susceptibility to castrating parasitic infections (Stroud, 2021). Local extinctions of plant-residue feeding earthworms influence the abundance of bird and animal species that depend on this food resource, the trophic transfer of pollutants and parasitic infections, plus soil erosion, crop emergence and crop pest and disease incidences.

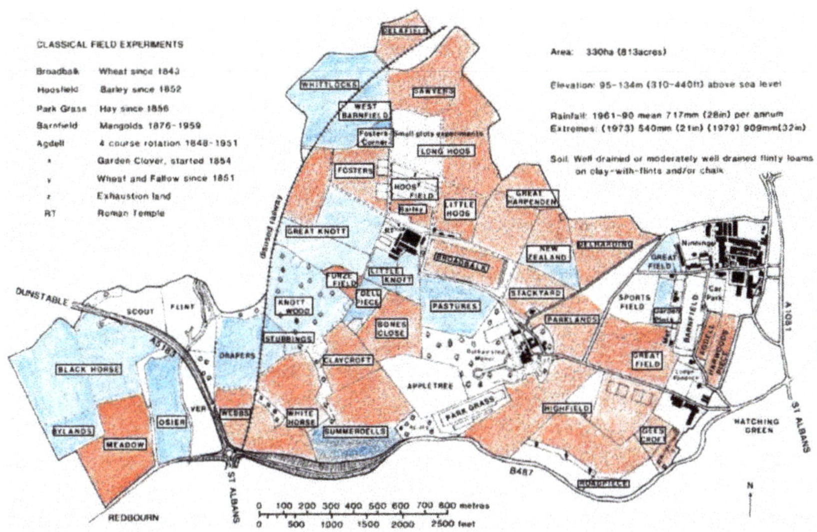

Figure 3 Whole-farm analysis at Rothamsted Research for #WorldWormWeek. Key: Blue-coloured fields were sites where all three ecological groups of earthworms were recorded, and orange-coloured fields were sites where there was at least one ecological group of earthworms absent.

6 Conclusion

No systematic earthworm observation network exists despite the trophic significance of earthworms. The Twitter-based initiative demonstrates that it is possible to co-ordinate biological sampling and farmers have proven capabilities in developing skills in earthworm identification to ecological group level. The widespread absences of epigeic and anecic earthworms are unexplained.

7 Where to look for further information

- Soil health (Baveye et al., 2016).
- Psychology of farmer behaviour (Moran et al., 2013).
- What is a 'good farmer'? Social sciences research (Burton, 2004).

8 References

Archer, M., Barraclough, D., Bone, J., Eggleton, P., Head, M., Jones, D. T. and Voulvoulis, N. 2009. *The OPAL Soil and Earthworm Survey Report*. Imperial College Press, Imperial.

Baldivieso-Freitas, P., Blanco-Moreno, J. M., Gutiérrez-López, M., Peigné, J., Pérez-Ferrer, A., Trigo-Aza, D. and Sans, F. X. 2018. Earthworm abundance response to conservation agriculture practices in organic arable farming under a Mediterranean climate. *Pedobiologia* 66, 58–64. http://dx.doi.org/10.1016/j.pedobi.2017.10.002.

Bastos, M., Piccardi, C., Levy, M., Mcroberts, N. and Lubell, M. 2018. Core-periphery or decentralized? Topological shifts of specialized information on Twitter. *Social Networks* 52, 282–293.

Baveye, P. C., Baveye, J. and Gowdy, J. 2016. Soil "Ecosystem" services and natural capital: critical appraisal of research on uncertain ground. *Frontiers in Environmental Science* 4, 41.

Bertrand, M., Barot, S., Blouin, M., Whalen, J., De Oliveira, T. and Roger-Estrade, J. 2015. Earthworm services for cropping systems: a review. *Agronomy for Sustainable Development* 35(2), 553–567.

BIOBIO 2012. Available at: http://www.biobio-indicator.org/factsheets/earthworms.pdf [Online].

Blouin, M., Hodson, M. E., Delgado, E. A., Baker, G., Brussaard, L., Butt, K. R., Dai, J., Dendooven, L., Peres, G., Tondoh, J. E., Cluzeau, D. and Brun, J.-J. 2013. A review of earthworm impact on soil function and ecosystem services. *European Journal of Soil Science* 64(2), 161–182.

Bouché, M. B. 1977. Stratégies lombriciennes. In: Lohm, U. and Persson, T. (Eds), *Soil Organisms as Components of Ecosystems*. Ecological Bulletin, Stockholm.

Briones, M. J. I. and Schmidt, O. 2017. Conventional tillage decreases the abundance and biomass of earthworms and alters their community structure in a global meta-analysis. *Global Change Biology* 23(10), 4396–4419.

Brown, G. G., Benito, N. P., Pasini, A., Sautter, K. D., Guimarães, de F. M. and Torres, E. 2003. No-tillage greatly increases earthworm populations in Paraná state, Brazil:

the 7th international symposium on earthworm ecology - Cardiff - Wales - 2002. *Pedobiologia*, 47, 764–771.

Brown, G. G., Da Silva, E., Thomazini, M., Niva, C., Decaëns, T., Cunha, L., Nadolny, H., Demetrio, W., Santos, A., Ferreira, T., Maia, L., Conrado, A., Segalla, R., Ferreira, A., Pasini, A., Bartz, M., Sautter, K., James, S., Baretta, D. and Lavelle, P. 2018. The role of soil fauna in soil health and delivery of ecosystem services. In: Reicosky, D. (Ed.), *Managing Soil Health for Sustainable Agriculture Volume 1: Fundamentals*. Burleigh Dodds Science Publishing, Cambridge, UK.

Brown, G. G., Edwards, C. A. and Brussaard, L. 2004. How earthworms affect plant growth: burrowing into the mechanisms. In: Edwards, C. (Ed.), *Earthworm Ecology*. CRC Press, Boca Raton, FL.

Bruyn, L. A. Ld and Abbey, J. A. 2003. Characterisation of farmers' soil sense and the implications for on-farm monitoring of soil health. *Australian Journal of Experimental Agriculture* 43(3), 285–305.

Burton, R. J. F. 2004. Seeing through the 'Good Farmer's' eyes: towards developing an understanding of the social symbolic value of 'Productivist' behaviour. *Sociologia Ruralis* 44(2), 195–215.

Castellanos-Navarette, A., Rodríguez-Aragonés, C., De Goede, R. G. M., Kooistra, M. J., Sayre, K. D., Brussaard, L. and Pulleman, M. M. 2012. Earthworm activity and soil structure changes under conservation agriculture in central Mexico. *Soil and Tillage Research* 123, 61–70.

Cristofari, H., Girard, N. and Magda, D. 2017. Supporting transition toward conservation agriculture: a framework to analyze the learning processes of farmers. *Hungarian Geographical Bulletin* 66(1), 65–76.

Cristofari, H., Girard, N. and Magda, D. 2018. How agroecological farmers develop their own practices: a framework to describe their learning processes. *Agroecology and Sustainable Food Systems* 42(7), 777–795.

Darwin, C. 1882. *The Formation of Vegetable Mould: Through the Action of Worms, with Observations on their Habits*. John Murray, London.

Edwards, C. A. and Bohlen, P. J. 1996. *Biology and Ecology of Earthworms*. Chapman and Hall Ltd., London, New York.

Goncharov, A. A., Glebova, A. A. and Tiunov, A. V. 2020. Trophic interactions between Fusarium species and soil fauna: a meta-analysis of experimental studies. *Applied Soil Ecology* 145, 103302.

Hickman, Z. A. and Reid, B. J. 2008. Earthworm assisted bioremediation of organic contaminants. *Environment International* 34(7), 1072–1081.

Hofgaard, I. S., Seehusen, T., Aamot, H. U., Riley, H., Razzaghian, J., Le, V. H., Hjelkrem, A. G., Dill-Macky, R. and Brodal, G. 2016. Inoculum potential of Fusarium spp. relates to tillage and straw management in Norwegian fields of spring oats. *Frontiers in Microbiology* 7, 556–556.

King, R. A., Tibble, A. L. and Symondson, W. O. 2008. Opening a can of worms: unprecedented sympatric cryptic diversity within British lumbricid earthworms. *Molecular Ecology* 17(21), 4684–4698.

Kladivko, E. J., Akhouri, N. M. and Weesies, G. 1997. Earthworm populations and species distributions under no-till and conventional tillage in Indiana and Illinois. *Soil Biology and Biochemistry* 29(3–4), 613–615.

Kuperman, R. G., Checkai, R. T., Garcia, M. V. B., Römbke, J., Stephenson, G. L. and Sousa, J. P. 2009. State of the science and the way forward for the ecotoxicological assessment of contaminated land. *Pesquisa Agropecuaria Brasileira* 44(8), 811-824.

Lahmar, R. 2010. Adoption of conservation agriculture in Europe: lessons of the KASSA project. *Land Use Policy* 27(1), 4-10.

Lazcano, C. and Dominguez, J. 2011. The use of vermicompost in sustainable agriculture. In: Miransari, M. (Ed.), *Soil Nutrients*. Nova Science Publishers, New York.

Lee, K. E. 1985. *Earthworms: Their Ecology and Relationships with Soils and Land Use*. Academic Press, Sydney and Orlando, FL.

López, A. and Martinson, S. A. 2017. Chapter 9. Respiratory system, mediastinum, and pleurae1. In: Zachary, J. F. (Ed.), *Pathologic Basis of Veterinary Disease* (6th edn.). Mosby, St Louis, MO.

Macdonald, D. W. 1983. Predation on earthworms by terrestrial vertebrates. In: Satchell, J. E. (Ed.), *Earthworm Ecology: From Darwin to Vermiculture*. Springer Netherlands, Dordrecht.

Mills, J., Reed, M., Skaalsveen, K. and Ingram, J. 2019. The use of Twitter for knowledge exchange on sustainable soil management. *Soil Use and Management* 35(1), 195-203.

Moran, D., Lucas, A. and Barnes, A. 2013. Mitigation win-win. *Nature Climate Change* 3(7), 611-613.

Nxele, T. C., Lamani, S., Measey, G. J., Armstrong, A. J., Plisko, J. D., Willows-Munro, S., Janion-Scheepers, C. and Wilson, J. R. U. 2015. Studying earthworms (Annelida: Oligochaeta) in South Africa. *African Invertebrates* 56(3), 779-806.

Pauli, N., Abbott, L. K., Negrete-Yankelevich, S. and Andrés, P. 2016. Farmers knowledge and use of soil fauna in agriculture: a worldwide review. *Ecology and Society* 21(3) 19.

Plaas, E., Meyer-Wolfarth, F., Banse, M., Bengtsson, J., Bergmann, H., Faber, J., Potthoff, M., Runge, T., Schrader, S. and Taylor, A. 2019. Towards valuation of biodiversity in agricultural soils: a case for earthworms. *Ecological Economics* 159, 291-300.

Sanchez-Hernandez, J. C., Aira, M. and Domínguez, J. 2014. Extracellular pesticide detoxification in the gastrointestinal tract of earthworms. *Soil Biology and Biochemistry* 79, 1-4.

Scheu, S. 2003. Effects of earthworms on plant growth: patterns and perspectives. *Pedobiologia* 47(5-6), 846-856.

Schneider, F., Steiger, D., Ledermann, T., Fry, P. and Rist, S. 2012. No-tillage farming: co-creation of innovation through network building. *Land Degradation and Development* 23(3), 242-255.

Stephenson, J. 1930. *The Oligochaeta*. Clarendon Press, Oxford.

Stroud, J. L. 2019. Soil health pilot study in England: outcomes from an on-farm earthworm survey. *PLoS ONE* 14(2), e0203909.

Stroud, J. L. 2021. Annelid diseases. In: Rowley, A. F., Coates, C. J. and Whitten, M. M. A. (Eds), *Invertebrate Pathology*. Oxford University Press, Oxford.

Stroud, J. L., Irons, D. E., Watts, C. W., White, R. P., Mcgrath, S. P. and Whitmore, A. P. 2016. Population collapse of *Lumbricus terrestris* in conventional arable cultivations and response to straw applications. *Applied Soil Ecology* 108, 72-75.

Targetti, S., Herzog, F., Geijzendorffer, I. R., Wolfrum, S., Arndorfer, M., Balàzs, K., Choisis, J. P., Dennis, P., Eiter, S., Fjellstad, W., Friedel, J. K., Jeanneret, P., Jongman, R. H. G., Kainz, M., Luescher, G., Moreno, G., Zanetti, T., Sarthou, J. P., Stoyanova, S., Wiley,

D., Paoletti, M. G. and Viaggi, D. 2014. Estimating the cost of different strategies for measuring farmland biodiversity: evidence from a Europe-wide field evaluation. *Ecological Indicators* 45, 434–443.

Van Groenigen, J. W., Lubbers, I. M., Vos, H. M. J., Brown, G. G., De Deyn, G. B. and Van Groenigen, K. J. 2014. Earthworms increase plant production: a meta-analysis. *Scientific Reports* 4, 6365.

Zúñiga, M. C., Feijoo, M. A., Quintero, H., Aldana, N. J. and Carvajal, A. F. 2013. Farmers' perceptions of earthworms and their role in soil. *Applied Soil Ecology* 69, 61-68.

Chapter 3

Characterisation of fungal communities and functions in agricultural soils

Andy F. S. Taylor, The James Hutton Institute and University of Aberdeen, UK; and Thomas Freitag, Lucinda J. Robinson and Duncan White, The James Hutton Institute, UK

1 Introduction

Fungi are a large, taxonomically and ecologically heterogenous group of organisms that are grouped into the Kingdom – Fungi (Stajich et al., 2009). There are an estimated 2–6 million species, of which only 2–6% have been formally described (Taylor et al., 2014; Hawksworth and Lüking, 2017). Morphologically they vary from microscopic, single-celled organisms (yeasts) up to vast complex structures, which are considered to be the largest individual organisms on the planet (Smith et al., 1992). Fungi, along with the plethora of other organisms inhabiting our soils, are essential components of all terrestrial ecosystems, including agricultural systems (Wagg et al., 2014). Functionally they range from ecosystem recyclers, as the main saprotrophic decomposers of recalcitrant organic materials, to those forming a great diversity of symbiotic associations with plants and animals. The latter form a gradient from devastating crop pathogens to essential mutualistic root symbionts. Despite a broad appreciation of their potential importance, it is only relatively recently that there has been a realisation of the possible benefits of supporting fungal communities in agricultural soils.

A number of recent articles have demonstrated different aspects of fungi in agricultural systems and potential benefits derived from their presence

http://dx.doi.org/10.19103/AS.2020.0079.09

(e.g. Ellouze et al., 2014; Gilbert and Johnson, 2017). Frąc et al. (2018), in particular, presented a useful broad overview of the roles of fungi in soil health in agricultural and horticultural ecosystems and support the view that the complexity of the research field merits a multidisciplinary approach. One of the main drivers for the recent interest in utilising fungi has been the development of molecular technologies which allow us to investigate fungal communities in detail. Prior to this, it has been effectively impossible to gain answers to even simple questions such as – how many fungi (species) are in this soil? Or perhaps more importantly – What are their functions? We have made massive steps forward in answering these questions, but are still a long way from having an in-depth understanding of what constitutes a healthy fungal community and how management practices impact their sustainability. If we are to make significant progress in harnessing fungal communities for improving the sustainability of our production systems, then we have to greatly improve our ability to accurately monitor species composition, abundance and interactions, and to vastly increase our understanding of functional traits. In this chapter, we will present some recent advances in the study of these ubiquitous and vital components of all terrestrial ecosystems.

The various roles of fungi in agricultural systems will only be discussed briefly to illustrate particular points, as these have been well covered by Frąc et al. (2018). Instead, the main focus will be on some technical advances that have enabled high throughput generation of sequence data from fungal communities. These advances are discussed in relation to aspects of fungal biology which combine to make addressing such a simple question as 'How diverse are my soil fungal communities?' so problematic. An example of an approach to a national scale survey of soil fungi is given to demonstrate the large-scale characterisation of fungi in soils. Some considerations are then given to knowledge gaps and how we may gain greater insights into fungal communities in order to increase our understanding of potential benefits.

2 Challenges in characterising fungal communities

2.1 Characterising species-rich communities

Measures of community diversity of any organism group comprise two components:

- species richness (the number of species present within the community); and
- the relative abundance of the species (the number of individuals of each species).

For communities of organisms forming discrete bodies (e.g. earthworms), the determination of both these parameters can be – given the appropriate taxonomic skills – relatively straightforward. Simple counts of the number of individual earthworms of each species present within a given space would give both species richness and relative abundance (Bertrand et al., 2015). In addition, drying and weighing the earthworms would also give a good measure of relative importance as biomass. Finally, since the ecological traits of many worm species are fairly well known, at least in temperate zones, a community functional trait profile could be generated using the relative abundance or biomass of each species present (Ernst and Emmerling, 2009). In short, a high resolution taxonomic and functional picture of an earthworm community could be obtained relatively simply and quickly (Pelosi et al., 2014). With fungi, it is currently impossible, practically, to achieve this kind of resolution. To date, there is not a single field-based study of a fungal community that has included measures of both species richness and the numbers of individuals of each species. The reasons for this relate to fundamental aspects of fungal biology: aspects that have developed over more than a billion years of evolutionary history (Wang et al., 1999; Loron et al., 2019).

2.2 Community structure, spatial distribution and sampling issues

Prior to the 1990s, virtually all our knowledge of fungal communities in agricultural systems came from studies that involved isolating and culturing fungi onto a range of artificial media (see Kjøller and Struwe, 1982). The fungal colonies which produced distinctive features (e.g. spores, specialised vegetative structures or growth on particular media) could then be identified. Unfortunately, this approach is very much biased towards those fungi that actually grow in culture, in particular, those that grow fast and readily in culture. These studies certainly provided evidence for the presence of particular groups of fungi (particularly the opportunistic moulds) but totally failed to detect the vast majority of other fungi which either do not grow in culture or those which would have been rapidly overgrown. The measures of abundance, which were based on the number of individual colonies of each taxon, were also strongly biased to species reproducing via millions of asexual spores and those which survived fragmentation of the preparation process.

It was not until the development of molecular approaches that allow the extraction of DNA from almost any environmental substrate and the selective amplification and sequencing of targeted organism groups, that we had the tools to start to tackle the complexities of fungal communities in soil. Using these approaches, we can simultaneously generate massive amounts of data from hundreds of samples, where each sample contains sequences from a species-rich community. In principle, the direct extraction of DNA from soil could recover all fungal species present, but the unique ecology of fungi means that there are many caveats associated with using these molecular approaches that need to be considered in order to obtain a reasonably comprehensive picture of the species present.

3 Molecular characterisation of fungal communities

The application of molecular approaches for assessing biological diversity has been termed as barcoding and is based on the use of short sections of DNA from a standardised region of the genome to identify species (Hebert et al., 2003; Hebert and Gregory, 2005). It is an approach that has been widely adopted as a taxonomic and conservation tool for assessing diversity. Where this involves generating barcodes for entire communities simultaneously, the approach is called metabarcoding. DNA metabarcoding is based upon the simple principles of generating a set or library of barcodes from the target community and then grouping or clustering these into groups of similar sequences, which are then 'identified' against a reference library of barcodes from known species. The field-generated barcodes are clustered together based on the degree of similarity and a single cluster is referred to as an operational taxonomic unit (OTU). The OTUs are compared with a reference set of barcode sequences of known identities and these names are then assigned to the field community OTUs based upon the similarity between the OTU and the reference barcode. An essential part of DNA barcoding is therefore the existence of a comprehensive sequence reference library that allows the correct assignment of unknown sequences to already known reference species.

Metabarcoding of species-rich communities, including fungi, has only really been possible due to the development of next-generation sequencing (NGS) platforms that can simultaneously sequence samples of mixed DNA fragments (Shendure and Ji, 2008). Prior to the development of these techniques, which were developed for human genome sequencing and adapted for ecological applications, community analyses were based on only a few dozen to hundreds of sequences generated either directly from fungal structures or from cloning of mixed samples and then subsequent Sanger sequencing of individual sequences (e.g. Helgason et al., 1998). The low numbers of sequences meant that it was possible to manually compare each sequence produced with reference sequences in a database in order to identify species. With the advent of NGS data sets containing millions of reads, bioinformatic processing using designated processing software pipelines became essential (pipelines are discussed further in the case study below).

International databases such as the National Center for Biotechnology Information (NCBI) contain millions of submitted sequences and these can act as reference sources for comparing environmental sequences. However, it became clear in the early 2000s that around 20% of sequences in some databases were taxonomically unreliable (Bridge et al., 2003; Nilsson et al., 2006). In an attempt to provide a more accurate reference set, the UNITE fungal sequence database was established (Kõljalg et al., 2005) and this has grown to become the main database dedicated to fungal sequence data (https://unite.ut.ee/index.php). The UNITE database is discussed further in the future research section below.

The official barcoding marker for fungi is the nuclear ribosomal internal transcribed spacer (ITS) region (Schoch et al., 2012), although other markers are preferred for certain groups (for arbuscular mycorrhizal (AM) fungi see https://maarjam.botany.ut.ee/?action). The ITS region is a multicopy region that shows low variation within species but has sufficient interspecific variation to enable identification across a very broad range of fungi. The slight variation within species is accounted for by clustering generated reads into OTUs with the level of clustering influenced on the group of fungi being analysed. However, the most commonly used cut-off is 2.5% dissimilarity. This means that all the sequences within a single OTU cluster will share 97.5% of the sequence.

Earlier studies of fungal communities using an NGS approach used 454 pyrosequencing (e.g. Sugiyama et al., 2010) but this has been superseded by other platforms capable of generating huge numbers of sequences or reads. One commonly used approach is the MySeq platform from Illumina, which can generate 25–50 million reads in a single run (https://emea.illumina.com/science/technology/next-generation-sequencing.html; Ravi et al., 2018). The huge number of sequences or reads can be spread across hundreds of samples, generating community profiles containing tens of thousands of reads recovered from each sample. This level of coverage is required for species-rich communities. The downside of huge numbers of reads is that they are short – each read is only a few hundred base pairs in length. This means that the whole of the ITS cannot be sequenced using this approach and studies are based on ITS1 or more commonly ITS2 (see Ihrmark et al., 2012 for discussion about primers for ITS 2).

There are other platforms, such as Pacbio (https://www.pacb.com/) and Oxford Nanopore Technologies (https://nanoporetech.com/), that can generate much longer reads, many kilobases in length, but at present, it is difficult to run multiple samples and the number of reads is low. The very long reads mean that these approaches are particularly useful for reconstructing genomes. Oxford Nanopore technologies produce the MinION, which is a self-contained sequencing device about the size of a memory stick that plugs into a laptop. They have already been utilised in generating whole genomes for fungi (Dutreux et al., 2018). The huge advantages of this technology are that it is rapid, sample preparation is relatively easy, there are no amplification steps involved and it can produce very long reads (5 kb–200 kb) in real time.

Further development of this technology could see a rapid, real-time generation of sequence data in the field.

NGS approaches have without doubt given us unprecedented power to investigate species-rich, cryptic communities, such as soil fungi. However, the vast increase in read numbers does not necessarily circumvent some fundamental aspects of fungal biology which need to be borne in mind when characterising fungi in soil – or any other opaque medium.

The fungi are extraordinarily diverse, with an estimated 2–6 million species (Taylor et al., 2014; Hawksworth and Lüking, 2017), although more recent estimates suggest the number may be close to 12 million (Wu et al., 2019). Given that a tiny fraction of these has been formally described, even if we distinguish between all the species in a community, a considerable proportion of them would be without a formal scientific name. This applies equally to tropical systems as it does to systems that have been studied to a much greater degree (Sommermann, et al., 2018). Many community studies using NGS approaches often have many OTUs simply identified as 'Unknown fungus' (e.g. Hardoim et al., 2015; Tedersoo et al., 2014). This means that the DNA sequence clearly belongs within the Kingdom Fungi, but we are ignorant of where it is taxonomically placed within the whole Kingdom. Some of these will represent new evolutionary lineages within the fungal kingdom (see Tedersoo et al., 2014) and this highlights one of the main misguided preconceptions about using DNA approaches to characterise communities (irrespective of which organism group) – that we can use DNA to identify everything. The paucity of reference sequences is discussed further in future research below.

Fungal communities are typically structured in a similar way to any other species-rich communities, with a few common species and many rare species (Taylor, 2002). The common taxa make up the bulk of the fungal structures and are often widespread in the system under study, while the often much larger number of rarer species can be very patchily distributed. This means that sampling strategies for characterising fungal communities must take into account both the very uneven relative abundance of species and the spatial ecology of fungi. In natural and seminatural systems, the mycelia of fungal species can show a high degree of spatial segregation, being stratified within the soil profile (Taylor et al., 2014) or are only associated with the roots of particular plant species (Öpik et al., 2009). This spatial segregation of species creates a considerable sampling problem when trying to obtain a comprehensive picture of a fungal community, as many species may exist only in small discrete areas and/or in low abundance and will only be recovered with intense sampling. These spatial issues may be less of a problem in agricultural soils which are regularly mixed and homogenised with repeated tillage. However, homogenisation may be restricted to the upper soil layers as Moll et al. (2015, 2016) found that communities were vertically structured in agricultural soils with communities in the ploughed soil (0–10 cm), rooted soil (40–50 cm), root-free soil (60–70 cm), soil depth and in plant roots having different taxonomic structures. At present, we do not know either the horizontal or vertical spatial sampling scales or the volumes of sampled soil that would be required to capture a comprehensive representation of fungal communities in agricultural systems.

Accounting for the spatial structuring of species is compounded by the very small amounts of soil from which DNA is usually extracted - commonly in the region of 250 mg. In terms of soil volume, this may be less than 1 cm^3 of soil. Various methods have been proposed for making this sample (or multiple samples of the same quantity) as representative as possible (see Tedersoo et al., 2014). But it is often still a small finite amount relative to the total volume sampled - which in turn will be a very small representation of the entire area under study. The species included in the sample will be those which are most common and widespread and there will be a proportion of the rarer taxa. It is the latter pool that will vary the most across space and consequently between replicate samples. Studies should ideally include an initial assessment of spatial dispersal of species, but time and costs usually preclude this.

The basic unit for most fungi is the hypha (hyphae - plural) (Fig. 1). Hyphae are effective tubes, made up of cells, which extend apically into the growing substrate. Individual hyphae range from 3 µm to 10 µm in diameter but can be many centimetres in length and collectively they make up the mycelium of the fungus - the vegetative body. The production of mycelia in the soil can be prodigious (Leake et al., 2004), with hundreds of meters of hyphae per gram of soil. It is problematic to extract hyphae from most soils and obtaining hyphal measurements is very time-consuming. In addition, with very few exceptions,

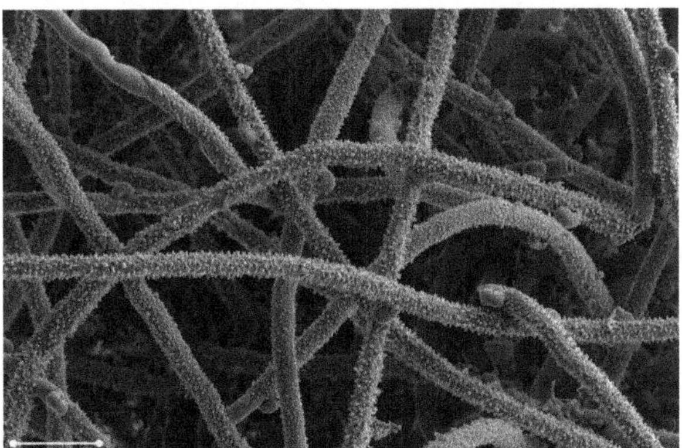

Figure 1 A scanning electron micrograph of the mycelium of a soil fungus composed of interwoven hyphae. The hyphal surface is covered in crystals. Scale bar = 10 µm. (Photo - Evelyne Delbos, The James Hutton Institute, UK).

it is not possible to differentiate between hyphae of different fungi. The small diameter of the hyphae means that they can explore very small pores in the soil for nutrients and organic carbon (C). This property of fungi is one reason why they are much more efficient at exploring and extracting nutrients from the soil than plant roots. In undisturbed substrates, mycelia can expand and form large connective systems that can move nutrients and organic C over considerable distances. The mycelia can also act as stores of nutrients, particularly N, and thus mycelia can reduce nutrient loss. These may be lost when a mycelium is damaged either naturally during a disruption by feeding animals or during soil cultivation.

Mycelial disruption also has an important consequence for community structure as the resulting fragments may survive and continue to grow. This means that individual, separate mycelia may be genetically identical – effectively clones of the original mycelium. This clearly creates problems when trying to estimate the numbers of genetically distinct individuals of a given species within a population. The study of population genetic structures of pathogens is a major factor in understanding and controlling disease (see Millgroom and Peever, 2003; Burdon and Laine, 2019). The lack of genetic structure in crop systems is a recognised driver enabling the development of low genetic diversity in pathogen populations (McDonald and Stukenbrock, 2016). There are as yet apparently no studies on how the low diversity in crop plants impacts on population structures of non-pathogenic soil fungi (see Section 5 for species richness in arable systems).

4 Proxies for fungal abundance

The difficulties associated with determining the abundance of individual fungi in soil has led to the development of a range of proxies for estimating fungal biomass. These may be direct measures of hyphal length (Shen et al., 2016), or indirect measures based on a fungal biomarker. Determining direct visual estimates of hyphal length can be very time-consuming to generate and it can be difficult to differentiate between live (active) mycelia and dead fragments. In addition, it is rarely possible to differentiate between the hyphae of individual species. The use of specific (or group) markers, such as monoclonal antibodies (Thornton, 2008), can be used to visualise the hyphae of individual species and can be extremely useful for tracking pathogen spread in soils but would have limited applicability for community-level studies.

The membrane-bound sterol ergosterol is one of the most commonly used fungal biomarkers and has been used extensively across many different systems and conditions (see Ruzicka et al., 2000). But it is not without its critics as the conversion factor used to convert ergosterol content to fungal biomass can be highly variable and context dependent (Chen et al., 2008; Nurika et al., 2018). In addition, the turnover time of ergosterol in soils may lag behind major perturbations to fungal communities suggesting that the turnover be not reflecting current active biomass (Mille-Lindblom et al., 2004; Högberg, 2006). Despite these concerns, ergosterol is used extensively as a proxy for total fungal biomass.

Ergosterol does not allow any resolution of individual groups of fungi. Greater resolution can be obtained using another set of biomarkers, phospholipid fatty acids (PLFA), which are produced as components of cellular membranes of bacteria and eukaryotes. These have been used extensively to generate combined microbial and fungal profiles (Willers et al., 2015; Quideau et al., 2016). Since they are rapidly degraded on cell death, PLFAs are often considered as biomarkers of an active or viable community (Quideau et al., 2016).

Some PLFAs can be used as a proxy for specific groups of fungi: the marker PLFA 16:1ω5c is used to estimate the arbuscular mycorrhizal fungal abundance (e.g. Helgason et al., 2010). However, Veum et al. (2019) demonstrated PLFA profiles are very sensitive to sample handling and preparation. In addition, the pioneers of utilising PLFA data for examining soil microbial communities (Frostegård et al., 1993), reiterated, in a revisit of their 1993 classic citation paper, that PLFA data must be interpreted with caution as applying diversity indices to PLFA profiles is inappropriate (Frostegård et al., 2011). A standard protocol for PLFA analysis is available (ISO/TS, 29843-1:2010).

Quantitative molecular techniques, such as quantitative PCR (qPCR), have been used to generate estimates of total fungal abundance (Liu et al., 2012), but is more commonly used for quantifying individual species (e.g. Gerin et al., 2018). This approach can give a very sensitive and potentially accurate estimate of the numbers of target DNA templates in a substrate. However, there is still the issue of converting this to a measure of biomass or abundance. Since the number of nuclei per given hyphal length is determined by cell length, which is highly variable between and within species (which is highly context dependent), estimates of biomass based on template number could be widely inaccurate. This is compounded where the template determined is a multi-copy gene or DNA region, which can vary in number between species (see comments on the ITS region used for barcoding fungi), making comparisons between species even more problematic.

In one of the very few studies to compare a range of proxies for estimating fungal biomass, Baldrian et al. (2013) used ergosterol contents, PLFA data and ITS copy number as estimates of fungal biomass in litter and soil. They found that the three measures gave quite different estimates of biomass and concluded that the three measures were not very comparable. Determining accurate estimates of the relative abundance of individual species within fungal communities is still very much an open area which will require much more effort to yield data that could be used with confidence. This field is more advanced in studies of microbiome (bacterial) communities (see Knight et al., 2018), including approaches linking qPCR with data produced from NGS sequencing approaches (Jian et al., 2020).

5 Case study: investigating soil fungal communities

The following is an example of a large-scale investigation into soil fungal communities, which illustrates one approach by which these complex communities can be investigated. We have included a simplified description as there are numerous detailed publications and protocols for each stage of this process. In this fast-moving field, the review by Lindahl et al. (2013) is an excellent 'Users guide' to high-throughput sequencing approaches to assess fungal communities. This review outlines many of the basic principles which go to ensure high-quality reliable data, and these principles apply to any study irrespective of the sequencing approach utilised. A more general guide to the sequence-based assessment of biodiversity is given by Creer et al. (2016).

Once we are in a position to characterise fungal communities in soil, we can start to ask questions relating to what constitutes a healthy community and what environmental factors and management practices promote healthy communities? Understanding the factors which determine community composition could ultimately enable us to manipulate communities to increase potential benefits. One way of approaching these issues is to investigate fungal communities across a range of habitats and determine the main drivers of community diversity. We have been taking this approach by studying the fungal communities in soils collected as part of the second National Soils Inventory of Scotland (NSIS). The survey methodology is described here: https://www.hutton.ac.uk/sites/default/files/files/soils/NSIS%202007-2009%20Field%20Protocols.pdf, and see Chapman et al. (2013) for more details. This survey was carried out between 2007 and 2010 and involved very detailed characterisation of soils across a 20-km grid across the whole of Scotland. During the survey, the habitat and soil at each sampling site were described in detail and soil samples were taken for chemical characterisation. The soil was also archived at −80°C and we have used this material to generate sequence data for fungal (and bacterial and archaeal) communities from 203 sites. The site-based metadata enables an investigation of relationships among the biotic and abiotic variables across the whole of the country. At each sampling site on the 20-km grid, central soil pits were excavated allowing samples to be retrieved from each soil horizon (Fig. 2). This would allow an investigation of the stratification of soil biology as fungi (and soil organisms in general) are known to be strongly spatially segregated

Figure 2 Characterisation of abiotic and biotic soil parameters in the Cairngorms, Scotland, with samples collected for chemical analysis and the determination of soil fungal and microbial communities. All tools are cleaned and disinfected between samples. (Photo - Dave Riach, The James Hutton Institute, UK).

in soils (Taylor et al., 2014). The data reported here are derived from only the upper horizon from the central pit.

The simplified flow chart in Fig. 3 shows the major steps used in generating sequence-based community profiles of fungi in the NSIS soil samples. Soil samples were kept cool and processed as quickly as possible. It is often forgotten that the average temperatures experienced by soil biota at northern and higher elevation sites may be similar to or even lower than those in standard cold rooms or fridges (3–4°C). Hence soil processes and activities will continue in soil samples in cold rooms, which can significantly impact on community structures even after only a few days (see Lindahl et al., 2010). In more recent studies, we have frozen soil samples at −20°C directly from the field and then cryo-milled them in liquid nitrogen before subsampling for storing at −80°C. This approach has worked well and has yielded good quality DNA and RNA. In the NSIS study, soils were sieved to remove larger stones and to create homogenised samples from which subsamples were placed in microcentrifuge tubes and stored at −80°C.

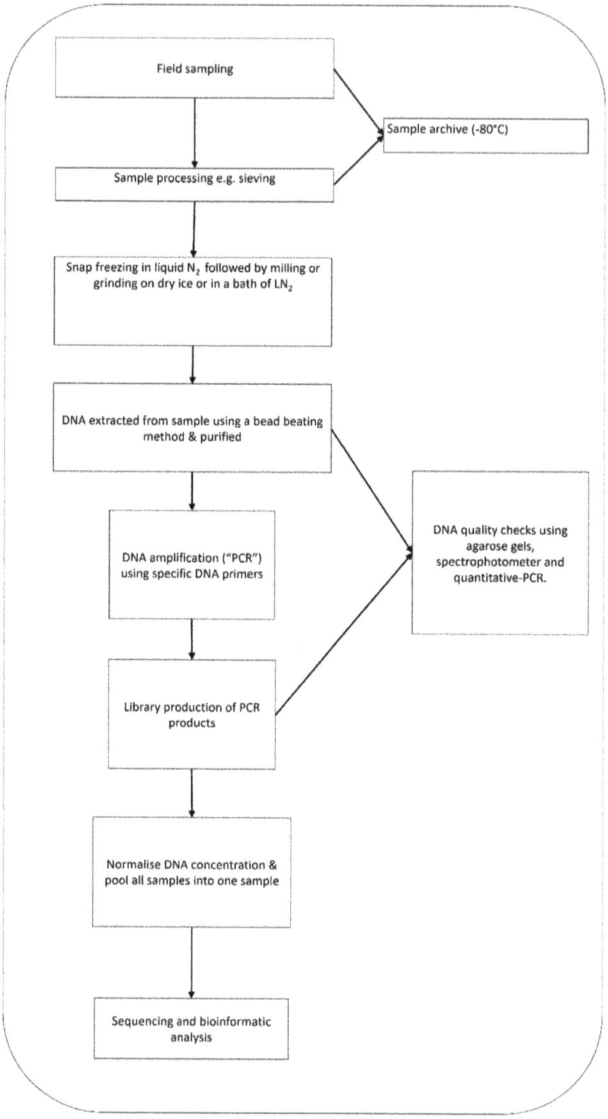

Figure 3 Simplified flow chart showing the major steps used in generating sequence-based community profiles of fungi from soil samples.

5.1 Extraction of DNA

The extraction of nucleic acids is a crucial step in the whole process and the approach may differ markedly between soil types (Fig. 2). Those with high organic contents (the majority of those in the NSIS samples) require particular care and post-extraction measures may need to be taken in order to remove contaminants that can inhibit the amplification step (Lindahl et al., 2013). Extraction involves a disruption step and then extraction of the released products. There are many different kits on the market for extraction involving maceration with beads ('bead beating') or hand grinding using a pestle and mortars followed by chemical lysis. The quantities of soil which are extracted are generally small, ca. 250 mg of soil, and it is common to extract multiple subsamples from a single larger sample in order to gain a greater representation of the communities present. For the NSIS samples, total nucleic acids were extracted from 250 ± 50 mg of cryo-milled soil in triplicate using a CTAB extraction buffer (Griffiths et al., 2000). Extraction is a time-consuming process and there are attempts to speed up the process via automation (Chiodi et al., 2020) and to extract from larger quantities of soil (7.5 g, see Dopheide et al., 2019). DNA extraction from soil, particularly those with high organic contents is still problematic, and further studies are required to optimise the process and to determine the number and volumes of samples required to give representative estimates of fungal communities in soil.

The quality and purity of the DNA is checked at various stages during the whole procedure and can be done using a combination of determining DNA/protein (260/280) ratios spectrophotometrically, analysis of agarose electrophoresis gels, quantification using fluorescent probes or using q-PCR. An examination of the DNA on an agarose gel at this stage should show clean high-molecular-weight bands (Fig. 4).

5.2 Preparation of amplicon libraries

Having obtained the necessary quantity and quality of DNA, the target barcode can then be amplified using the appropriate primer set (Fig. 2). For fungi, the ITS2 PCR primers fITS7 and ITS4 (Ihrmark et al., 2012) were used for the

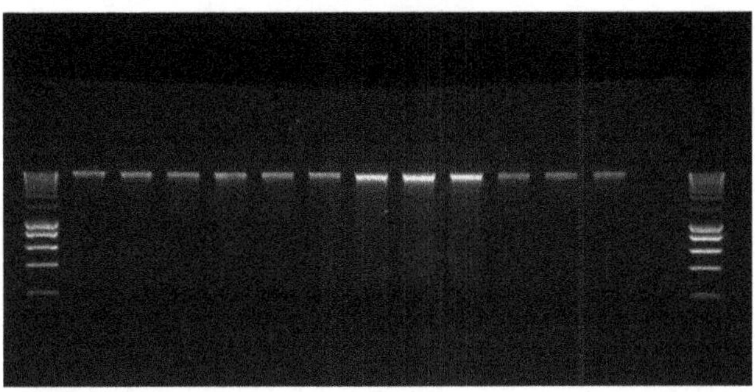

Figure 4 Agarose gel of total DNA extracted from soil. The clear single bands show high-molecular-weight DNA with little or no smearing due to the shearing of DNA into smaller fragments.

NSIS samples. One of the great advantages that NGS approaches have is that multiple samples can be sequenced in parallel simultaneously. This is achieved by multiplexing samples by creating libraries of barcodes (or amplicons) from each sample that are uniquely identified. This is done by attaching sample identifiers or short sequence tags to the amplicons. This can be done either directly at the stage of amplification by using uniquely tagged primers for each sample or by adding the tags to the amplicons post-amplification. These tags are sequenced along with the amplicon which allows post-sequencing identification of the sample origin of the barcode sequence.

5.3 Normalising and pooling libraries

The final steps prior to sequencing are the normalisation of the concentration of DNA in each library and then pooling all into one sample for delivery to the sequencer (Fig. 2). The normalisation is critical to ensure that the amplicons from each sample are equally represented in the final mix and hence have an equal probably of being sequenced. If one library contained a much higher concentration than the rest, then most of the returned sequences would be from this sample.

Controls were included throughout the whole procedure in order to check for contamination, for recovery and to assess the effectiveness of post-sequencing data processing.

5.4 Bioinformatic analysis

The bioinformatic analysis of NGS metabarcoding sequence data is a huge and rapidly evolving field, with new procedures and processing pipelines constantly being developed (Fig. 2). The basic aims are invariably that of quality control and assigning sequences to samples and hopefully named taxa. Each sequencing platform has an associated level of sequencing error and a large part of the post-sequencing bioinformatic analysis of the sequencing data is in quality control where sequences that fall below certain quality levels are removed. Unless the primers are truly specific to the target group, quality control also includes removing sequences from non-target organism groups (for fungi see Bengtsson-Palme et al., 2013). Two common packages for processing fungal high-throughput amplicon sequencing data have been the mothur pipeline (Schloss et al., 2009; https://mothur.org/wiki/mothur_v.1.30.0/) and the DADA2 packages (Callahan et al., 2016, version 1.6). In mothur, sequences are clustered at a user-defined degree of similarity (in fungi this is usually 97.5% similarity), and each cluster is then considered as an operational taxonomic unit (OTU). Whereas in DADA2, each unique sequence is considered as a separate entity - an assigned sequence variant (ASV). The DADA2 approach was designed for bacterial 16S data, where the detection of single nucleotide variants is important. For fungal ITS, where intragenomic and intraspecific variation in the ITS is common (Nilsson et al., 2008; Lindner et al., 2013), clustering is necessary to avoid over-inflation of taxon richness estimates. We, therefore, carry out a further clustering step of the ASVs produced by DADA2 to produce OTUs clustered at 97.5% similarity. The OTUs produced by either mothur or DADA2 are then classified taxonomically by comparing them with the latest reference set available from the UNITE database (Abarenkov et al., 2010; https://unite.ut.ee/repository.php). Finally, in order to gain functional information on the fungi identified in the communities, the OTU table can be parsed into the online FUNGuild database (Nguyen et al., 2016) which assigns each, where possible, to an ecologically meaningful group or functional guild.

The community data acquired using a metabarcoding approach such as that briefly outlined above effectively produces a matrix of the occurrence of OTUs in each sample in the study. The presence of each OTU is denoted by the number of sequences or reads retrieved for that OTU in each sample - seemingly typical community data. However, there are numerous publications on the interpretation of the read counts from these studies. It is common that the very rare OTUs are removed, particularly those OTUs represented by only single reads (singletons). Unless rare taxa are the focus of the study, even removing OTUS with < 10 reads is unlikely to have a major impact on the study outcomes. The use of read numbers as measures of abundance is a point of contention as there are many caveats in their interpretation (see Amend et al., 2010; Lindahl et al., 2013). It is questionable given these caveats how reliable read numbers reflect the actual abundance of the fungi in the sample, but it is a common practice. It should also be borne in mind that these measures of abundance ideally be a reflection of the numbers of individuals of the taxa present. A situation which, as already pointed out, is currently unattainable for fungal communities at present. Read numbers are at best an uncertain proxy of abundance as there are limited possibilities to calibrate these data against the independent measure of biomass. Using the data simply as presence/absence avoids these issues and can be used to generate community richness estimates as well for examining patterns of community structure in relation to multiple environmental variables in multivariate approaches.

5.5 A nationwide analysis of fungal communities

The sequencing of the NSIS samples produced ca. 36 million usable sequences or reads, from which we recognised an overall fungal community richness of 5670 OTUs. The average community richness in a sample was 144 OTUs. Somewhat surprisingly, the more managed systems appeared to support higher numbers of taxa (Fig. 5). This could be interpreted as the frequent disruption of the soil in these systems favouring species with smaller mycelia or indeed

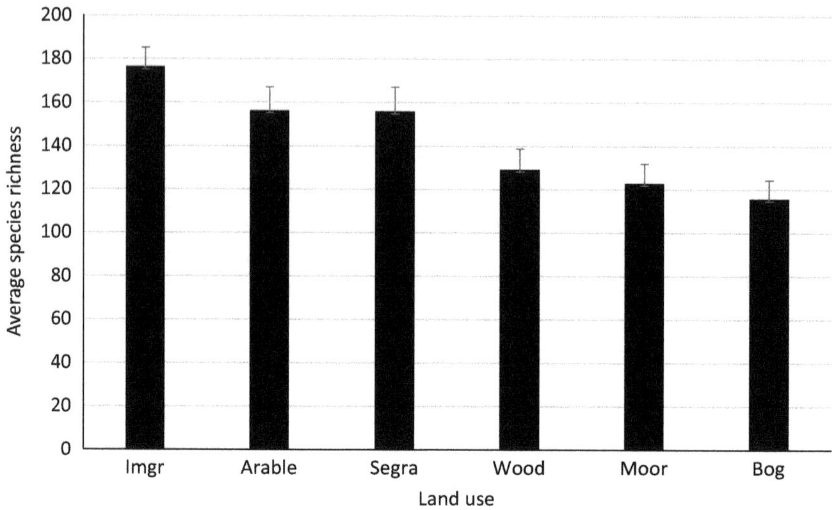

Figure 5 The average community richness of fungal communities in soils under different land use types sampled across Scotland. Imgr - improved grasslands; Arable; Segra - seminatural grasslands; Wood - woodlands; Moor - moorlands; bog - wet bogs.

unicellular fungi. This may be coupled with disruption preventing the build-up of large, dominant mycelia which could restrict the number of available niches. Without spatial and population data for individual species, we cannot address these questions. Large, well-developed mycelial are known to store nutrients, which would be a very useful trait to promote in our agricultural systems.

Analyses of the community data from the NSIS study has demonstrated that land use also impacts on the species present. This means that the soil fungal communities associated with the different land uses are different both quantitatively (species richness) and most likely qualitatively (functional traits), due to different species being present. However, we still know very little about the functions of most species in nature. The lack of trait data is a major hurdle to building community functional profiles, which would be a requirement for developing approaches for designing or engineering community structures for maximum benefit. In addition to the lack of trait, data are the problem of the poor resolution of identification of the OTUs. Around 20% of the OTUs in the study could only be assigned to the fungal kingdom: that is, the sequence is of fungal origin, but we cannot assign it to any known group within the kingdom. This is not unusual and is a reflection of the very small proportion of fungi that have actually received a formal description, including a sequence of the type specimen, and of the overall poor coverage of reference sequences in databases. This does not mean that species names could be assigned to the remaining 80% of the OTUs. In fact, only 14% of the OTUs could actually be identified to species level, with the rest identified to various levels of the taxonomic hierarchy (i.e. genus, family, etc.). Since species even within a single genus may differ markedly in their functional capabilities, having 86% of the OTUs identified above the species level means that there are severe restrictions of what can be inferred about community function. This is clearly a research field that requires considerable work and/or new approaches in generating community functional profiles.

6 Conclusion

In a recent publication, Chen et al. (2020) demonstrated links between plant richness and productivity with the composition of the fungal communities in soils. They suggested that the interactions between fungal guilds and the plants contributed to the overall productivity and diversity above ground. Of particular interest is the negative relationship they found between beneficial symbiotic fungi and plant pathogens. It is studies such as this which demonstrate the potential importance of fungal communities in agricultural systems. However, despite the huge advances in our abilities to detect and potentially identify soil fungal communities that have been outlined above, we are still a long way from being able to answer the question posed at the start of this review - what constitutes a healthy fungal community and how do management practices impact its sustainability?

We can now extract DNA from soil, amplify portions of fungal DNA and theoretically use these to distinguish between most of the fungal species present within the sample. Fungal communities are not hugely specious like bacterial communities and the massive numbers of reads which can be generated from single sequence runs of NGS platforms can easily accommodate processing high numbers of samples simultaneously for fungal communities. High-quality generated reads can be sorted or clustered, based on similarity, to generate OTUs (species), which in turn gives richness data for fungal communities. This can be done for any substrate from which we can extract DNA. This is a huge step forward in characterising fungal communities.

However, we will be lucky to identify more than a small proportion of the OTUs recovered to the level of species (in the case of the national survey of Scotland, this was only 14%) due to the lack of reference sequences and the tiny proportion of fungi which have actually been named and have associated sequence data. Linking OTUs to names means that any published information on that species, including distribution and trait information, can be used to build up a functional profile of the community. The finer the resolution there is in naming OTUs, the more accurate this information will be. Populating databases with more reference sequences to identify unknowns is an active area of work (see Torres-Cruz et al., 2017). Many of the OTUs which are only classified as far as the Kingdom is actually widespread and common across the globe, yet the only information available about them is that they are fungi. Nilsson et al. (2016)

published a list of these fungi as the 'Top 50 most wanted fungi' to highlight this issue and to generate interest in identifying these common and, presumably, important unknown species.

While great advances have been made, and are continually being made, in determining taxon richness within communities, we are still unable to answer the second part of the community diversity definition, namely, what is the relative abundance of each taxon present? The use of proxies of abundance, such as ergosterol and PLFAs, give some measure of fungal-related variables but the relationship with biomass can be context dependent and there may be little agreement among different measures. In addition, the level of resolution at which proxies relate to specific fungi is either negligible or only to broad groups (e.g. AM fungi).

The use of qPCR to quantify species can give very accurate estimates of DNA template numbers of individual fungal species. But here there is the issue of how template numbers relate to biomass, due to cell length being context dependent (each cell will contain a set number of templates) and also the actual number of copies of the template in each genome will not be known. The latter may vary by two orders of magnitude for the ITS region in fungi (Lofgren et al., 2018). In addition, qPCR approaches are usually limited to detecting and assessing single species.

Read numbers of individual OTUs generated during NGS studies of fungal communities are commonly used as measures of abundance but there are many caveats relating to their use, such as primer bias giving uneven amplification of particular taxa (Tedersoo et al., 2015). Issues such as this will lead to false representation of taxon template abundance and hence to false read numbers. Even if read numbers were an accurate reflection of the original template number, there is still the issue of how these can be related to abundance as discussed above in respect to qPCR.

Overlying all of these attempts at measuring abundance is the fact that we are not measuring community relative abundance in the traditional sense of numbers of individuals within a population. Instead, they are measures of a population without regard for the numbers of individuals within that population and the genetic diversity contained within these individuals. Until we can incorporate population structure into our assessments, we are missing a crucial component of what constitutes a soil fungal community, and how we can maximise the utility of these communities to support the sustainability and resilience of agricultural systems.

7 Future trends in research

Without massive investment, traditional taxonomic approaches cannot hope to produce descriptions and reference sequences for all or even a significant fraction of the undescribed fungal species. Recent estimates actually place the number of fungal species closer to 12 million (Wu et al., 2019). A number of alternative approaches have been proposed which would enable OTUs to be identified where possible, but which also allow unnamed OTUs to be compared across studies based upon fixed reference sequences. Kõljalg et al. (2013) proposed such a system that is based upon the concept that each cluster or OTU is considered as a 'species hypotheses' (SH). Each SH is given a reference sequence and a unique DOI tag which allows each SH to be tracked across time and space. Any new fungi which are identified, sequenced and given a validly published name can then be added to the relevant unnamed species hypotheses - if it already exists. This is the approach that is utilised in the already mentioned public-access UNITE database (https://unite.ut.ee/), which serves as the hub for sequence-based identification of fungi. The database provides regular updates of reference data sets that are used in the processing of NGS to assign taxonomic affiliations for generated OTUs. As of January 2020, the number of species hypotheses recognised at 97.5% similarity is 106 906 from a total of 1 799 133 ITS sequences included in the analysis. Even with these impressive numbers, there is still a long way to go before we have reference sequences for a significant proportion of the possible global fungal diversity.

The advent of NGS approaches has enabled immense amounts of sequence data to be generated from soil samples and the field of soil metagenomics where the entire bacterial genomic material in samples is sequenced has massively advanced soil microbiome research (see Nesme et al., 2016). This has enabled thousands of new genomes of unculturable bacteria to be generated, greatly expanding our knowledge of bacterial taxonomic and functional diversity (Delmont et al., 2015; Anantharaman et al., 2016). The greater complexity of fungal genomes has so far been restrictive in taking this approach to fungi but this is an area with huge potential to enable new insights into fungal communities.

Metagenomic studies will also be able to better predict community functional diversity as the number of annotated fungal genomes available continues to increase (https://mycocosm.jgi.doe.gov/programs/fungi/1000fungalgenomes.jsf). Such approaches have been available for bacterial communities for some years (Langille et al., 2013). Falconer et al. (2015) proposed a trait-based approach for modelling resilience of fungal communities and metagenomic data could contribute to such an approach.

The intergenic spacer (IGS) region of rDNA has been used successfully in population studies of a number of fungal species (see Gherbi et al., 1999; Guidot et al., 2001) as it allows recognition of genetically distinct individuals to be recognised. Identification of the IGS (or other individual-level markers) in metagenomic studies coupled with ITS data could potentially overcome the issues associated with determining the relative abundance of numbers of individuals in fungal populations simultaneously. Reconstruction of fungal genomes from environmental samples would open up whole new fields of research but the amounts of data involved, and the computing power required would be prodigious.

Fungi are often grouped with bacteria as part of the soil 'microbial' diversity and assumed to be subject to the same chemical and physical environment and conditions. There may be some logical support for this assumption being applied to unicellular fungi like yeasts, but even these are commonly an order of magnitude greater in size than bacteria. A greater appreciation is required that fungi are not bacteria and that they can, if allowed to develop mycelia, integrate across the heterogenous soil environment, moving nutrients and carbon from sources to sinks. With greater emphasis on sustainability and reduced inputs, there is a need for more fundamental research into how changes in soil management, in particular, reduced tillage, will impact on mycelial development and the retention of nutrients in agricultural systems. This seems a simple question, but we need to be able to recognise the intact mycelia of individuals of the same species, which as outlined above is difficult to achieve.

So far, we have been considering fungi largely in isolation without the myriad of interactions that will be with the plethora of other soil organisms. Interactions with soil bacteria, in particular, have been shown to be prevalent at both small scales (Deveau et al., 2018) and at global scales (Bahram et al., 2018) and are likely to have been significant drivers of evolution in both groups (Boer et al., 2010). Deveau et al. (2018) present an excellent review of fungal/bacterial interactions and highlight their importance across diverse research fields and ecosystems. There is currently great interest in engineering beneficial microbiomes and fungi for crop plants (see Sergaki et al., 2018; Martínez-Hidalgo et al., 2019) but studies are often done in isolation without consideration for how engineered communities will interact in situ with existing soil biology. A more holistic approach, although certainly more complicated and difficult, would most likely lead to more stable outcomes under field conditions.

Frąc et al. (2018) gave a comprehensive review of some potential benefits of fungi in agricultural systems. While many of these could be predicted based on our current knowledge, the possibility that plants can activate defence mechanisms in response to their neighbours being attacked by aphids through inter-plant networks of soil mycelia highlights wonderfully how little we actually know about the complex interactions even in relatively simple agricultural systems (Gilbert and Johnson, 2017). Without doubt there will be many other surprises waiting to be discovered with similar potential to enhance sustainable agriculture.

8 Where to look for further information

Across the globe, there are a huge number of initiatives and organisations involved in research on soil biodiversity, including fungi. The Global Soil Biodiversity Initiative (https://www.globalsoilbiodiversity.org/) is an excellent forum for promoting the importance of soil biodiversity and for raising awareness about the need for soil biodiversity to be considered in land-use policies. The web site also provides access to the Global Soil Biodiversity Atlas which is an excellent resource for topics relating to soil health. The European Soil Data Centre (ESDAC) hosts a number of themes including soil biodiversity and has a useful source of links and references related to soils and soil biodiversity (https://esdac.jrc.ec.europa.eu/themes/soil-biodiversity).

9 References

Abarenkov, K., Henrik Nilsson, R., Larsson, K. H., Alexander, I. J., Eberhardt, U., Erland, S., Høiland, K., Kjøller, R., Larsson, E., Pennanen, T., Sen, R., Taylor, A. F. S., Tedersoo, L., Ursing, B. M., Vrålstad, T., Liimatainen, K., Peintner, U. and Kõljalg, U. (2010). The UNITE database for molecular identification of fungi–recent updates and future perspectives. *New Phytologist* 186(2): 281–285.

Amend, A. S., Seifert, K. A. and Bruns, T. D. (2010). Quantifying microbial communities with 454 pyrosequencing: does read abundance count? *Molecular Ecology* 19(24): 5555–5565. 10.1111/j.1365-294X.2010.04898.x.

Anantharaman, K., Brown, C. T., Hug, L. A., Sharon, I., Castelle, C. J., Probst, A. J., Thomas, B. C., Singh, A., Wilkins, M. J., Karaoz, U., Brodie, E. L., Williams, K. H., Hubbard, S. S. and Banfield, J. F. (2016). Thousands of microbial genomes shed light on interconnected biogeochemical processes in an aquifer system. *Nature Communications* 7: 13219. 10.1038/ncomms13219.

Bahram, M., Hildebrand, F., Forslund, S. K., Anderson, J. L., Soudzilovskaia, N. A., Bodegom, P. M., Bengtsson-Palme, J., Anslan, S., Coelho, L. P., Harend, H., Huerta-Cepas, J., Medema, M. H., Maltz, M. R., Mundra, S., Olsson, P. A., Pent, M., Põlme, S., Sunagawa, S., Ryberg, M., Tedersoo, L. and Bork, P. (2018). Structure and function of the global topsoil microbiome. *Nature* 560(7717): 233–237. 10.1038/s41586-018-0386-6.

Baldrian, P., Větrovský, T., Cajthaml, T., Dobiášová, P., Petránková, M., Šnajdr, J. and Eichlerová, I. (2013). Estimation of fungal biomass in forest litter and soil. *Fungal Ecology* 6(1): 1–11. 10.1016/j.funeco.2012.10.002.

Bengtsson-Palme, J., Ryberg, M., Hartmann, M., Branco, S., Wang, Z., Godhe, A., De Wit, P., Sanchez-Garcia, M., Ebersberger, I., de Sousa, F., Amend, A. S., Jumpponen, A., Unterseher, M., Kristiansson, E., Abarenkov, K., Bertrand, Y. J. K., Sanli, K., Eriksson, K. M., Vik, U., Veldre, V. and Nilsson, R. H. (2013). Improved software detection and extraction of ITS1 and ITS2 from ribosomal ITS sequences of fungi and other eukaryotes for analysis of environmental sequencing data. *Methods in Ecology and Evolution* 4: 914–919.

Bertrand, M., Barot, S., Blouin, M., Whalen, J., de Olivera, T. and Roger-Estrade, J. (2015). Earthworm services for cropping systems: a review. *Agronomy for Sustainable Development* 35(2): 553–567. 10.1007/s13593-014-0269-7.

Boer, Wd, Folman, L. B., Summerbell, R. C. and Boddy, L. (2010). Living in a fungal world: impact of fungi on soil bacterial niche development. *FEMS Microbiology Reviews* 29(4): 795–811. 10.1016/j.femsre.2004.11.005.

Bridge, P. D., Roberts, P. J., Spooner, B. M. and Panchal, G. (2003). On the unreliability of published DNA sequences. *New Phytologist* 160(1): 43–48. 10.1046/j.1469-8137.2003.00861.x.

Burdon, J. and Laine, A. (2019). Sources and patterns of variation in plant pathogens. In: *Evolutionary Dynamics of Plant-Pathogen Interactions* (pp. 91–122). Cambridge: Cambridge University Press. 10.1017/9781108625517.005.

Callahan, B. J., McMurdie, P. J., Rosen, M. J., Han, A. W., Johnson, A. J. A. and Holmes, S. P. (2016). DADA2: high-resolution sample inference from Illumina amplicon data. *Nature Methods* 13(7): 581–583. 10.1038/nmeth.3869.

Chapman, S. J., Bell, J. S., Campbell, C. D., Hudson, G., Lilly, A., Nolan, A. J., Robertson, A. H. J., Potts, J. M. and Towers, W. (2013). Comparison of soil carbon stocks in Scottish soils between 1978 and 2009. *European Journal of Soil Science* 64(4): 455–465. 10.1111/ejss.12041.

Chen, J. Y. W., Lau, A. P. S. and Fang, M. (2008). Assessment of the atmospheric fungal prevalence through field ergosterol measurement II: establishing the conversion factor. *Atmospheric Environment* 42(22): 5534–5541.

Chen, W., Wang, J., Meng, Z., Xu, R., Chen, J., Zhang, Y. and Hu, T. (2020). Fertility-related interplay between fungal guilds underlies plant richness–productivity relationships in natural grasslands. *New Phytologist* 226(4): 1129–1143. 10.1111/nph.16390.

Chiodi, C., Moro, M., Squartini, A., Concheri, G., Occhi, F., Fornsaier, F., Cagnin, M., Bertoldo, G., Broccanello, C. and Stevanato, P. (2020). High-throughput isolation of nucleic acids from soil. *Soil Systems* 4(1): 3. 10.3390/soilsystems4010003.

Creer, S., Thomas, W. K., Deiner, K., Frey, S., Porazinska, D., Taberlet, P., Potter, C. and Bik, H. M. (2016). The ecologist's field guide to sequence-based identification of biodiversity. *Methods in Ecology and Evolution* 7: 1008–1018. 10.1111/2041-210X.12574.

Delmont, T. O., Eren, A. M., Maccario, L., Prestat, E., Esen, Ö. C., Pelletier, E., Le Paslier, D., Simonet, P. and Vogel, T. M. (2015). Reconstructing rare soil microbial genomes using in situ enrichments and metagenomics. *Frontiers in Microbiology* 6: 358. 10.3389/fmicb.2015.00358.

Deveau, A., Bonito, G., Uehling, J., Paoletti, M., Becker, M., Bindschedler, S., Hacquard, S., Hervé, V., Labbé, J., Lastovetsky, O. A., Mieszkin, S., Millet, L. J., Vajna, B., Junier, P., Bonfante, P., Krom, B. P., Olsson, S., van Elsas, J. D. and Wick, L. Y. (2018). Bacterial-fungal interactions: ecology, mechanisms and challenges. *FEMS Microbiology Reviews* 42(3): 335–352, 10.1093/femsre/fuy008.

Dopheide, A., Xie, D., Buckley, T. R., Drummond, A. J. and Newcomb, R. D. (2019). Impacts of DNA extraction and PCR on DNA metabarcoding estimates of soil biodiversity. *Methods in Ecology and Evolution* 10(1): 120–133. 10.1111/2041-210X.13086.

Dutreux, F., Da Silva, C., d'Agata, L., Couloux, A., Gay, E. J., Istace, B., Lapalu, N., Lemainque, A., Linglin, J., Noel, B., Wincker, P., Cruad, C., Rouxel, T., Balesdent, M. H. and Aury, J. M. (2018). De novo assembly and annotation of three *Leptosphaeria* genomes using Oxford nanopore MinION sequencing. *Scientific Data* 5: 180235 10.1038/sdata.2018.235.

Ellouze, W., Esmaeili Taheri, A., Bainard, L. D., Yang, C., Bazghaleh, N., Navarro-Borrell, A., Hanson, K. and Hamel, C. (2014). Soil fungal resources in annual cropping systems and their potential for management. *BioMed Research International* 2014: 531824. 10.1155/2014/531824.

Ernst, G. and Emmerling, C. (2009). Impact of five different tillage systems on soil organic carbon content and the density, biomass, and community composition of earthworms after a ten-year period. *European Journal of Soil Biology* 45(3): 247–251.

Falconer, R. E., Otten, W. and White, N. A. (2015). Toward modelling the resistance and resilience of "Below-ground" fungal communities: a mechanistic and trait-based approach. *Advances in Applied Microbiology* 93: 1–44.

Frąc, M., Hannula, S. E., Bełka, M. and Jędryczka, M. (2018). Fungal biodiversity and their role in soil health. *Frontiers in Microbiology* 9: 707. 10.3389/fmicb.2018.00707.

Frostegård, Å., Bååth, E. and Tunlid, A. (1993). Shifts in the structure of soil microbial communities in limed forests as revealed by phospholipid fatty acid analysis. *Soil Biology and Biochemistry* 25(6): 723–730.

Frostegård, Å., Tunlid, A. and Bååth, E. (2011). Use and misuse of PLFA measurements in soils. *Soil Biology and Biochemistry* 43(8): 1621–1625.

Gerin, D., Pollastro, S., Raguseo, C., De Miccolis Angelini, R. M. and Faretra, F. (2018). A ready-to-use single- and duplex-Taqman-qPCR assay to detect and quantify the biocontrol agents *Trichoderma asperellum* and *Trichoderma gamsii*. *Frontiers in Microbiology* 9: 2073. 10.3389/fmicb.2018.02073.

Gherbi, H., Delaruelle, C., Selosse, M.-A. and Martin, F. (1999). High genetic diversity in a population of the ectomycorrhizal basidiomycete *Laccaria amethystina* in a 150-year-old beech forest. *Molecular Ecology* 8(12): 2003 –2013.

Gilbert, L. and Johnson, D. (2017). Plant-plant communication through common mycorrhizal networks. *Advances in Botanical Research* 82: 83–97.

Griffiths, R. I., Whiteley, A. S., O'Donnell, A. G. and Bailey, M. J. (2000). Rapid method for coextraction of DNA and RNA from natural environments for analysis of ribosomal DNA- and rRNA-based microbial community composition. *Applied and Environmental Microbiology* 66(12): 5488–5491.

Guidot, A., Debaud, J. C. and Marmeisse, R. (2001). Correspondence between genet diversity and spatial distribution of above- and below-ground populations of the ectomycorrhizal fungus *Hebeloma cylindrosporum*. *Molecular Ecology* 10(5): 1121–1131.

Hardoim, P. R., van Overbeek, L. S., Berg, G., Pirttilä, A. M., Company, S., Campisano, A., Döring, M. and Sessitsch, A. (2015). The hidden world within plants: ecological and evolutionary considerations for defining functioning of microbial endophytes. *Microbiology and Molecular Biology Reviews* 79(3): 293–320. 10.1128/MMBR.00050-14.

Hawksworth, D. L. and Lüking, R. (2017). Fungal diversity revisited: 2.2 to 3.8 million species. *Microbiology Spectrum* 5(4): FUNK-0052-2016.

Hebert, P. D. N., Cywinska, A., Ball, S. L. and deWaard, J. R. (2003). Biological identifications through DNA barcodes. *Proceedings. Biological Sciences* 270(1512): 313–321.

Hebert, P. D. N. and Gregory, T. R. (2005). The promise of DNA barcoding for taxonomy. *Systematic Biology* 54(5): 852–859.

Helgason, B. L., Walley, F. L. and Germida, J. J. (2010). No-till soil management increases microbial biomass and alters community profiles in soil aggregates. *Applied Soil Ecology* 46(3): 390–397. 10.1016/j.apsoil.2010.10.002.

Helgason, T., Daniell, T. J., Husband, R., Fitter, A. H. and Young, J. P. W. (1998). Ploughing up the wood-wide web? *Nature* 394(6692): 431.

Högberg, M. N. (2006). Discrepancies between ergosterol and the phospholipid fatty acid 18:2ω6,9 as biomarkers for fungi in boreal forest soils. *Soil Biology and Biochemistry* 38(12): 3431–3435.

Ihrmark, K., Bödeker, I. T. M., Cruz-Martinez, K., Friberg, H., Kubartova, A., Schenck, J., Strid, Y., Stenlid, J., Brandström-Durling, M., Clemmensen, K. E. and Lindahl, B. D. (2012). New primers to amplify the fungal ITS2 region–evaluation by 454-sequencing of artificial and natural communities. *FEMS Microbiology Ecology* 82(3): 666–677.

ISO/TS, 29843-1 (2010). Soil quality-determination of soil microbial diversity. Geneva, Switzerland: ISO/TC. http://www.iso.org/iso/home.htm.

Jian, C., Luukkonen, P., Yki-Järvinen, H., Salonen, A. and Korpela, K. (2020). Quantitative PCR provides a simple and accessible method for quantitative microbiota profiling. *PLoS ONE* 15(1): e0227285. 10.1371/journal.pone.0227285.

Kjøller, A. and Struwe, S. (1982). Microfungi in ecosystems: fungal occurrence and activity in litter and soil. *Oikos* 39(3): 391–422.

Knight, R., Vrbanac, A., Taylor, B. C., Aksenov, A., Callewaert, C., Debelius, J., Gonzalez, A., Kosciolek, T., McCall, L. I., McDonald, D., Melnik, A. V., Morton, J. T., Navas, J., Quinn, R. A., Sanders, J. G., Swafford, A. D., Thompson, L. R., Tripathi, A., Xu, Z. Z., Zaneveld, J. R., Zhu, Q., Caporaso, J. G. and Dorrestein, P. C. (2018). Best practices for analysing microbiomes. *Nature Reviews. Microbiology* 16(7): 410–422. 10.1038/s41579-018-0029-9.

Kõljalg, U., Larsson, K. H., Abarenkov, K., Nilsson, R. H., Alexander, I. J., Eberhardt, U., Erland, S., Høiland, K., Kjøller, R., Larsson, E., Pennanen, T., Sen, R., Taylor, A. F., Tedersoo, L., Vrålstad, T. and Ursing, B. M. (2005). UNITE–a database providing web-based methods for the molecular identification of ectomycorrhizal fungi. *New Phytologist* 166(3): 1063–1068.

Kõljalg, U., Nilsson, R. H., Abarenkov, K., Tedersoo, L., Taylor, A. F. S., Bahram, M., Bates, S. T., Bruns, T. D., Bengtsson-Palme, J., Callaghan, T. M., Douglas, B., Drenkhan, T., Eberhardt, U., Dueñas, M., Grebenc, T., Griffith, G. W., Hartmann, M., Kirk, P. M., Kohout, P., Larsson, E., Lindahl, B. D., Lücking, R., Martín, M. P., Matheny, P. B., Nguyen, N. H., Niskanen, T., Oja, J., Peay, K. G., Peintner, U., Peterson, M., Põldmaa, K., Saag, L., Saar, I., Schüßler, A., Scott, J. A., Senés, C., Smith, M. E., Suija, A., Taylor, D. L., Telleria, M. T., Weiss, M. and Larsson, K. H. (2013). Towards a unified paradigm for sequence-based identification of fungi. *Molecular Ecology* 22(21): 5271–5277.

Langille, M. G., Zaneveld, J., Caporaso, J. G., McDonald, D., Knights, D., Reyes, J. A., Clemente, J. C., Burkepile, D. E., Vega Thurber, R. L., Knight, R., Beiko, R. G. and Huttenhower, C. (2013). Predictive functional profiling of microbial communities using 16S rRNA marker gene sequences. *Nature Biotechnology* 31(9): 814–821. 10.1038/nbt.2676.

Leake, J. R., Johnson, D., Donnelly, D., Muckle, G., Boddy, L. and Read, D. J. (2004). Networks of power and influence: the role of mycorrhizal mycelium in controlling plant communities and agroecosystem functioning. *Canadian Journal of Botany* 82(8): 1016–1045.

Lindahl, B. D., de Boer, W. and Finlay, R. D. (2010). Disruption of root carbon transport into forest humus stimulates fungal opportunists at the expense of mycorrhizal fungi. *The ISME Journal* 4(7): 872–881.

Lindahl, B. D., Nilsson, R. H., Tedersoo, L., Abarenkov, K., Carlsen, T., Kjøller, R., Kõljalg, U., Pennanen, T., Rosendahl, S., Stenlid, J. and Kauserud, H. (2013). Fungal community analysis by high-throughput sequencing of amplified markers–a user's guide. *New Phytologist* 199(1): 288–299.

Lindner, D. L., Carlsen, T., Nilsson, R. H., Davey, M., Schumacher, T. and Kauserud, H. (2013). Employing 454 amplicon pyrosequencing to reveal intragenomic divergence in the internal transcribed spacer rDNA region in fungi. *Ecology and Evolution* 3(6): 1751–1764.

Liu, C. M., Kachur, S., Dwan, M. G., Abraham, A. G., Aziz, M., Hsueh, P. R., Huang, Y. T., Busch, J. D., Lamit, L. J., Gehring, C. A., Keim, P. and Price, L. B. (2012). FungiQuant: a broad-coverage fungal quantitative real-time PCR assay. *BMC Microbiology* 12: 255 10.1186/1471-2180-12-255.

Lofgren, L. A., Uehling, J. K., Branco, S., Bruns, T. D., Martin, F. and Kennedy, P. G. (2018). Genome-based estimates of fungal rDNA copy number variation across phylogenetic scales and ecological lifestyles. *Molecular Ecology* 28(4): 721–730. 10.1111/mec.14995.

Loron, C. C., François, C., Rainbird, R. H., Turner, E. C., Borensztajn, S. and Javaux, E. J. (2019). Early fungi from the Proterozoic era in Arctic Canada. *Nature* 570(7760): 232–235. 10.1038/s41586-019-1217-0.

Martínez-Hidalgo, P., Maymon, M., Pule-Meulenberg, F. and Hirsch, A. M. (2019). Engineering root microbiomes for healthier crops and soils using beneficial, environmentally safe bacteria. *Canadian Journal of Microbiology* 65(2): 91–104, 10.1139/cjm-2018-0315.

McDonald, B. A. and Stukenbrock, E. H. (2016). Rapid emergence of pathogens in agro-ecosystems: global threats to agricultural sustainability and food security. *Philosophical Transactions of the Royal Society of London Series B* 371. 10.1098/rstb.2016.0026: 20160026.

Mille-Lindblom, C., Wachenfeldt, E. and Tranvik, L. J. (2004). Ergosterol as a measure of living fungal biomass: persistence in environmental samples after fungal death. *Journal of Microbiological Methods* 59(2): 253–262. 10.1016/j.mimet.2004.07.010.

Millgroom, M. G. and Peever, T. L. (2003). Population biology of plant pathogens: the synthesis of plant disease epidemiology and population genetics. *Plant Disease* 87(6): 608–617. 10.1094/PDIS.2003.87.6.608.

Moll, J., Goldmann, K., Kramer, S., Hempel, S., Kandeler, E., Marhan, S., Ruess, L., Krüger, D. and Buscot, F. (2015). Resource type and availability regulate fungal communities along arable soil profiles. *Microbial Ecology*70(2): 390–399.

Moll, J., Hoppe, B., König, S., Wubet, T., Buscot, F. and Krüger, D. (2016). Spatial distribution of fungal communities in an arable soil. *PLoS ONE* 11(2): e0148130. 10.1371/journal.pone.0148130

Nesme, J., Achouak, W., Agathos, S. N., Bailey, M., Baldrian, P., Brunel, D., Frostegård, Å, Heulin, T., Jansson, J. K., Jurkevitch, E., Kruus, K. L., Kowalchuk, G. A., Lagares, A., Lappin-Scott, H. M., Lemanceau, P., Le Paslier, D., Mandic-Mulec, I., Murrell, J. C., Myrold, D. D., Nalin, R., Nannipieri, P., Neufeld, J. D., O'Gara, F., Parnell, J. J., Pühler, A., Pylro, V., Ramos, J. L., Roesch, L. F., Schloter, M., Schleper, C., Sczyrba, A., Sessitsch, A., Sjöling, S., Sørensen, J., Sørensen, S. J., Tebbe, C. C., Topp, E., Tsiamis, G., van Elsas, J. D., van Keulen, G., Widmer, F., Wagner, M., Zhang, T., Zhang, X., Zhao, L., Zhu, Y. G., Vogel, T. M. and Simonet, P. (2016). Back to the future of soil metagenomics. *Frontiers in Microbiology* 7: 73. 10.3389/fmicb.2016.00073.

Nguyen, N. H., Song, Z. W., Bates, S. T., Branco, S., Tedersoo, L., Menke, J., Schilling, J. S. and Kennedy, P. G. (2016). FUNGuild: an open annotation tool for parsing fungal community datasets by ecological guild. *Fungal Ecology* 20: 241–248.

Nilsson, R. H., Kristiansson, E., Ryberg, M., Hallenberg, N. and Larsson, K.-H. (2008). Intraspecific ITS Variability in the kingdom fungi as expressed in the international sequence databases and its implications for molecular species identification. *Evolutionary Bioinformatics* 4: 193–201.

Nilsson, R. H., Ryberg, M., Kristiansson, E., Abarenkov, K., Larsson, K.-H., Kõljalg, U. (2006). Taxonomic reliability of DNA sequences in public sequence databases: a fungal perspective. *PLoS ONE* 1(1): e59. 10.1371/journal.pone.0000059.

Nilsson, R. H., Wurzbacher, C., Bahram, M., R. M. Coimbra, V., Larsson, E., Tedersoo, L., Eriksson, J., Duarte, C., Svantesson, S., Sánchez-García, M., Ryberg, M. K., Kristiansson, E. and Abarenkov, K. (2016). Top 50 most wanted fungi. *MycoKeys* 12: 29–40. 10.3897/mycokeys.12.7553.

Nurika, I., Eastwood, D. C. and Barker, G. C. (2018). A comparison of ergosterol and PLFA methods for monitoring the growth of ligninolytic fungi during wheat straw solid state fermentation. *Journal of Microbiological Methods* 148: 49–54. 10.1016/j.mimet.2018.03.006.

Öpik, M., Metsis, M., Daniell, T. J., Zobel, M. and Moora, M. (2009). Large-scale parallel 454 sequencing reveals host ecological group specificity of arbuscular mycorrhizal fungi in a boreonemoral forest. New Phytologist 184: 424–437.

Pelosi, C., Pey, B., Hedde, M., Caro, G., Capowiez, Y., Guernion, M., Peigné, J., Piron, D., Betrand, M. and Cluzeau, D. (2014). Reducing tillage in cultivated fields increases earthworm functional diversity. *Applied Soil Ecology* 83: 79–87.

Quideau, S. A., McIntosh, A. C., Norris, C. E., Lloret, E., Swallow, M. J. and Hannam, K. (2016). Extraction and analysis of microbial phospholipid fatty acids in soils. *Journal of Visualized Experiments: JoVE* 114(114): e54360, 10.3791/54360.

Ravi, R. K., Walton, K. and Khosroheidari, M. (2018). MiSeq: a next generation sequencing platform for genomic analysis. *Methods in Molecular Biology* 1706: 223–232. 10.1007/978-1-4939-7471-9_12.

Ruzicka, S., Edgerton, D., Norman, M. and Hill, T. (2000). The utility of ergosterol as a bioindicator of fungi in temperate soils. *Soil Biology and Biochemistry* 32(7): 989–1005.

Schloss, P. D., Westcott, S. L., Ryabin, T., Hall, J. R., Hartmann, M., Hollister, E. B., Lesniewski, R. A., Oakley, B. B., Parks, D. H., Robinson, C. J., Sahl, J. W., Stres, B., Thallinger, G. G., Van Horn, D. J. and Weber, C. F. (2009). Introducing mothur: open-source, platform-independent, community-supported software for describing and comparing microbial communities. Applied and Environmental Microbiology 75: 7537–7541. 10.1128/AEM.01541-09.

Schoch, C. L., Seifert, K. A., Huhndorf, S., Robert, V., Spouge, J. L., Levesque, C. A., Chen, W., Fungal Barcoding Consortium and Fungal Barcoding Consortium Author List (2012). Nuclear ribosomal internal transcribed spacer (ITS) region as a universal DNA barcode marker for Fungi. *Proceedings of the National Academy of Sciences of the United States of America* 109(16): 6241–6246. 10.1073/pnas.1117018109.

Sergaki, C., Lagunas, B., Lidbury, I., Gifford, M. L. and Schäfer, P. (2018). Challenges and approaches in microbiome research: from fundamental to applied. *Frontiers in Plant Science* 9: 1205. 10.3389/fpls.2018.01205.

Shen, Q., Kirschbaum, M. U. F., Hedley, M. J. and Arbestain, M. C. (2016). Testing an alternative method for estimating the length of fungal hyphae using photomicrography and image processing. *PLoS ONE* 11(6): e0157017.

Shendure, J. and Ji, H. (2008). Next-Generation DNA sequencing. *Nature Biotechnology* 26(10): 1135–1145.

Smith, M. L., Bruhn, J. N. and Anderson, J. B. (1992). The fungus *Armillaria bulbosa* is among the largest and oldest living organisms. *Nature* 356(6368): 428–431.

Sommermann, L., Geistlinger, J., Wibberg, D., Deubel, A., Zwanzig, J., Babin, D., Schlüter, A. and Schellenberg, I. (2018). Fungal community profiles in agricultural soils of a long-term field trial under different tillage, fertilization and crop rotation conditions analyzed by high-throughput ITS-amplicon sequencing. *PLoS ONE* 13(4): e0195345. 10.1371/journal.pone.0195345.

Stajich, J. E., Berbee, M. L., Blackwell, M., Hibbett, D. S., James, T. Y., Spatafora, J. W. and Taylor, J. W. (2009). The fungi. *Current Biology* 19(18): R840–R845.

Sugiyama, A., Vivanco, J. M., Jayanty, S. S. and Manter, D. K. (2010). Pyrosequencing assessment of soil microbial communities in organic and conventional potato farms. *Plant Disease* 94(11): 1329–1335. 10.1094/PDIS-02-10-0090.

Taylor, A. F. S. (2002). Fungal diversity in ectomycorrhizal communities: sampling effort and species detection. *Plant and Soil* 244(1/2): 19–28.

Taylor, D. L., Hollingsworth, T. N., McFarland, J. W., Lennon, N. J., Nusbaum, C. and Ruess, R. W. (2014). A first comprehensive census of fungi in soil reveals both hyperdiversity and fine-scale niche partitioning. *Ecological Monographs* 84(1): 3–20.

Tedersoo, L., Anslan, S., Bahram, M., Põlme, S., Riit, T., Liiv, I., Kõljalg, U., Kisand, V., Nilsson, H., Hildebrand, F., Bork, P. and Abarenkov, K. (2015). Shotgun metagenomes and multiple primer pair-barcode combinations of amplicons reveal biases in metabarcoding analyses of fungi. *MycoKeys* 10: 1–43. 10.3897/mycokeys.10.4852.

Tedersoo, L., Bahram, M., Põlme, S., Kõljalg, U., Yorou, N. S., Wijesundera, R., Ruiz, L. V., Vasco-Palacios, A. M., Thu, P. Q., Suija, A., Smith, M. E., Sharp, C., Saluveer, E., Saitta, A., Rosas, M., Riit, T., Ratkowsky, D., Pritsch, K., Põldmaa, K., Piepenbring, M., Phosri, C., Peterson, M., Parts, K., Pärtel, K., Otsing, E., Nouhra, E., Njouonkou, A. L., Henrik Nilsson, R., Morgado, L. N., Mayor, J., May, T. W., Majuakim, L., Lodge, D. J., Lee, S. S., Larsson, K.-H., Kohout, P., Hosaka, K., Hiiesalu, I., Henkel, T. W., Harend, H., Guo, L.-d., Greslebin, A., Grelet, G., Geml, J., Gates, G., Dunstan, W., Dunk, C., Drenkhan, R., Dearnaley, J., De Kesel, A., Dang, T., Chen, X., Buegger, F., Brearley, F. Q., Bonito, G., Anslan, S., Abell, S. and Abarenkov, K. (2014). Global diversity and geography of soil fungi. Science 346: 1078.

Thornton, C. R. (2008). Tracking fungi in soil with monoclonal antibodies. *European Journal of Plant Pathology* 121(3): 347–353. 10.1007/s10658-007-9228-3.

Torres-Cruz, T. J., Billingsley Tobias, T. L., Almatruk, M., Hesse, C. N., Kuske, C. R., Desirò, A., Benucci, G. M. N., Bonito, G., Stajich, J. E., Dunlap, C., Arnold, A. E. and Porras-Alfaro, A. (2017). Bifiguratus adelaidae, gen. et sp. nov., a new member of Mucoromycotina in endophytic and soil-dwelling habitats. *Mycologia* 109(3): 363–378. 10.1080/00275514.2017.1364958.

Veum, K. S., Lorenz, T. and Kremer, R. J. (2019). Phospholipid fatty acid profiles of soils under variable handling and storage conditions. *Agronomy Journal* 111(3): 1090–1096.

Wagg, C., Bender, S. F., Widmer, F. and Van der Heijden, M. G. A. (2014). Soil biodiversity and soil community composition determine ecosystem multifunctionality. *Proceedings of the National Academy of Sciences of the United States of America* 111(14): 5266–5270.

Wang, D. Y. C., Kumar, S. and Hedges, S. B. (1999). Divergence time estimates for the early history of animal phyla and the origin of plants, animals and fungi. *Proceedings. Biological Sciences* 266(1415): 163–171. 10.1098/rspb.1999.0617.

Willers, C., Jansen van Rensburg, P. J. and Claasans, S. (2015). Phospholipid fatty acid profiling of microbial communities–a review of interpretations and recent applications. *Journal of Applied Microbiology* 119(5): 1207–1218.

Wu, B., Hussain, M., Zhang, W., Stadler, M., Liu, X. and Xiang, M. (2019). Current insights into fungal species diversity and perspective on naming the environmental DNA sequences of fungi. *Mycology* 10(3): 127–140. 10.1080/21501203.2019.1614106.

Part 2

Measuring soil physical and chemical properties

Chapter 4

Advances in visual soil evaluation techniques

Mansonia Pulido-Moncada, Aarhus University, Denmark; Bruce C. Ball, formerly Scotland's Rural College (SRUC), UK; and Wim M. Cornelis, Ghent University, Belgium

1 Introduction

The Food and Agriculture Organization of the United Nations (FAO) defines soil degradation as 'a change in the soil health status resulting in a diminished capacity of the ecosystem to provide goods and services for its beneficiaries'. Therefore, detrimental changes in soil health/quality boost the vulnerability of the soils as a limited resource, directly affecting the food production and rise of people's needs worldwide (Osman, 2014). From technically developed to underdeveloped countries, soil health threats are a consequence of poor management practices in either forestry or agricultural soils (Bennett *et al.*, 2010; Cardoso *et al.*, 2013).

The broad aspects that affect soil health are most readily recognised through their influence on soil structure. A poor soil structure is the common cause/result of the degradation problems that can co-occur in the soil (Dexter, 2004; Pagliai *et al.*, 2004). The soil structure controls the interaction between the liquid, gaseous and solid phases in soil (Pulido-Moncada *et al.*, 2017)

http://dx.doi.org/10.19103/AS.2020.0079.10

and is thus the central pillar that supports soil functions (Eswaran *et al.*, 2001; Osman, 2014).

According to Horn and Smucker (2005) soil structure is 'the arrangement of single mineral particles and organic substances into greater units known as aggregates and the corresponding inter-aggregate pore system'. Dexter (1988) mentioned that soil structure involves the different properties that are revealed at many different size scales in soil, ranging from the level of clay particles and clay-organic matter complexes to the spatial arrangement of aggregates/clods in the soil profile. Additionally, soil structure-related processes are impacted by natural pedogenesis, but in agricultural soils these processes are also affected by soil management practices (Carter, 2004). For example, the use of heavy machinery in modern agriculture worldwide affects soil properties from the surface down to the subsoil (Arvidsson and Keller, 2007; Etana *et al.*, 2013).

The wide range of scales in which soil structure can be assessed, together with its dynamic nature, is thus the principal reason for the complexity of soil structure (Lal and Shukla, 2004). Due to the complexity of soil structure, there is no unique method for soil structure characterisation and the evaluation of its changes. A complete assessment of soil structure in agricultural soils might involve the direct evaluation of structural form (morphological aspect), structural stability and structural resilience (Kay, 1990).

Soil structure highly affects physical and hydraulic soil properties that determine crop yield and the soil's vulnerability to degradation, including water infiltrability, water and air permeability, water holding and air capacity, among others (Pagliai *et al.*, 2004; Reynolds *et al.*, 2009; Kodešová *et al.*, 2011). Traditional quantitative methods that measure such properties in the laboratory or field are vice versa an indirect way to assess and monitor soil structure (Rabot *et al.*, 2018), and these soil properties are thus often considered as soil health/quality indicators. Indirect evaluation of soil structure is also possible using modelling techniques such as neural networks, pedotransfer functions, fractal theory and network models (Young *et al.*, 2001).

Direct methods involve measuring of aggregate size and stability (Lal and Shukla, 2004); observation of morphological, structural features by microscopy or analysis of images, for example, X-ray computed tomography scans, electrical resistivity tomography, or thin sections (Young *et al.*, 2001; Lal and Shukla, 2004; Pagliai *et al.*, 2004); and examination of structural form by naked-eye using visual soil evaluation techniques (Ball *et al.*, 2013).

Focus on the use of direct methods for visual soil evaluation has increased in the last two decades. Popularisation of the use of visual soil evaluation techniques has spread to several countries, in areas of a temperate, subtropical and tropical climate. Worldwide studies applying visual examinations have been conducted on agricultural fields, and plot experiments (Munkholm and

Holden, 2015), but also the application of these methods has extended to remote areas in the Amazon (Guimarães *et al.*, 2017a; Cherubin *et al.*, 2019) and rural areas in Africa (Cornelis *et al.*, 2019a; Obour *et al.*, 2020). Franco *et al.* (2019) suggest that this is a reflection of a global need for reliable and low-cost methods that can provide an evaluation of soil structure status complementary to direct evaluation for research and extension service purposes.

These methods require less money and labour costs than direct evaluation and can be used by farmers and advisors after training (Rabot *et al.*, 2018; van Leeuwen *et al.*, 2018). The visual soil evaluation methods provide an assessment of the effect of soil management on soil structure in a soil volume larger than the core samples typically used for soil physical characterisation allowing better capture of soil variability (Guimarães *et al.*, 2017b). Visual techniques are applicable at the site and farm level (Sonneveld *et al.*, 2014) for a quick assessment of the topsoil, but also allow a more complete, comprehensive evaluation of the soil structure changes of the soil profile and their link to soil management practices. This is of importance to farmers and consultants in soil management decisions, as well as to soil scientists for planning sampling in timely assessments or monitoring campaigns (Ball *et al.*, 2007).

The development of visual soil assessment methods covers two aspects, one relates to the use of comprehensive description of the soil structure and related causative factors that required scientific knowledge, and the other that implies accurate but simpler assessment schemes to be applicable by farmers. These two aspects are the bases that have contributed to the development of new methods or the refinement of pre-existing ones. To date, visual soil assessment techniques either focus on or integrate indicators such as ease of fragmentation, fragment size distribution, outer and inner morphological description of aggregates, aggregate stability, and biological activity. The latest advances of these techniques respond to the need for them to be applicable to local conditions or to include additional parameters for a broader capture of agronomic information that can be used for decision-making on soil management practices.

2 Assessing soil structural quality by visual soil evaluation techniques

Visual techniques to assess soil structure have been shown to detect changes in soil structural-related properties and to relate well to laboratory measurements (Mueller *et al.*, 2009a; Guimarães *et al.*, 2013; Pulido-Moncada *et al.*, 2014c). Estimates of soil structural quality determined in the field visually and by handling the soil have an important part to play in soil physical analysis because they allow the assessment of the overall condition of the soil in three dimensions (Batey *et al.*, 2015). Soil structural changes assessed by visual assessment in the field have even been validated by X-ray computed tomography images analysis (Garbout *et al.*, 2013; Carducci *et al.*, 2017).

However, the generation of useful visual soil structure data depends on the skill and experience of the user of the method, and the type of examination or description of the soil structure made (McKeague *et al.*, 1986). To avoid subjectivity when conducting visual soil evaluation, new operators need training with experts (van Leeuwen *et al.*, 2018). During training or evaluation per se forest and natural un-degraded conditions are critical as a reference for people to conduct an accurate assessment or to distinguish between natural or anthropogenic soil structure degradation (Ball *et al.*, 2017). Additionally, users of visual examinations of soil structure have to identify the purpose for describing soil structure, in order to define where and how to conduct the assessment, for example, monitoring of soil compaction risk in grazing systems at the field level by farmers, or assessing management impact on soil quality at the regional/national level by policymakers/researchers. This involves the selection of the method, as well as establishing the number of replicates or the dimensions of a profile pit to assess the variability (McKeague *et al.*, 1986). Definition of the purpose and procedure to be followed will determine the duration of the test and degree of detail of the assessment (Batey, 2000; Ball and Douglas, 2003; Boizard *et al.*, 2005). Hence, the selection of the visual technique is likely to depend on both the principles involved and how well these suit the purpose of the assessment required by the stakeholder and the resources available.

Visual techniques can be applied any time; however, the time of assessment is closely related to its purpose. Emphasis on the need to conduct the visual assessment when the soil moisture is near field capacity and when the soil is readily accessible has been highlighted by several authors (Ball and Douglas, 2003; Ball *et al.*, 2007; Guimarães *et al.*, 2011). Soil moisture conditions affect the breaking up of soil fragments, the reproducibility of the visual soil structure assessment, and the relation between visual examination and laboratory measurements (van Leeuwen *et al.*, 2018).

The different visual methods used to assess soil structural quality directly in the field can be grouped into methods based on topsoil examination and those based on soil profile examination. To decide the operations to follow in the assessment of soil structure, it is necessary to know the differences or the principal advantages between the methods developed for this purpose.

The working group F 'Visual soil examination and evaluation' of the International Soil Tillage Research Organisation (ISTRO) met in 2005 and compared 10 visual soil evaluation methods in the field (Boizard *et al.*, 2005), that is, Whole Profile Assessment, SOILpak, Profil Cultural, Peerklamp method, Central West Soil Structure Classification, Visual Soil Assessment (VSA), Soil Quality Scoring Procedure (SQSP), The Spade Analysis, FAL method and Guide to Better Structure. The working group stated that each of these methods had been developed in response to a specific need and that their potential applicability from one area to another stimulated further studies in other countries under different conditions.

Mueller *et al.* (2009b) later contributed a comparison of the degree of agreement among seven topsoil methods and their relationships with measured parameters of soil structure. Their study confirmed the feasibility of visual techniques on soil health/quality assessment, and that methods such as the Werner method (Werner and Thaemert, 1989), the VSA (Shepherd, 2000) and the FAO guidelines (Jahn *et al.*, 2006) were capable of comparisons of soil structure of different sites. Nevertheless, Mueller *et al.* (2009b) pointed out that the time of evaluation and biological activity in relation to structure dynamics influenced the choice of the method, that is, the type and size of aggregates is an important diagnostic indicator for capturing structural changes over the season, but also the rotation dependency of biological features call for the evaluation of biological indicators at different periods of the growing cycle to capture the interpretation of soil and vegetation dynamic relationship. Additionally, although visual features correlated well with measured parameters, site-specific relationships were found, especially under soil conditions where the compaction-induced pore morphology drives the transport of fluids (Mueller *et al.*, 2009b).

In 2016, Emmet-Booth *et al.* reviewed the majority of visual soil evaluation techniques that have been published in English, German and French. Their review provides a critical comparison of the methodological approaches, strengths and weaknesses of the techniques. The authors highlighted the dependency on soil moisture, soil texture and sampling procedure on the outputs of the visual techniques.

Although there are several visual techniques of local application or international recognition, the following two sections summarise selected visual techniques, which are ranked in chronological order of development, for the purpose of showing differences in methodological approaches.

3 Methods based on topsoil examination (spade methods)

Methods related to topsoil examination are generally focussed on the topsoil layer or one spade depth (0 – 0.2/0.3 m), and describe the status or condition of specific topsoil stemming from relatively recent past land use and management. The topsoil examination methods call for the evaluation of a block of soil extracted with a spade from the soil surface. In general, these methods involve the assessment of different soil features by breaking up the extracted soil block by hand, or by a drop-shatter test. Several topsoil examination methods, also called spade methods, have been proposed, differing in aspects such as depth of the soil under consideration (from the superficial top layer of 0.15 m to also including the transition layer to the subsoil down to 0.50 m), handling the soil before the assessment, criteria employed and emphasis placed on particular features of soil structure, and scoring system (Mueller *et al.*, 2009b; Emmet-Booth *et al.*, 2016). Topsoil examination methods have been well accepted because they are quick, straightforward and widely applicable (Mueller *et al.*, 2009b; Guimarães *et al.*, 2011).

3.1 From 'The Spade Diagnosis' to 'The Spade Analysis'

Development of the method:

Görbing (1947) presented a method that is based on the collection of 25 years knowledge and 50 000 single spade diagnoses in Germany. The Spade Diagnosis method was later reintroduced for the evaluation or 'assessment of closeness to a' desirable structural state (crumb structure or soil tilth) in organic farming (Diez, 1982; Preuschen, 1983; Suhr et al., 1995). Modification of the Görbing method by Preuschen (1983) allowed a comprehensive examination of the relation between soil structure, root growth, faunal activity, decomposition of soil organic matter and fragmentation of the topsoil. Sobelius (1995) further modified the Görbing method by a description of soil structure and root growth, including root morphology following the FAO guidelines for soil description.

Munkholm (2000) later developed The Spade Analysis method by integrating the criteria used by the previously mentioned authors with well-known standardised guidelines. This intended the derivation of a complete description of the actual state of the soil tilth with minimal subjectivity by the operators. The guidelines used by Munkholm are those of Madsen and Jensen (1988) for describing layering and boundaries, soil texture, grade and type of structural units; the Munsell colour chart for soil colour description; aggregate size and soil consistency following the FAO guidelines; root growth characteristics, characterisation of pores and earthworm activity according to the description of Petersen and Møberg (1987) and Greve *et al.* (1999); and description of organic matter decomposition based on Preuschen (1983).

The method has been mainly used in Denmark and Germany for extension services. Though its use has not extended internationally, it represents a good example of a comprehensive evaluation of soil tilth, including morphological aspects of the soil structure, for which operators need background knowledge in soil description, though it is time-consuming.

Assessment protocol (key criteria and scale of scoring):

The Spade Analysis by Munkholm can be used at any time of the year for soil tilth assessment, except when the soil is frozen. Briefly, the extracted soil block is evaluated on a flat spade standing at the operator eye level (Fig. 1). The method involves the description of general soil features (soil surface, layering,

Figure 1 Example of a soil block examination with The Spade Analysis method. Photo courtesy of Lars Munkholm.

boundaries, colour, moisture, texture); soil structure (aggregate type and size, structural grade, soil consistency); macropores characterisation (number and size, pore distribution, pore continuity, pore orientation); internal surface features (clay coating and organic matter); concentration or concretion of minerals (type and quantity); root growth (root number and size-classes, branching, distribution, impediments, deformation features, number and distribution of root nodules); soil fauna; and decomposition of soil organic matter. The visual soil evaluation technique provided by Munkholm (2000) is qualitative, and no numerical translation of the soil tilth is given. The observation and classification of each of the indicators are used to provide a general evaluation of the soil tilth, distinguishing between good and poor soil tilth. The Spade Analysis manual is available at the following link: https://dcapub.au.dk/pub/djfrap_markbrug_28.pdf.

3.1.1 Soil Quality Scoring Procedure

The soil quality scoring procedure (SQSP) was developed by Ball and Douglas (2003) in Scotland. The authors' objective was to simplify and improve the Munkholm's system by ranking the topsoil's quality in terms of soil surface condition, soil structure and the network of roots, using a scoring scheme similar to that of Beste (1999). The topsoil score reflects changes in soil quality properties and the effects of soil management and residue incorporation practices. Ball and Douglas (2003) suggest either direct field evaluation or transport of the soil block to a laboratory for further assessment. The SQSP method provides a simplified version of Munkholm's explanatory notes to identify and describe the factors under assessment, which should suit soils of all texture. However, the evaluation of soil structure is noted to be mainly applicable to sandy and loamy soils. Additionally, the SQSP provides a scoring system with five categories, where 1 represents an ideal crumb structure and desirable rooting, and 5 reveals poor structure and restricted rooting. When applied in tropical conditions, the SQSP was found to overestimate the soil structural quality and was not sensitive enough when roots were absent during the evaluation (Pulido-Moncada *et al.*, 2014b). Therefore, the method should be used preferably when the root system is well developed. As advancement in the development of The Spade Analysis, the simplified SQSP method provides a semi-quantitative visual assessment that can be conducted by farmers after a short training.

3.2 From 'The Peerlkamp method' to 'VESS' method – including VESS variants

Development of the method:

The 'Peerlkamp' Visual Method of soil-structure evaluation developed in the Netherlands is a modified version of the visual evaluation of soil structure introduced by Ferwerda in the 1940s for arable fields. The method involves characterisation of aggregate size, soil particle cohesion and aggregate porosity, which is summarised in a St-number based on a scale of 10, that is, where 1 refers to a very poor structure and 10 to an ideal soil structure (Peerlkamp, 1959). The St-number was found to be strongly correlated with measurements of aggregate size, porosity, consistency and crop yield on soils at the Institute for Soil Fertility in the Netherlands (Peerlkamp, 1959). Peerlkamp (1959) stated that for his method, the soil structure description and the score assigned differs according to soil texture. The author also stated that before cultivation and after harvesting are the preferable times for conducting the assessment of changes due to mechanical stress in soil structure. The Peerlkamp method, as it is known in soil science societies, has been used as the base protocol for the development of other methods.

The Working Group F of ISTRO in its meeting held in 2005 (Boizard *et al.*, 2005) concluded the need for a revised version of the Peerlkamp method with a summarised criteria description and a shorter scoring scale while retaining accuracy and the speed of assessment (Ball *et al.*, 2007). In response to that, the Visual Evaluation of Soil Structure (VESS) was developed in Scotland by Ball *et al.* (2007) to provide a quick 'look-see' spade method for assessing soil structure to aid management decisions.

The VESS method was criticised for possible subjectivity in terms of breaking up the aggregates by hand (Giarola *et al.*, 2009). Based on this concern, Guimarães *et al.* (2011) conducted a comparison of the aggregate break-up technique with the drop shatter test and demonstrated that both resulted in the same soil quality score. Additionally, Guimarães *et al.* (2011) improved the original VESS by introducing the evaluation of shape and internal porosity of small aggregates (1.5 cm in diameter) following Roger-Estrade *et al.* (2004). The improved VESS method by Guimarães *et al.* (2011) has been amply used by ISTRO researchers and other colleagues, demonstrating its capability to correlate well with several measurements of soil physical quality (Pulido-Moncada *et al.*, 2014c; Ball *et al.*, 2017; Cornelis *et al.*, 2019a), macro-invertebrates density (Franco *et al.*, 2017) and crop yield (Munkholm *et al.*, 2013). This showed its relevance as an indicator of soil health/quality for a range of soil types; also to detect compaction damage (Ball *et al.*, 2017), and changes in soil quality across a land area (Giarola *et al.*, 2013; Leopizzi *et al.*, 2018); as well as to estimate the minimum number of tests needed to characterise a field (Guimarães *et al.*, 2017a; Leopizzi *et al.*, 2018).

The main advantage of VESS technique is that it provides quick and easy identification of soil structure status (Batey *et al.*, 2015), with the aid of close-up images of structural features that are helpful tools for illustration of the soil block description. The influence of soil water content and texture on results of structure evaluation has been reported for this method (Giarola *et al.*, 2010; Batey *et al.*, 2015). Variants of the VESS method have been developed in response to specific limitations for its applicability (briefly mentioned below).

Assessment protocol (key criteria and scale of scoring):

A guide for the block extraction and examination, block break-up and soil scoring is given in Table 1 in Ball *et al.* (2007). The improved description of the breaking up of the aggregates and the method chart by Guimarães *et al.* (2011) are the key guidelines for the examination indicators and scoring. The detailed description for the break-up of fragments by single hand and examination of the internal structural features of small aggregates are key to reduce subjectivity by operators. A score, so-called Sq, is given after evaluation of size and shape of aggregates, visible porosity and roots, distinguishing feature in the aggregates (anaerobic colours), and appearance and description of small aggregates (Fig. 2). Similar to the SQSP by Ball and Douglas (2003), the number of scores was set as five. However, for the VESS method, Sq 1 is best, and 5 is worst, in order to agree with the generically used concept of class 1 being the highest quality. Scores 1–3 indicate good and adequate structures and 4–5 indicate structures that require improvement by management. As suggested by Peerlkamp (1959), layering is considered by scoring each identified topsoil layer separately and calculating the overall soil score by multiplying the score of each layer by its thickness and dividing the product by the overall depth. The use of VESS in different countries has been helped by translation of the reference card into local languages. These and a pictures-only version are all available at Scotland's Rural College

Figure 2 Example of drop/shatter test for the Visual Soil Assessment (VSA) structure method (left) and manual fragmentation for Visual Evaluation of Soil Structure (VESS) method (right) of a silt loam soil under pasture in Belgium. VSA score was 2 (moderate) for soil structure indicator and 31 (moderate) for soil quality, whereas overall VESS scores were 3 (firm or moderate).

(SRUC) website: https://www.sruc.ac.uk/info/120625/visual_evaluation_of_soil_structure.

3.2.1 VESS variant for core samples: CoreVESS

Johannes *et al.* (2017) proposed an adaptation of the VESS method to soil core samples of 150 cm^3 (CoreVESS) in order to investigate the relationship between measurements of soil physical properties and the visual examination of soil structure on the same soil volume. The protocol of CoreVESS suggests the equilibration of the moisture of the soil samples to at specific matric potential before the examination. This improves the VESS method by reducing subjectivity and any influence of varying moisture conditions of the soil in the field. The CoreVESS also allows covering a larger number of replicates that can improve the estimation of spatial variability in a field (Johannes *et al.*, 2017). Another advantage of CoreVESS is the easy assessment of the subsoil by using a dedicated core sampler (Cornelis *et al.*, 2019a). However, the volume of the sample is considerably smaller than a spade-extracted soil block. The small size of core samples used by Johannes *et al.* (2017) limited the range of criteria usable during the visual-tactile examination, leaving the diagnosis of the soil quality to be based only on the difficulty of core break-up, aggregate shape, and visible porosity. Yet, the CoreVESS has been validated by correlations with soil physical properties such as soil organic carbon, bulk density and porosity in temperate soils (Johannes *et al.*, 2017) and air capacity, air permeability, saturated hydraulic conductivity and aggregate stability in weathered tropical soils (Cornelis *et al.*, 2019a).

The first author of this chapter had previously performed the VESS method in the laboratory on a much larger soil core of *ca.* 1200 cm^3 (Fig. 3), and tested the correspondence in soil quality score between spade soil block and soil core when using the full criteria of the standard VESS (data no published). Although the relationship between visual score and measured soil physical properties in the same soil volume of 150 cm^3 was demonstrated by Johannes *et al.* (2017), it is recommended that larger core samples could also be used in the laboratory. Nevertheless, this is subjected to the restriction of the need to adapt sample holders for measurement apparatus such as air or water permeameters. Alternatively, Cornelis *et al.* (2019a) demonstrated that sample size is of no consequence when the 'number and distribution of roots' is included as part of the evaluation criteria of CoreVESS.

3.2.2 VESS variant for grasslands: GrassVESS

The VESS method has been conducted under different land uses; however, in grasslands, the presence of a root-mat has been considered to influence the applicability of the method (Cui *et al.*, 2014). In response, Emmet-Booth *et al.* (2018) developed the GrassVESS as a variant of the VESS method that takes into account the evaluation of root-mat on pasture soils. The GrassVESS protocol of Emmet-Booth *et al.* (2018) suggests evaluation of the root-mat layer separately from the underlying proportion of the soil block denoted as the lower portion (Fig. 4). Therefore, the GrassVESS provides separate flow

Figure 3 Example of CoreVESS procedure on samples of 1200 cm^3 taken after harvesting from a sandy loam under winter wheat in Belgium. From left to right, core in holder, intact core, core after break-up. VESS score was Sq2 (intact, good quality) in the upper 10 cm over Sq4 (compact, poor quality) in the lower 5 cm.

Figure 4 Following careful extraction of an intact soil sample block, GrassVESS requires the separate assessment and scoring of the root-mat layer and the lower portion. The root-mat is the surface soil layer tightly bound by grass roots and typically ~ 5 cm in depth. Assessment involves the manual break-up of soil layers, exposure of aggregates and evaluation of various soil properties according to the GrassVESS flowchart. Photos and drawing courtesy of Jeremy Emmet-Booth.

charts for each classification procedure. Briefly, for the evaluation of the lower portion of the soil block, a decision structure flow chart with reference pictures was constructed to guide the operator through a four score classification. For this, the adjustment of the shape of the aggregates for Sq2 and root growth and aggregates/clods size for Sq4 was needed. The flow chart for the root-mat layer, illustrated with reference pictures, provides three score levels defined by aggregates size, shape and porosity, together with root growth pathway. At 30 different fields in Ireland, GrassVESS was found to be more accurate for visual assessment of grasslands than VESS, and the assigned Sq score was better correlated with the bulk density and porosity of each soil layer (Emmet-Booth *et al.*, 2018). Further studies under different pasture species, soil type and climate conditions are needed.

3.2.3 VESS variant for transition layer evaluation across topsoil and subsoil: Double Spade method

Emmet-Booth *et al.* (2019) proposed the use of VESS down to 0.4 m in order to identify soil management requirements based on the structural condition of both topsoil and plough pan layer (transition layer to subsoil). The authors named this variant of the method the Double Spade method, as it evaluates two spade depths. The Double Spade represents an intermediate method between VESS and SubVESS (described below) by taking principles of both. In brief, the authors' protocol consists of manually excavating a mini trench of ~0.45 m deep, in which the distinguishing layers are identified by knife insertion on the exposed soil wall (Fig. 5). The evaluation of each layer involves as criteria the penetration resistance, redox evidence, aggregate size and shape, intra-aggregate porosity, aggregate rupture resistance and rooting. An overall score for the top ~0.4 m is then given by taking into account the thickness of the identified layers. The Double Spade method has been demonstrated as capable of evaluating traffic-induced changes in soil structure in temperate soils (Emmet-Booth *et al.*, 2020).

Alternatively, Cornelis *et al.* (2019a) used the standard VESS as a double block method on hand-hoed soils, which showed a more shallow transition layer. They extracted two consecutive soil blocks with a spade and evaluated the blocks rather than the pit wall is in the Double Spade method.

Figure 5 The Double Spade Method involves excavation of a soil pit to ~ 0.4 m depth by spade and the separate assessment of structural layers identified on one of the exposed soil pit walls. Different soil properties are evaluated and scored separately. Drawing and description courtesy of Jeremy Emmet-Booth.

3.3 Visual Soil Assessment (VSA)

Development of the method:

For New Zealand soils, Shepherd (2000) developed the detailed visual soil assessment (VSA) method, which involves the visual assessment of both key soil state and plant indicators of soil health/quality, presented in a field guide. The method was developed as a tool to assess and monitor soil health and plant performance to help farmers in New Zealand to develop and implement appropriate sustainable land management practices. In effect, land management centred on soil conditions.

In 2008, the VSA field guides were adapted by the FAO in collaboration with the University of Teramo as a manual consisting of a series of booklets for soil quality assessment for vineyards, olive orchards, orchards, wheat, maize, annual crops and pasture. The adapted VSA manual was published with open access at: http://www.fao.org/3/i0007e/i0007e00.htm.

The VSA by Shepherd (2000) was also selected by the FAO as a method for soil health evaluation in the land degradation assessment in drylands (LADA) project (2006–2010). The adapted version of VSA for the LADA project was applied as part of the assessment of land degradation and sustainable land management at a local level in Argentina, China, Cuba, Senegal, South Africa and Tunisia (Bunning *et al.*, 2011). The final report of this project provides a manual with a soil assessment section as a guideline for scoring soil health and soil erosion, which is of open access at: http://www.fao.org/nr/kagera/tools-and-methods/lada-local-level-assessment-manuals/en/.

In 2009, Shepherd launched updated field guides for pasture and cropping, which also provide a quick protocol to estimate the potential for nutrient loss, carbon sequestration and emission of greenhouse gases. The 2009 field guides of the VSA method are thus tools for an integrated evaluation of the soil health and sustainability of pastoral and cropping lands. The evaluations of soil, plant and environmental concerns are conducted as individual evaluations, but the comparison of their results allows making cause-and-effect links.

The VSA has been successfully applied for soil health/quality examination in temperate agricultural soils of northern Europe (Newell-Price *et al.*, 2013; Sonneveld *et al.*, 2014), Canada and China (Mueller *et al.*, 2009b). Murphy *et al.* (2013) showed that the VSA scores of soil indicators have strong relationships to conventional soil physical measures, and scores from other comprehensive systems such as the SOILpak (described below). They also found the VSA to apply to Australian soil monitoring programmes. However, the VSA, when used in agricultural tropical soils from Venezuela, local adjustment to the expected rating of number of earthworms was needed and calculated based on the density of earthworm of the studied soils (Pulido-Moncada *et al.*, 2014b). The need of further validation of the method in different soil types was mentioned by van Leeuwen *et al.* (2018) after finding VSA score depended on soil type in a study conducted in the Netherlands.

Assessment protocol (key criteria and scale of scoring):

Following the field guide manuals, the soil blocks are extracted and described according to various criteria (Shepherd, 2009). The soil indicators' section of the field guide manuals launched by Shepherd in 2009 provides a detailed description of the criteria of each of the following indicators: soil texture, soil structure, soil porosity, number and colour of soil mottles, soil colour, earthworms (number, size and species), soil smell, potential rooting depth, surface ponding, soil surface (surface cover and crusting, and soil erosion for soils under cropping, whereas soil relief is for pastoral soils).

Different from the rest of the topsoil methods mentioned here, the main soil health assessment in VSA uses a drop shatter technique for fractionation of a soil block into component structural units. The VSA drop test requires coarse soils to be dropped from 0.5 m to 1.0 m for medium- and fine-textured soil (Shepherd, 2009). Soil structure is evaluated by arranging the resulting aggregates after the drop test by size and observing the different proportions (Fig. 2), comparing these with reference photographs and then allocating a score. Additional fresh soil samples are used for porosity and smell evaluation. Each indicator is assigned a score of 2 (good condition), 1 (moderate condition) or 0 (poor condition) comparing the soil with the criteria description and the reference photographs given in the field guides. Each indicator (key diagnostic factor) used by the VSA is considered as a separate entity. In this way, each is a useful early indicator of changes in soil conditions.

To arrive at an overall score of the soil condition each indicator is weighed by a factor of importance suggested by Shepherd (2009) based on expert knowledge and given in the provided scorecard of the field guides. Determination of these soil indicators or overall soil quality condition, when compared with cropping indicators, gives a good means of assessing the agronomic value of the land.

4 Methods based on soil profile examination

Soil profile examination methods broadly follow classical soil survey principles (Ball *et al.*, 2015) of evaluating soil structure morphology and its variation with depth, modified to allow systematic assessments of anthropogenic impact on soil structure. The soil profile assessment is conducted on a large soil pit of~1.5 m depth with varying width depending on the method criteria. Soil layers and horizons are distinguished from top to bottom of the profile, and individual characterisation of soil structure status is conducted per soil layer. Diagnosis of the structural condition of the zone between topsoil and subsoil - the transition layer - and compacted subsoil layers are the main emphasis of the profile examination methods. Detailed examination of visual porosity and aggregate morphology are conducted by evaluation of soil fragments extracted with a knife. Although all the soil profile methods focus on the evaluation of soil compaction, they mainly differ in the dimension of the profile and the type of morphological characterisation of the soil structure applied. Compared to the topsoil methods, soil profile methods provide more detailed information on the footprint of agricultural activities in-depth, as well as on intrinsic soil characteristics and natural soil structure development. However, replication is limited by the destructive and labour-intensive nature of the test.

4.1 The 'Profil Cultural' including Mini 3D Soil Profile variant

Development of the method:

Manichon developed the Profil Cultural method in the 1980s based on the approach for the analysis of tillage implemented by Hénin and his collaborators in the 1960s at the INRA Laboratory of Cultural Techniques in France. In its origins, the method was used as an educational tool, but it was later introduced as a research tool in experimental and survey protocols (Manichon, 1993). The evolution of the Profil Cultural involves the definition of different levels and stratification of structural organisation, which allows identification of the way clods are assembled and the internal state of clods in the topsoil (Manichon, 1987). The method delineates contrasting zones of soil structure resulting from tillage and wheel actions allowing a 'map' of the vertical (tillage horizons) and lateral (wheel tracks) variation of the soil structure.

Roger-Estrade *et al.* (2004) described this method in English. They also successfully used the morphological data from the Profil Cultural with soil physical measurements as the inputs of a model that estimates soil structural dynamics in tilled soils. Peigné *et al.* (2013) introduced a modified version of the Profil Cultural that included the diagnosis of the subsoil with an emphasis on the examination of the number of earthworm burrows per m^2 in a horizontal plane and the quantification of cracking in the transition layer, as factors that can explain the root development in compacted subsoils. Several attempts to improve and adapt the method to specific conditions include a description of platy structure (Boizard *et al.*, 2013; Sasal *et al.*, 2017), and characterisation of earthworm activities (Piron *et al.*, 2017). Boizard *et al.* (2017) unified those different attempts and proposed a new approach to assess degradation and regeneration of soil structure under tillage and no-till. They introduced an improved description of crack type; the addition of platy structures and their horizontal cracking to the system; and linking porosity to a biological activity criterion when present in the different type of structural forms.

The Profil Cultural method has been widely used in France by extension services, and research programmes up to date. In addition, earlier in 1999, the

Profil Cultural was adapted by Tavares Filho *et al.* (1999) for use as a tool for the assessment of changes in soil structure of cultivated clay soils under different management systems in a tropical area.

Assessment protocol (key criteria and scale of scoring):

Well-described protocols are given by Peigné *et al.* (2013) and Boizard *et al.* (2017). Briefly, a pit of ~ 2–3 m wide and 0.8–1.5 m deep is dug to represent the spatial variation of soil structure and is vertically partitioned in layers. In each layer, the structural state is defined in a lateral orientation and compartments of homogeneous soil structure form are delineated. In each compartment, the internal structural state of clods or zones is evaluated focussing on the type of porosity within the soil fragments. The structural state of each identified compartment in the soil profile is qualitatively classified by Greek letters and subscripts that summarised the spatial arrangement of aggregates/clods and type of porosity (Fig. 6). Boizard *et al.* (2013) proposed that the proportion of each structural type can be calculated as the ratio of the area of each

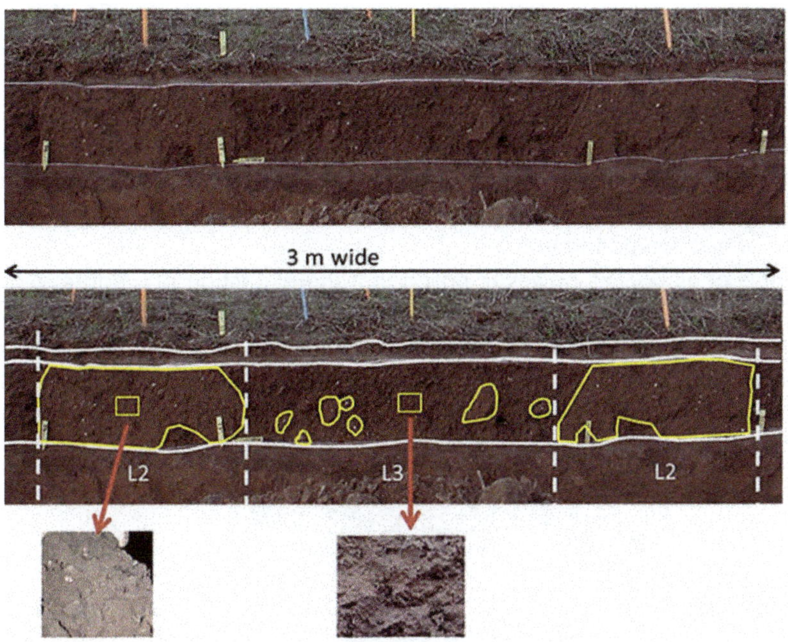

Figure 6 The observation face of a soil pit before and after delineation with the Profil Cultural method. Vertical stratification: H1, seed bed; H5, part of the top layer untilled since the last deep tillage; P1, first untilled layer. Lateral stratification: L1, part of the profile located under the wheel tracks of field operation(s) done after secondary tillage (visible at soil surface); L2, part of the profile located under the wheel tracks of field operation(s) at secondary tillage; L3, part of the profile un-wheeled since ploughing. Zones are delimited by yellow lines. Structural states are: CΔb0 in L2 compartment and CΓb2 in L3. Photos and drawing courtesy of Hubert Boizard.

structural type to the total area of the horizon. Compared to the other methods discussed in this chapter, the Profil Cultural method requires a higher degree of scientific knowledge for soil structure interpretation. This is a reflection of its comprehensive, thorough and time-consuming morphological assessment of soil structure, which integrates changes with time resulting from natural and anthropogenic factors.

4.1.1 Mini 3D Soil Profile

The Mini 3D soil profile was developed by Tomis *et al.* (2019) as a simplified version of the Profil Cultural method, which provides a time-saving and less soil-disturbing alternative to the extensive digging and replacing of the soil, and the limited replicates in a given field. The Mini 3D Soil Profile involves the extraction of a 1 m^3 soil block and its examination in situ while on the loader of a tractor (Fig. 7). Tomis *et al.* (2019) demonstrated that the new approach, Mini 3D Soil Profile, provides similar results to that of Boizard *et al.* (2017), and that variation between the two methods might be related to the location of the soil profiles. Three soil blocks are required as the minimum to represent the spatial variability and to correspond with the ~3 m wide pit of the Profil Cultural. The soil block is partitioned by eye on the vertical and lateral orientation, and each compartment delineated is visually examined following the same criteria as the Profil Cultural (Boizard *et al.*, 2017), that is, type of morphology, type of porosity and biological activity. The structural state of the soil block is a comparative analysis of each compartment.

Tomis *et al.* (2019), however, noted that the extraction procedure might cause slight overestimation of *phi* structure types (fragments with sharp edges, separated by flat cracks oriented in all spatial directions) by creating new cracks or revealing pre-existing cracks that are not detectable by the Profil Cultural. The dimension of the soil block, taken for the Mini 3D Soil Profile, limits the

Figure 7 Example of a soil profile extraction and observation face with the mini 3D Soil Profile method. Photos courtesy of Vincent Tomis, Agro-Transfert-RT.

mapping of each cultural operation of the evaluated area. This variant does not intend to replace the Profil Cultural method; however, Tomis *et al.* (2019) considered their new approach as more practical for extension services with farmers. Information obtained from the method allows examination of the causes of differences between the identified soil structure zones or compartments, and decision-making regarding the loosening of the soil.

4.2 SOILpak Scoring Procedure

Development of the method:

The SOILpak is the Soil Management Package for cotton production on cracking clay soils (Vertisols) designed originally by Daniells and Larsen (1991) in Australia. The SOILpak system was developed to assess soil compaction of Vertisols under irrigated cotton, and to benefit the Australian cotton growers in their decisions on soil management (Daniells *et al.*, 1996). The SOILpak system is described in a manual of five sections, from which the section 'Diagnosis of Soil Structure' is the key in the soil structural form evaluation process. This SOILpak section is an all-embracing structural description that covers the evaluation of structural form, structural stability in water and structural resilience (McKenzie, 2013).

A refined edition of the cotton SOILpak manual by McKenzie (1998) was extensively used by The Cotton Research and Development Corporation in Australia to conduct hands-on training in soil assessment for cotton growers, consultants and advisers. Expert knowledge from researchers, agronomists and crop growers has been the source base for dissemination, but also for further adaptations of this numeric system for assessing structural form. After extensive use of the SOILpak system, that is, 1500 times, McKenzie (2001a) provided a revised SOILpak scoring procedure that takes into account (i) contradictory component scores or unavailable factors by using an equation in which each factor is weighted by a factor of importance for root growth, (ii) features such as continuous vertical macropores and smears layers, and (iii) adaptation of the assessment system for loam soil. The modified SOILpak scores proved to be strongly correlated to shear strength and air-filled porosity for Vertisols (McKenzie, 2001b). The SOILpak was hence modified by McKenzie (2001a) to be used on a wide range of soils for cotton root growth relevance. Further adaptations of the SOILpak procedure are applicable to cropping systems such as cereals and vegetables (McMullen, 2000; Anderson *et al.*, 2007).

The evaluation of aggregate stability in water is proposed by McKenzie (2013) to complement SOILpak 'soil structural form' in the assessment of slaking and dispersion of waterlogged soils (Field *et al.*, 1997). The aggregate stability test also includes the evaluation of the dispersion of remoulded aggregates after being moistened. The manually remoulded aggregates simulate the impact of machinery on wet soil.

The SOILpak method has been used in Australian cotton industry by farmers and advisers since the 1980s and was reported to help to boost productivity (McKenzie, 2013). The SOILpak method has been recognised as providing a relatively quick understanding and evaluation of soil structure variation in medium- and fine-textured soils, but an adaptation for its use in sandy soils is needed (Batey *et al.*, 2015).

Assessment protocol (key criteria and scale of scoring):

SOILpak is a scoring technique applicable to the top 0.9 m of soil. Features for deriving the SOILpak 'soil structural form' score (listed in descending order of importance for root growth) for cotton in firm Vertisols are the width of primary clods, ease of breakage of the soil sample, behaviour of fresh cotton roots, shape of clods produced by moderate hand pressure, features of fracture faces, the proportion of primary clods within the compound clods, internal porosity of primary clods and the colour of the interior of primary clods. In the case of loose Vertisols, the previously listed factors are applicable, except for root morphology, features of fracture faces and colour of clods (McKenzie, 2001a). For loamy soils and other crop systems, the factors used are the size of natural subunits of aggregates, ease of fracture, aggregate shape, fracture faces, the proportion of small aggregates within larger aggregates and internal porosity of smallest aggregates (McMullen, 2000; Anderson *et al.*, 2007). As an additional factor for very dry soil, supplementary assessment procedures for ease of soil fragments fracture are considered.

The SOILpak 'soil structural form' scoring procedure is based on the scheme of Batey (1988), and includes three categories on a scale of 0 to 2, that is, 0 (poor structure), 1 (moderate structure) and 2 (good structure), for severely compacted to excellent structure for root growth, respectively. When the operator of the method becomes an expert, he/she may apply a 20 point scale for the scoring procedure, that is, 0.0–2.0 in steps of 0.1 (McKenzie, 2001a). As part of the measurement package for 'soil structural form', the aggregate stability score is given after 10 min and again after 2 h (Fig. 8) after the application of the dispersion treatment. Dispersion scores range from 0 (nil dispersion) to 4 (complete dispersion) for both undisturbed and remoulded aggregates (McKenzie, 2013). Interpretation of aggregate form, stability and resilience are provided for each layer of the soil profile by classifying the soil structure as 'terrible, moderate or excellent' and providing suggestions on management strategies based on the soil structure classification.

4.3 The Numeric Visual Evaluation of Subsoil Structure (SubVESS)

Development of the method:

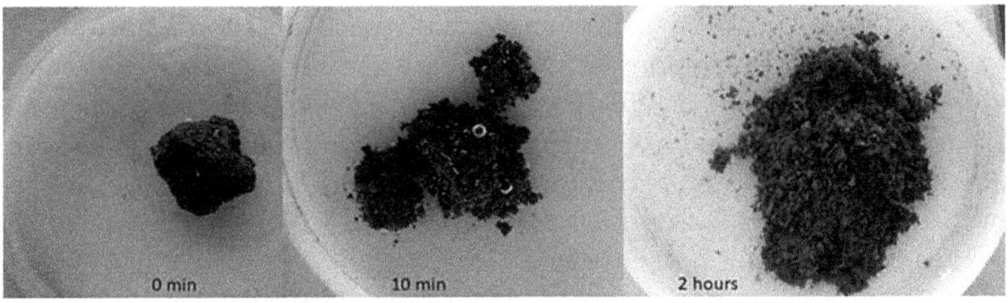

Figure 8 Example of the water aggregate stability test in the field. The soil sample was taken from a silt loam under mono-cropping and conventional tillage in Venezuela.

The Working Group F of ISTRO in its meeting in 2005 (Boizard *et al.*, 2005) also recognised the need for a rapid numerical assessment of subsoil structure. Later in its meeting in 2011, an adaptation of the VESS method was developed for the numerical assessment of any anthropic structural deformation of the transition layer and identification of subsoil compaction. The subsoil structural quality assessment, called SubVESS, was then published by Ball *et al.* (2015). The main purpose of the SubVESS was to identify the requirement for subsoil compaction mitigation practices and the depth at which these are needed. The key criteria used in SubVESS include drainage status, as well as cracking and biopores as essential soil characteristics for gas transport and pathways for roots in the subsoil.

Ball *et al.* (2015) applied the method in soils of contrasting compaction status under a temperate and tropical climate, as well as in soils of contrasting types in a temperate climate. The authors showed that SubVESS could identify subsoil quality under diverse conditions. Pulido-Moncada *et al.* (2019) found that SubVESS detected signs of increased compaction after multiple passes of heavy traffic on temperate sandy loam soil, reflecting a diminution of effectiveness of macroporosity found from pore characterisation conducted in the laboratory. However, Ball *et al.* (2015) recognised that the method did not assess the water-holding capacity, which is a factor of importance in stony soils and Vertisols, for which systems such as the SOILpak or the Profile Cultural are more suitable.

Assessment protocol (key criteria and scale of scoring):

As in the case of VESS, it is preferable to conduct the assessment when the soil is moist and when roots are fully developed. On the soil profile, layers of contrasting structure in the subsoil are identified by prodding with the point of a knife and also by recognition of distinct soil colours. Examination of each soil layer and its soil fragments involves mottling, strength, porosity, roots, and aggregate size and shape using a diagnostic flowchart provided by Ball *et al.* (2015). The SubVESS scoring system assigns a separate score to each indicator or key factor. The SubVESS does not distinguish the level of importance among the indicators by a weighting factor and leads the user to decide an overall score for each soil layer based on the most frequently occurring score of the five factors or indicators. The final score given to each individual layer is assigned with greater emphasis on porosity and aggregates. The SubVESS in common with VESS comprises 5 scores, where 1 is best, and 5 is worst. However, the description criteria are not interchangeable with VESS scoring. The VESS and SubVESS can be applied together for a full description of the soil profile.

5 Dissemination of visual soil evaluation techniques and their future trends in research

The use of visual soil evaluation techniques is not exclusive to those mentioned here, or in Boizard *et al.* (2005) and Emmet-Booth *et al.* (2016). Local publications, which are not necessarily known by the international research community, also provide refined or novel techniques for soil health/quality assessment that have sensitivity and suitability to local conditions and cropping.

Examples of these are the Rapid Diagnosis of Soil Structure (DRES, for its acronym in Portuguese) (Ralisch *et al.*, 2017) and the Practical Guide for Participative Evaluation of Soil Quality (PGPE) (Comin *et al.*, 2016), which were developed for applicability in tropical and subtropical areas in Brazil. DRES is a spade method based on both VESS and refined version of the Profil Cultural for tropical soils (Tavares Filho *et al.*, 1999). DRES is a topsoil spade method with a focus on the assessment of soil degradation and recovery. Examination criteria are root development and distribution and pathways; the shape, roughness, and size of aggregates; relative proportion (in volume) of the different aggregate sizes and their morphological characteristics; transitional level of structural condition between layers; compaction; and biological activity. By contrast, PGPE is a profile method that was developed for soil quality assessment on no-tillage vegetable systems on tropical soils (Comin *et al.*, 2016). This involves the evaluation of soil organic matter, rooting, soil structure, compaction and infiltration, erosion, water retention, macrofauna activity and surface cover. When comparing these two techniques with the analytical Soil Management Assessment Framework developed by Andrews *et al.* (2004), Pereira Valani *et al.* (2020) suggested that both DRES and PGPE have the potential to distinguish soil quality within management systems under local conditions.

The previous examples in connection with the case studies presented in this chapter provide evidence that the use and the advancement of the visual soil evaluation techniques have been disseminated to technically underdeveloped, newly industrialised and developed countries. Additionally, as visual soil evaluation techniques are already important for evaluating and monitoring soil degradation by compaction and erosion, they also have potential to be extended to other forms of degradation such as desertification in different geographical areas (Ball and Munkholm, 2015).

There is also a potential for the use of the visual scores of crop and soil, and accompanying soil fertility measurements to form a network of indicators of fertility that can be part of a management package to maximise crop productivity and/or to optimise environmental qualities (Shepherd, 2009; Guimarães *et al.*, 2017a).

Morphological observations of soil obtained from visual examination have also been shown to be applicable as potential independent variables for predicting other structure-related properties (Lilly *et al.*, 2008; Vereecken *et al.*, 2010; Nguyen *et al.*, 2014) and to develop soil health/quality frameworks (Pulido-Moncada *et al.*, 2014a; Bondi *et al.*, 2018; Johannes *et al.*, 2019). The structure indicators used as criteria of the visual soil evaluation techniques have also been used and proposed as quick tools for assessing the capacity of the soil structure status to function as a regulator of carbon storage, greenhouse gas emissions and nutrient leaching (Cloy *et al.*, 2015). Hence, the use of visual techniques for in situ soil data collection and their inclusion in the development of soil health/quality frameworks is an area of significance in future research (Karlen *et al.*, 2019).

Advanced techniques for soil structural assessment such as remote and proximal sensing, and the well-established laboratory analyses of soil samples, have been proposed for integration with visual soil evaluation data (Guimarães *et al.*, 2017b). The use of visual techniques can provide a useful first approximation of the variation in soil structure in the field, which helps to define boundaries of areas of contrasting structures and areas of the poor structure that would benefit from further detailed assessment (McKenzie, 2013). Additionally, visual soil data can be used to reveal degraded areas that need more detailed evaluation for a specific recommendation of soil management practices (Guimarães *et al.*, 2017b). Furthermore, visual soil evaluation methods may be useful for validation of the development of advanced soil profile visualisation such as using portable X-ray fluorescence spectrometry (Sun *et al.*, 2020).

Perhaps the greatest benefit of visual techniques for soil structure assessment is the provision of a ready understanding of the soil as a basic resource vulnerable to management that is simply derived from feeling, observation and smell. Awareness of the value of soil as a resource can bring a deeper perception in land users of their shared, inherited values (Wahlhütter *et al.*, 2016). These approaches would provide valuable support to agroecological approaches with a focus on sustaining and regenerating soil health/quality.

Visual soil evaluation techniques continue to be refined as some researchers seek to adapt the criteria of assessment or the scoring system in order to cover the complexity of different soil systems or local needs. As an illustration of this, the VESS method was presented in a tutorial video produced by the Swiss Confederation's Centre of Agroscope in 2020, with a modified diagnostic chart that gives more emphasis in the evaluation of inner characteristics of the aggregates/clods after fragmentation including aspects such as roughness of the face of the aggregates (https://youtu.be/BWUeERE-wJw). Another example is an upcoming variant of the VESS method, currently under development by Joanna Cloy and colleagues, that intends to provide an extended scoring system with improving the evaluation of existing properties and new properties. Their goal is to provide a tool that offers agronomic information for soil management and crop productivity decisions.

6 Case study I: VESS and sustainable agricultural assessment and management

The VESS method has been applied extensively in the UK, Scandinavia and Brazil and has been particularly valuable as an indicator of the sensitivity of soil to management such as tillage (Giarola *et al.*, 2013; Guimarães *et al.*, 2017a).VESS is now applied in general farming practice with the scores used to determine options to maintain or improve soil quality in both arable land and grassland (Ball *et al.*, 2017; Emmet-Booth *et al.*, 2018). In Scotland, VESS has been incorporated into practical agricultural guidance literature such as 'Valuing your soils' where it forms part of general soil fertility assessments (Cloy *et al.*, 2016). Here the application of VESS as a tool for soil management and environmental assessment is illustrated for medium-heavy textured soils in South Scotland, an area with moist soils (annual rainfall ~1000–1360 mm in the previous decade) (Fig. 9).

Along with other semi-quantitative visual soil evaluation techniques, VESS has also become popular with scientists for making decisions on sampling and soil management for cropping (Batey *et al.*, 2015; Emmet-Booth *et al.*, 2016). VESS can be used to detect zones of differing soil structure and has proved particularly useful in identifying compact layers both in topsoil and subsoil (Ball *et al.*, 2017; Emmet-Booth *et al.*, 2019). This makes it suitable for choosing the extent, type and depth of use of soil loosening equipment such as subsoilers and for identifying the need for controlled traffic (Godwin and Spoor, 2015) (Fig. 9).

VESS can be used in the derivation of environmental diagnostics. The ability to identify layers helps to specify layers that either restrict infiltration, increase runoff or are anaerobic, possibly requiring drainage (Fig. 9). Such layers can also contain zones of restricted porosity that increase the risk of the production and emission of greenhouse gases, N_2O and CH_4 (Ball *et al.*, 2017). Evidence of this can be inferred from the correlation of VESS scores with the shape of the water release curve and with hydraulic conductivity in weathered soils in

Environmental diagnostics		Temporal change, comparing A and B at the same location in a compaction experiment under grass	
Carbon accumulation in the upper 0.10 m in forest soil	Poor aeration due to prolonged flooding after extreme rainfall	A) In 2011 before annual compaction	B) In 2014 after 3 annual compaction treatments and with higher winter rainfall than average

Soil management actions			
Soil structure improving under organic grass plus clover	Well-structured soil suitable for no-tillage	Soil with compact layer above 0.1-0.15 m depth that requires shallow loosening before cropping	Soil with compact layer below ~0.1 m that requires loosening to 0.25-0.30 m depth before cropping

Figure 9 Illustrations of the use of VESS for environmental diagnostics, monitoring of temporal change associated with management and weather and for specifying management decisions. Compact layers are shown loosened to allow comparison. These examples are from medium-heavy textured soils in South Scotland.

Uganda (Cornelis *et al.*, 2019a). Soil colour, readily identified using VESS, can be used as an indicator of soil organic matter/soil carbon content (Wills *et al.*, 2007), with darker colours indicating higher organic matter contents (Fig. 9). VESS scores also correlated with soil C in tropical soils (Pulido-Moncada *et al.*, 2014c; Guimarães *et al.*, 2017a). Visual soil evaluation techniques can be used to develop a greenhouse gas emission index, an index of nutrient leaching and a soil carbon storage index though, to achieve this, extra visual assessments and measurement of some soil properties are required (Shepherd, 2009).

With good sensitivity to detection of compaction damage, regular repetitions of VESS can be used to monitor changes in compaction status as may occur during weather extremes associated with climate change (Ball *et al.*, 2017). Heavy tractor compaction of grassland in unusually wet autumn in South Scotland resulted in a structure (VESS score 4.2), considerably poorer than the good structure (VESS score 2.9) in the non-compacted reference (Fig. 9). Taking photographs of the soil at different times helps to illustrate and record these changes in soil structure.

Visual evaluation of soil has proved useful in fostering the connection to the soil of a wide range of users from farmers and gardeners to scientists and students through demonstrations and teaching (Ball *et al.*, 2018). It is valuable for raising awareness of soil and soil management at farmers' meetings where handling and inspection of soil samples can lead to sharing of ideas and experiences. This can help the transition away from intensive farming by participation in groups on farming focussed on soil conditions and regeneration of soil fertility such as organic farming or no-till (Ball *et al.*, 2018). A study in Ghana showed how the VESS method could help integrate farmers' knowledge with a science-based assessment of soil qualities (Obour *et al.*, 2020). One of the properties of soil evaluation during visual soil evaluation that is most appreciated by farmers and general users is smell; this provides a useful, intuitive indicator of the function of living organisms within the soil (Ball *et al.*, 2017). This, along with the physical contact with the soil in VESS can increase awareness of the soil-food connection and develop respect for the soil and for others who work it (Ball *et al.*, 2018).

7 Case study II: visual techniques to assess soil structure application and contribution to agriculture in Africa

Agriculture, which is and will be the backbone of the African economy in the near future, is rainfed in > 95% of Africa (Wani *et al.*, 2009). As in other regions in the world, rainfall is becoming more erratic with later rainy seasons, more and longer dry spells and less rainy days, while on the other hand extreme precipitation might lead to an upsurge in flood incidents (Sillmann *et al.*, 2013). Building land resilience to droughts and floods is therefore key to secure food in Africa, particularly when water is limiting. Soil management aiming at improving the soil's infiltrability, and water holding and air capacity, is an important strategy to improve rainwater partitioning between green and blue water flows at the field scale thus optimising the root zone water balance (Cornelis *et al.*, 2019b), especially in regions with blue water shortage, but green water freely available (Rockström *et al.*, 2009). Blue water refers to stored run-off water in dams, lakes, and rivers, and recharged water in aquifers, while green water is the water stored in the soil as soil moisture generated from infiltrated rainwater (Falkenmark, 1995).

Suboptimal soil conditions regularly result from land degradation (Stroosnijder, 2009) and soil structural degradation in particular. In Africa, several national and international initiatives have been undertaken to improve soil structural quality and increase crop production. However, many institutes in Africa lack the means to measure soil structure-related properties and thus evaluate soil structural quality. Visual soil evaluation techniques have great potential as alternatives for or to complement more expensive, labour-intensive quantitative methods of indirect soil structure analysis. Moreover, they can be used by scientists, extension workers, farmers and their advisors, that is, all parties interested in assessing and monitoring soil structural quality. Non-governmental organisation collaborators working in Africa testified that visual soil evaluation methods could be a very low-cost way to track soil health in their projects. They can also effectively measure sustainable improvements in soil health that they often claim in their project proposals but hardly test before and after the project interventions, leaving them with very little evidence on which to base their claims.

When evaluating the impact of soil management on soil physical quality, most studies in resource-poor countries rely on measuring bulk density. Despite their particular potential for evaluating soil structural quality, visual soil evaluation methods are at present rarely applied in Africa (Franco *et al.*, 2019) or at least very few uses are reported in the scientific literature (Obour *et al.*, 2020). In one of such studies, Cornelis *et al.* (2019a) explored the potential of the VSA method (Shepherd, 2009), the VESS method (Guimarães *et al.*, 2011), VESS variants like CoreVESS (Johannes *et al.*, 2017) or with alternative scoring (i.e. arithmetic mean of individual scores given to the criteria considered as also done in Double Spade VESS, Emmet-Booth *et al.*, 2019), to assess soil structural quality of highly weathered ferralitic soils in Uganda. They applied these methods in a study aiming at the evaluation of two soil-improving maize cropping systems (i.e. planting pits and rip lines) that started three years earlier, *vis-a-vis* conventional tillage and natural forest (Fig. 10).

All visual soil evaluation approaches tested were able to detect changes in the structural quality of the highly weathered soil resulting from differences in land use. The Sq score given per approach matched rather well with each other (with Pearson r ranging between 0.66 and 0.92; see Fig. 11 for an example). Natural forests resulted in the best soil quality (with Sq scores ranging from 1.3 to 1.9 for VESS and its variants; $p<0.05$), but differences between the three cropping systems (Sq=2.3–2.9) were not significant. The soil-improving systems, after three years, did not significantly improve the visually estimated quality of the highly weathered soil when compared with conventional tillage. Other studies that used traditional quantitative methods to evaluate soil quality, though conducted under different agroecological conditions and soils, demonstrated that soil-improving systems need more than three years to become significantly effective (Araya *et al.*, 2016; Moreira *et al.*, 2016). The 'fair' Sq scores under all cropping systems might be due to the inherently good physical quality of highly weathered soils (WRB, 2015). The Sq scores under natural forest indicated 'good' soil structural quality, as expected.

Figure 10 Extraction of soil block with spade for visual soil evaluation and examination of soil structural quality of highly weathered soil in Uganda using VESS method. Photo courtesy of David Akodi.

The Ugandan study further showed that individual VESS Sq scores correlated well with a range of indirect lab-derived soil quality indicators (SQi's), including organic carbon content, bulk density/porosity, air capacity, air permeability, saturated hydraulic conductivity and mean-weighted aggregate diameter after having exposed aggregates to fast wetting (Cornelis *et al.*, 2019b). The correlation with plant available water capacity, which is less determined by soil structure but rather by clay content and mineralogy was, as expected, low. Interestingly, a VESS Sq score of 3 (which is the threshold for 'poor' quality) converged almost exactly with SQi threshold values often reported in the literature and relevant for highly weathered soils. For instance, Sys *et al.* (1993) considered 8 g SOC kg^{-1} as critical for soils in the tropics. This value coincided with a Sq score of 3 on the regression line relating both variables (Cornelis *et al.*, 2019a). Figure 12 shows an example of the relation between Sq scores (in this case from VESS) averaged per treatment and an overall soil physical quality index SQI. The latter was calculated from five significant (physical) SQis that

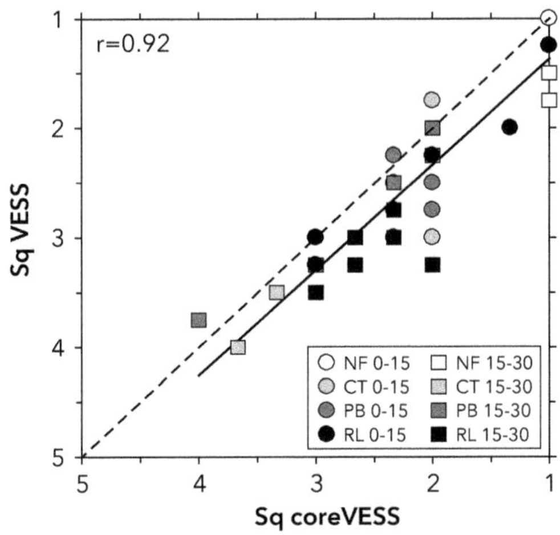

Figure 11 Example of comparison between visual soil evaluation and examination methods applied on highly weathered soil in Uganda: VESS vs. CoreVESS Sq scores (N=44). NF is natural forest, CT is conventional ploughing, PB is permanent beds, RL is rip lines, numbers refer to soil depth in cm. Sq is soil quality score calculated as the arithmetic mean of the individual scores given to the criteria considered. A score of 1 is 'good', 5 is 'poor'. Full line is regression line, dashed line is 1:1 line, *r* value shown is Pearson correlation coefficient. Note that several symbols overlap. (Reprinted from Cornelis et al. (2019), Copyright (2020), with permission from Elsevier).

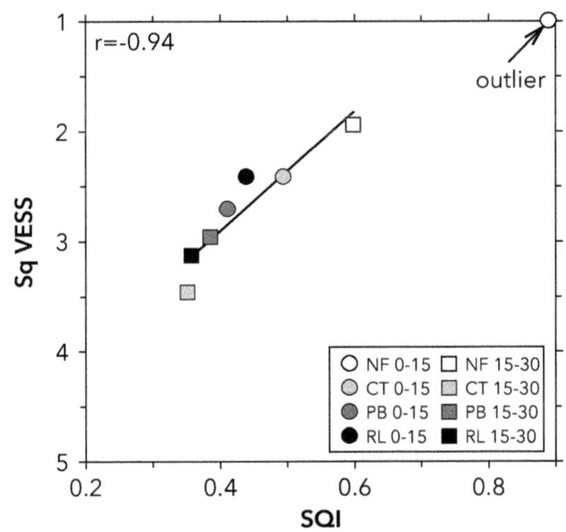

Figure 12 Example of relation between visual soil evaluation and examination-based Sq scores (VESS) and soil physical quality index SQI averaged per treatment and depth of highly weathered soil in Uganda (N=7; the outlier was excluded from the regression). NF is natural forest, CT is conventional ploughing, PB is permanent beds, RL is rip lines. Sq is soil quality score calculated as the arithmetic mean of the individual scores given to the criteria considered. Full line is regression line, r value shown is Pearson correlation coefficient (Reprinted from Cornelis et al. (2019), Copyright (2020), with permission from Elsevier).

characterise a different aspect of soil structure using linear scoring of 'more is better' and an (non-weighted) additive method. The high correlation suggests that VESS (but also variants/alternatives, not shown) captures very well the integrated aspects of soil structural quality.

Based on these promising results, there is need for further promote visual soil evaluation methods in Africa and to develop guidelines and other learning materials (e.g. video, apps) in local languages. VSEE methods should be given due attention in workshops and training courses given to a range of end-users, including, among others, scientists, students, surveyors, extension workers, farmers (including urban farming communities) and policymakers. It is encouraging that some non-governmental organisations are picking up these ideas. The methods might also be used at primary and secondary schools. Learning how to evaluate soils in a simple, intuitive manner can stimulate a better appreciation of soils by society. This can be best achieved by examining soils in the field; after all, a Swahili proverb says that *Mambo ya shamba, huishia shambani* (field stuff is to be discussed in the field). It can then add to a better comprehension of how soils can contribute to build resilience to climate change, to food and water security, to deliver ecosystem services, and thus to, directly and indirectly, achieve several United Nations' Sustainable Development Goals, which is of particular relevance to Africa.

8 Summary

Soil degradation caused by anthropogenic damage restricts soil functions and soil health/quality. The development of well-established protocols of methods to assess changes in soil health/quality, mainly related to agricultural activities, has been one of the significant achievements in soil science in the twentieth century. Changes of the soil under agricultural systems can be indirectly assessed through soil structure-related parameters. Among the several methods for soil structural quality assessment developed, direct visual soil evaluation methods are mainly related to visual and tactile examination of soil fragments and aggregate characteristics such as shape, size, stability, internal porosity and strength. Visual evaluation of soil structure made directly in the field generates a rapid and immediate assessment of soil health/quality that relates well to companion measurements in the laboratory. The visual soil evaluation methods have been developed in temperate regions, yet their use by agronomists, farmers and soil scientists has spread worldwide.

The wide range of visual soil evaluation techniques developed to date varies from those that involve general soil structure from description to more comprehensive systems whereby soil structure-related properties, root development, soil organic decomposition, faunal activity, aggregate stability, among other land and environmental indicators, are integrated. Topsoil visual methods are quick and accurate for qualitative or semi-quantitative evaluation of soil tilth and topsoil compaction in the field. Spade methods for topsoil examination have the advantage of being rapid enough for sufficient replication in experimental plots or fields. Visual methods focussed on subsoil or profile description, although limited in the number of replicates, being destructive, and more time-consuming, provide more information on the horizontal and vertical variability of soil structure that is valuable for cause-effect interpretation.

The development of new techniques or the improvement of old visual techniques has involved improved specification of the degree of detail for soil structural form description, soil type, land use, faunal activity and time taken for assessment. Further improvement of the visual techniques might focus on consideration of special soil conditions such as very sandy or very clayey soils, and stony soils, as well as on the influence of soil moisture content on soil structure evaluation.

Visual techniques have been successfully applied in different countries, and their applicability, feasibility and reliability have been proven through the relationship between morphological descriptions from the field and soil physical properties from the laboratory. The acceptable performance of the visual methods, and their variants derived from improvement, supports the applicability of them for timely assessment and monitoring of soil structure status in extension services and research projects.

The assessment of soil structural quality and selection of protocols can be considered as an agronomic basis for soil conservation/regeneration and sustainable agriculture frameworks. Researcher-farmer collaborations can generate data related to soil structure dynamics in agricultural soils, which can guide policymaking with an emphasis on addressing key research needs in both technically underdeveloped and developed countries.

9 Where to look for further information

9.1 The following articles and book provide a good overview of the subject:

- Ball, B. C. and Munkholm, L. J. (Eds). (2015). Textbook: *Visual Soil Evaluation: Realising Potential Crop Production with Minimum Environmental Impact*. CABI. ISBN-13: 978 1 78064 470 7.
- Ball, B. C., *et al.* (2017). Visual soil evaluation: a summary of some applications and potential developments for agriculture. *Soil and Tillage Research* 173: 114–124.
- Emmet-Booth, J., *et al.* (2016). "A review of visual soil evaluation techniques for soil structure." *Soil Use and Management* 32(4): 623–634.

- Guimarães, R., Lamandé M., Munkholm, L. J. and Keller, T. (Eds). (2017). Special issue: Visual soil evaluation and soil compaction research. *Soil and Tillage Research* 173: 1–124.
- Munkholm, L. J., Ball, B. C, and Batey, T. (Eds). (2013). Special issue: Applications of visual soil evaluation. *Soil and Tillage Research* 127: 1–100.

9.2 Key conference

- The triennial ISTRO conference (www.istro.org).

9.3 Key research in this area can be found at the following organisations:

- Scotland's Rural College (SRUC): https://www.sruc.ac.uk/info/120625/visual_evaluation_of_soil_structure.
- Bioagrinomics, Agricultural consultancy: https://www.bioagrinomics.com/visual-soil-assessment.
- Agro-transfert: http://www.agro-transfert-rt.org/.

10 References

Anderson, A., Kelly, J. and McKenzie, D. 2007. *Healthy Soils for Sustainable Vegetable Farms: Ute Guide*. Arris Pty Ltd, Australia .

Andrews, S. S., Karlen, D. L. and Cambardella, C. A. 2004. The soil management assessment framework. *Soil Science Society of America Journal* 68(6), 1945–1962.

Araya, T., Nyssen, J., Govaerts, B., Deckers, J., Sommer, R., Bauer, H., Gebrehiwot, K. and Cornelis, W. M. 2016. Seven years resource-conserving agriculture effect on soil quality and crop productivity in the Ethiopian drylands. *Soil and Tillage Research* 163, 99–109.

Arvidsson, J. and Keller, T. 2007. Soil stress as affected by wheel load and tyre inflation pressure. *Soil and Tillage Research* 96(1–2), 284–291.

Ball, B. C., Batey, T. and Munkholm, L. J. 2007. Field assessment of soil structural quality – a development of the Peerlkamp test. *Soil Use and Management* 23(4), 329–337.

Ball, B. C., Batey, T., Munkholm, L. J., Guimarães, R. M. L., Boizard, H., McKenzie, D. C., Peigné, J., Tormena, C. A. and Hargreaves, P. 2015. The numeric visual evaluation of subsoil structure (SubVESS) under agricultural production. *Soil and Tillage Research* 148, 85–96.

Ball, B. C. and Douglas, J. T. 2003. A simple procedure for assessing soil structural, rooting and surface conditions. *Soil Use and Management* 19(1), 50–56.

Ball, B. C., Guimarães, R. M. L., Cloy, J. M., Hargreaves, P. R., Shepherd, T. G. and McKenzie, B. M. 2017. Visual soil evaluation: a summary of some applications and potential developments for agriculture. *Soil and Tillage Research* 173, 114–124.

Ball, B. C., Hargreaves, P. R. and Watson, C. A. 2018. A framework of connections between soil and people can help improve sustainability of the food system and soil functions. *Ambio* 47(3), 269–283.

Ball, B. C., Munkholm, L. J. and Batey, T. 2013. Applications of visual soil evaluation. *Soil and Tillage Research* 127, 1-2.

Batey, T. 1988. *Soil Husbandry: A Practical Guide to the Use and Management of Soils*. Soil and Land Use Consultants Ltd., Aberdeen, UK.

Batey, T. 2000. Soil profile description and evaluation. In: Smith, K. A. and Mullins, C. E. (Eds), *Soil and Environmental Analysis, Physical Methods* (2nd edn). Marcel Dekker, New York, NY.

Batey, T., Guimarães, R. M., Peigné, J., Boizard, H., Ball, B. and Munkholm, L. 2015. Assessing structural quality for crop performance and for agronomy (VESS, VSA, SOILpak, Profil Cultural, SubVESS). In: Ball, B. C. and Munkholm, L. J. (Eds), *Visual Soil Evaluation: Realising Potential Crop Production with Minimum Environmental Impact*. CABI, Wallingford, pp. 15-30.

Bennett, L. T., Mele, P. M., Annett, S. and Kasel, S. 2010. Examining links between soil management, soil health, and public benefits in agricultural landscapes: an Australian perspective. *Agriculture, Ecosystems and Environment* 139(1-2), 1-12.

Beste, A. 1999. An applicable field method for the evaluation of some ecologically significant soil-function-parameters in science and agricultural consulting practice. In: *International Soil Conservation Organisation (ISCO)-Conference*, West Lafayette, IN.

Boizard, H., Batey, T., McKenzie, D., Richard, G., Roger-Estrade, J., Ball, B., Bradley, I., Cattle, S., Hasinger, G. and Munkholm, L. 2005. *Field Meeting "Visual Soil Structure Assessment"*. Detailed Report. INRA, France.

Boizard, H., Peigné, J., Sasal, M. C., de Fátima Guimarães, M., Piron, D., Tomis, V., Vian, J.-F., Cadoux, S., Ralisch, R., Tavares Filho, J., Heddadj, D., De Battista, J., Duparque, A., Franchini, J. C. and Roger-Estrade, J. 2017. Developments in the "profil cultural" method for an improved assessment of soil structure under no-till. *Soil and Tillage Research* 173, 92-103.

Boizard, H., Yoon, S. W., Leonard, J., Lheureux, S., Cousin, I., Roger-Estrade, J. and Richard, G. 2013. Using a morphological approach to evaluate the effect of traffic and weather conditions on the structure of a loamy soil in reduced tillage. *Soil and Tillage Research* 127, 34-44.

Bondi, G., Creamer, R., Ferrari, A., Fenton, O. and Wall, D. 2018. Using machine learning to predict soil bulk density on the basis of visual parameters: tools for in-field and post-field evaluation. *Geoderma* 318, 137-147.

Bunning, S., McDonagh, J., Rioux, J. and Woodfine, A. 2011. *Manual for Local Level Assessment of Land Degradation and Sustainable Land Management*. LADA Technical Report.

Cardoso, E. J. B. N., Vasconcellos, R. L. F., Bini, D., Miyauchi, M. Y. H., dos Santos, C. A., Alves, P. R. L., de Paula, A. M., Nakatani, A. S., de Moraes Pereira, J. and Nogueira, M. A. 2013. Soil health: looking for suitable indicators. What should be considered to assess the effects of use and management on soil health? *Scientia Agricola* 70(4), 274-289.

Carducci, C. E., Zinn, Y. L., Rossoni, D. F., Heck, R. J. and Oliveira, G. C. 2017. Visual analysis and X-ray computed tomography for assessing the spatial variability of soil structure in a cultivated Oxisol. *Soil and Tillage Research* 173, 15-23.

Carter, M. 2004. Researching structural complexity in agricultural soils. *Soil and Tillage Research* 79(1), 1-6.

Cherubin, M. R., Chavarro-Bermeo, J. P. and Silva-Olaya, A. M. 2019. Agroforestry systems improve soil physical quality in northwestern Colombian Amazon. *Agroforestry Systems* 93(5), 1741-1753.

Cloy, J. M., Audsley, R., Hargreaves, P. R., Ball, B. C. and Griffiths, B. S. 2016. *Valuing Your Soils: Practical Guidance for Scottish Farmers*. Centre for Expertise for Waters (CREW), Scottish Environment Protection Agency (SEPA) and Scotland's Rural College (SRUC), Edinburgh. Available at: https://www.sruc.ac.uk/downloads/file/2989/cloy_et_al_valuing_your_soils_practical_guidance_for_scottish_farmers.

Cloy, J. M., Ball, B. C., Shepherd, T. G. and Munkholm, L. 2015. Evaluating land quality for carbon storage, greenhouse gas emissions and nutrient leaching. In: Ball, B. C. and Munkholm, L. J. (Eds), *Visual Soil Evaluation: Realising Potential Crop Production with Minimum Environmental Impact*. CABI, Wallingford, pp. 103–121.

Comin, J., Fayad, J., Kurtz, C., Mafra, A. and Curmi, P. 2016. *Guia prático de avaliação participativa da qualidade do solo em Sistema de Plantio Direto de Hortaliças (SPDH)*. Open Brasil Gráfica e Editora, Brazil.

Cornelis, W. M., Akodi, D., Komutunga, E., Agaba, C., Ahumuza, E. and Oratungye, K. 2019a. Exploring visual soil evaluation and examination methods on highly-weathered tropical soil. *Soil and Tillage Research* 195, 104360.

Cornelis, W. M., Waweru, G. and Araya, T. 2019b. Building resilience against drought and floods: the soil-water management perspective. In: Lal, R. and Francaviglia, R. (Eds), *Sustainable Agriculture Reviews*, Vol. 29. Springer, Cham, pp. 125–142.

Cui, J., Askari, M. S. and Holden, N. M. 2014. Visual evaluation of soil structure under grassland management. *Soil Use and Management* 30(1), 129–138.

Daniells, I. and Larsen, D. 1991. *SOILpak: A Soil Management Package for Cotton Production on Cracking Clays*. NSW Agriculture, Narrabri.

Daniells, I., Larsen, D., McKenzie, D. and Anthony, D. 1996. SOILpak: a successful decision support system for managing the structure of Vertisols under irrigated cotton. *Soil Research* 34(6), 879–889.

Dexter, A. R. 1988. Advances in characterisation of soil structure. *Soil and Tillage Research* 11(3–4), 199–238.

Dexter, A. R. 2004. Soil physical quality: Part I. Theory, effects of soil texture, density, and organic matter, and effects on root growth. *Geoderma* 120(3–4), 201–214.

Diez, T. 1982. Spatendiagnostische Untersuchungen an der Krume von Lössboden. *Mitteilungen der Deutschen Bodenkundlichen Gesellschaft* 34, 145–148.

Emmet-Booth, J. P., Bondi, G., Fenton, O., Forristal, P. D., Jeuken, E., Creamer, R. E. and Holden, N. M. 2018. GrassVESS: a modification of the visual evaluation of soil structure method for grasslands. *Soil Use and Management* 34(1), 37–47.

Emmet-Booth, J. P., Forristal, P. D., Fenton, O., Ball, B. C. and Holden, N. M. 2016. A review of visual soil evaluation techniques for soil structure. *Soil Use and Management* 32(4), 623–634.

Emmet-Booth, J. P., Forristal, P. D., Fenton, O., Bondi, G. and Holden, N. M. 2019. Visual soil evaluation – spade vs. profile methods and the information conveyed for soil management. *Soil and Tillage Research* 187, 135–143.

Emmet-Booth, J. P., Holden, N. M., Fenton, O., Bondi, G. and Forristal, P. D. 2020. Exploring the sensitivity of visual soil evaluation to traffic-induced soil compaction. *Geoderma Regional* 20, e00243.

Eswaran, H., Lal, R. and Reich, P. 2001. Land degradation: an overview. *Responses to Land Degradation*. Proceedings of the 2nd International Conference on Land Degradation and Desertification, Khon Kaen. Oxford Press, New Delhi, pp. 20–35.

Etana, A., Larsbo, M., Keller, T., Arvidsson, J., Schjønning, P., Forkman, J. and Jarvis, N. 2013. Persistent subsoil compaction and its effects on preferential flow patterns in a loamy till soil. *Geoderma* 192, 430–436.

Falkenmark, M. 1995. Land-water linkages: a synopsis. In: Mather, T. (Ed.), *Land and Water Bulletin: Land and Water Integration and River Basin Management*, Vol. 1. Food and Agriculture Organization (FAO) of the United Nations, Rome, pp. 15–17.

Field, D. J., McKenzie, D. C. and Koppi, A. J. 1997. Development of an improved Vertisol stability test for SOILpak. *Soil Research* 35(4), 843–852.

Franco, A. L. C., Cherubin, M. R., Cerri, C. E. P., Guimarães, R. M. L. and Cerri, C. C. 2017. Relating the visual soil structure status and the abundance of soil engineering invertebrates across land use change. *Soil and Tillage Research* 173, 49–52.

Franco, H. H. S., Guimarães, R. M. L., Tormena, C. A., Cherubin, M. R. and Favilla, H. S. 2019. Global applications of the Visual Evaluation of Soil Structure method: a systematic review and meta-analysis. *Soil and Tillage Research* 190, 61–69.

Garbout, A., Munkholm, L. J. and Hansen, S. B. 2013. Tillage effects on topsoil structural quality assessed using X-ray CT, soil cores and visual soil evaluation. *Soil and Tillage Research* 128, 104–109.

Giarola, N. F. B., da Silva, A. P., Tormena, C. A., Ball, B. and Rosa, J. A. 2010. Visual soil structure quality assessment on Oxisols under no-tillage system. *Scientia Agricola* 67(4), 479–482.

Giarola, N. F. B., da Silva, Á. P., Tormena, C. A., Guimarães, R. M. L. and Ball, B. C. 2013. On the visual evaluation of soil structure: the Brazilian experience in Oxisols under no-tillage. *Soil and Tillage Research* 127, 60–64.

Giarola, N. F. B., Tormena, C. A., da Silva, A. P. and Ball, B. 2009. Método de avaliação visual da qualidade da estrutura aplicado a Latossolo Vermelho Distroférrico sob diferentes sistemas de uso e manejo. *Ciência Rural* 39(8), 2531–2534.

Godwin, R. J. and Spoor, G. 2015. Choosing and evaluating soil improvements by subsoiling and compaction control. In: Ball, B. C. and Munkholm, L. J. (Eds), *Visual Soil Evaluation: Realising Potential Crop Production With Minimum Environmental Impact*. CABI, Wallingford, pp. 66–85.

Görbing, J. 1947. *Die grundlagen der gare im praktischen ackerbau*, Vols I and II. Landbuch Verlag GMBH, Hannover.

Greve, M. H., Sperstad, R. and Nyborg, Å. 1999. *Retningslinjer for beskrivelse av jordprofil* (1st edn). Norsk Institutt for jord- og skogkartleggning, Ås.

Guimarães, R. M. L., Ball, B. C. and Tormena, C. A. 2011. Improvements in the visual evaluation of soil structure. *Soil Use and Management* 27, 395–403.

Guimarães, R. M. L., Ball, B. C., Tormena, C. A., Giarola, N. F. B. and da Silva, Á. P. 2013. Relating visual evaluation of soil structure to other physical properties in soils of contrasting texture and management. *Soil and Tillage Research* 127, 92–99.

Guimarães, R. M. L., Junior, A. F. N., Silva, W. G., Rogers, C. D., Ball, B. C., Montes, C. R. and Pereira, B. F. F. 2017a. The merits of the Visual Evaluation of Soil Structure method (VESS) for assessing soil physical quality in the remote, undeveloped regions of the Amazon basin. *Soil and Tillage Research* 173, 75–82.

Guimarães, R. M. L., Lamandé, M., Munkholm, L. J., Ball, B. C. and Keller, T. 2017b. Opportunities and future directions for visual soil evaluation methods in soil structure research. *Soil and Tillage Research* 173, 104–113.

Horn, R. and Smucker, A. 2005. Structure formation and its consequences for gas and water transport in unsaturated arable and forest soils. *Soil and Tillage Research* 82(1), 5–14.

Jahn, R., Blume, H. P., Asio, V. B., Spaargaren, O. and Schad, P. 2006. *Guidelines for Soil Description* (4th edn). Food and Agriculture Organization, Rome.

Johannes, A., Weisskopf, P., Schulin, R. and Boivin, P. 2017. To what extent do physical measurements match with visual evaluation of soil structure? *Soil and Tillage Research* 173, 24–32.

Johannes, A., Weisskopf, P., Schulin, R. and Boivin, P. 2019. Soil structure quality indicators and their limit values. *Ecological Indicators* 104, 686–694.

Karlen, D. L., Veum, K. S., Sudduth, K. A., Obrycki, J. F. and Nunes, M. R. 2019. Soil health assessment: past accomplishments, current activities, and future opportunities. *Soil and Tillage Research* 195, 104365.

Kay, B. D. 1990. Rates of change of soil structure under different cropping systems. In: Stewart, B. A. (Ed.), *Advances in Soil Science*, Vol. 12. Springer, New York, pp. 1–52.

Kodešová, R., Jirků, V., Kodeš, V., Mühlhanselová, M., Nikodem, A. and Žigová, A. 2011. Soil structure and soil hydraulic properties of Haplic Luvisol used as arable land and grassland. *Soil and Tillage Research* 111(2), 154–161.

Lal, R. and Shukla, M. K. 2004. *Principles of Soil Physics*. Marcel Dekker, New York.

Leopizzi, S., Gondret, K. and Boivin, P. 2018. Spatial variability and sampling requirements of the visual evaluation of soil structure in cropped fields. *Geoderma* 314, 58–62.

Lilly, A., Nemes, A., Rawls, W. J. and Pachepsky, Y. A. 2008. Probabilistic approach to the identification of input variables to estimate hydraulic conductivity. *Soil Science Society of America Journal* 72(1), 16–24.

Madsen, H. B. and Jensen, N. 1988. *Vejledning til beskrivelse af jordbundsprofiler*. Landbrugsministeriet, Denmark.

Manichon, H. 1987. Observation morphologique de l'état structural et mise en évidence d'effets de compactage des horizons travaillés. In: *Workshop on Soil Compaction: Consequences and Structural Regeneration Processes*, Avignon, France, pp. 39–52.

Manichon, H. 1993. Le profil cultural: une perspective nouvelle pour l'analyse du travail du sol. In: *Mélanges Stéphane Hénin: Sol, Agronomie, Environnement*. ORSTOM, INRA, ITCF, Paris, pp. 75–82.

McKeague, J., Wang, C. and Coen, G. 1986. *Describing and Interpreting the Macrostructure of Mineral Soils–A Preliminary Report*. Land Resource Research Institute, Ottawa.

McKenzie, D. C. 1998. *SOILpak for Cotton Growers*. NSW Agriculture, Orange.

McKenzie, D. C. 2001a. Rapid assessment of soil compaction damage I. The SOILpak score, a semi-quantitative measure of soil structural form. *Australian Journal of Soil Research* 39(1), 117–125.

McKenzie, D. C. 2001b. Rapid assessment of soil compaction damage II. Relationships between the SOILpak score, strength and aeration measurements, clod shrinkage parameters, and image analysis data on a Vertisol. *Australian Journal of Soil Research* 39(1), 127–141.

McKenzie, D. C. 2013. Visual soil examination techniques as part of a soil appraisal framework for farm evaluation in Australia. *Soil and Tillage Research* 127, 26–33.

McMullen, B. 2000. *SOILpak for Vegetable Growers*. NSW Agriculture, Orange.

Moreira, W. H., Tormena, C. A., Karlen, D. L., da Silva, Á. P., Keller, T. and Betioli, E. Jr. 2016. Seasonal changes in soil physical properties under long-term no-tillage. *Soil and Tillage Research* 160, 53–64.

Mueller, L., Kay, B. D., Deen, B., Hu, C., Zhang, Y., Wolff, M., Eulenstein, F. and Schindler, U. 2009a. Visual assessment of soil structure: Part II. Implications of tillage, rotation and traffic on sites in Canada, China and Germany. *Soil and Tillage Research* 103(1), 188–196.

Mueller, L., Kay, B. D., Hu, C., Li, Y., Schindler, U., Behrendt, A., Shepherd, T. G. and Ball, B. C. 2009b. Visual assessment of soil structure: evaluation of methodologies on sites in Canada, China and Germany. Part I: Comparing visual methods and linking them with soil physical data and grain yield of cereals. *Soil and Tillage Research* 103(1), 178–187.

Munkholm, L. J. 2000. *The Spade Analysis–A Modification of the Qualitative Spade Diagnosis for Scientific Use*. DIAS Report, Plant Production. CABI, Denmark, 73 pp.

Munkholm, L. J., Heck, R. J. and Deen, B. 2013. Long-term rotation and tillage effects on soil structure and crop yield. *Soil and Tillage Research* 127, 85–91.

Munkholm, L. J. and Holden, N. M. 2015. Visual evaluation of grassland and arable management impacts on soil quality. In: Ball, B. C. and Munkholm, L. J. (Eds), *Visual Soil Evaluation: Realising Potential Crop Production With Minimum Environmental Impact*. CABI, Wallingford, pp. 49–65.

Murphy, B. W., Crawford, M. H., Duncan, D. A., McKenzie, D. C. and Koen, T. B. 2013. The use of visual soil assessment schemes to evaluate surface structure in a soil monitoring program. *Soil and Tillage Research* 127, 3–12.

Newell-Price, J. P., Whittingham, M. J., Chambers, B. J. and Peel, S. 2013. Visual soil evaluation in relation to measured soil physical properties in a survey of grassland soil compaction in England and Wales. *Soil and Tillage Research* 127, 65–73.

Nguyen, P. M., Van Le, K. V. and Cornelis, W. M. 2014. Using categorical soil structure information to improve soil water retention estimates of tropical delta soils. *Soil Research* 52(5), 443–452.

Obour, P. B., Dadzie, F. A., Arthur, E., Munkholm, L. J., Saba, C. K. S., Rubæk, G. H. and Owusu, K. 2020. Integration of farmers' knowledge and science-based assessment of soil quality for peri-urban vegetable production in Ghana. *Renewable Agriculture and Food Systems* 35(2), 128–139.

Osman, K. T. 2014. *Soil Degradation, Conservation and Remediation*. Springer, Dordrecht.

Pagliai, M., Vignozzi, N. and Pellegrini, S. 2004. Soil structure and the effect of management practices. *Soil and Tillage Research* 79(2), 131–143.

Peerlkamp, P. 1959. A visual method of soil structure evaluation. *Meded. vd Landbouwhogeschool en Opzoekingsstations van de Staat te Gent* 24(24), 216–221.

Peigné, J., Vian, J.-F., Cannavacciuolo, M., Lefevre, V., Gautronneau, Y. and Boizard, H. 2013. Assessment of soil structure in the transition layer between topsoil and subsoil using the profil cultural method. *Soil and Tillage Research* 127, 13–25.

Pereira Valani, G., Machado Vezzani, F. and Cavalieri-Polizeli, K. M. V. 2020. Soil quality: evaluation of on-farm assessments in relation to analytical index. *Soil and Tillage Research* 198, 104565.

Petersen, L. and Møberg, J. P. 1987. *Vejledning til jordbundsprofilbeskrivelse*. Landbohøjskolen, København.

Piron, D., Boizard, H., Heddadj, D., Pérès, G., Hallaire, V. and Cluzeau, D. 2017. Indicators of earthworm bioturbation to improve visual assessment of soil structure. *Soil and Tillage Research* 173, 53–63.

Preuschen, G. 1983. Die Spatendiagnose und ihre Auswertung. In: Böhm, W., Kutschera, L. and Lichtenegger, E. (Eds), *Root Ecology and Its Pracitical Application*. International

Symposium Gumpenstein, 1982, Bundesanstalt Gumpenstein, A-8952 Irdning, pp. 355–368.

Pulido-Moncada, M., Gabriels, D. and Cornelis, W. M. 2014a. Data-driven analysis of soil quality indicators using limited data. *Geoderma* 235–236, 271–278.

Pulido-Moncada, M., Gabriels, D., Lobo, D., Rey, J. C. and Cornelis, W. M. 2014b. Visual field assessment of soil structural quality in tropical soils. *Soil and Tillage Research* 139, 8–18.

Pulido-Moncada, M., Helwig Penning, L., Timm, L. C., Gabriels, D. and Cornelis, W. M. 2017. Visual examination of changes in soil structural quality due to land use. *Soil and Tillage Research* 173, 83–91.

Pulido-Moncada, M., Munkholm, L. J. and Schjønning, P. 2019. Wheel load, repeated wheeling, and traction effects on subsoil compaction in northern Europe. *Soil and Tillage Research* 186, 300–309.

Pulido-Moncada, M., Penning, L. H., Timm, L. C., Gabriels, D. and Cornelis, W. M. 2014c. Visual examinations and soil physical and hydraulic properties for assessing soil structural quality of soils with contrasting textures and land uses. *Soil and Tillage Research* 140, 20–28.

Rabot, E., Wiesmeier, M., Schlüter, S. and Vogel, H.-J. 2018. Soil structure as an indicator of soil functions: a review. *Geoderma* 314, 122–137.

Ralisch, R., Debiasi, H., Franchini, J., Tomazi, M., Hernani, L., Melo, A. d. S., Santi, A., Martins, A. d. S. and de Bona, F. 2017. Diagnóstico rápido da estrutura do solo (DRES). Embrapa Solos-Capítulo em livro técnico (INFOTECA-E).

Reynolds, W. D., Drury, C. F., Tan, C. S., Fox, C. A. and Yang, X. M. 2009. Use of indicators and pore volume-function characteristics to quantify soil physical quality. *Geoderma* 152(3–4), 252–263.

Rockström, J., Falkenmark, M., Karlberg, L., Hoff, H., Rost, S. and Gerten, D. 2009. Future water availability for global food production: the potential of green water for increasing resilience to global change. *Water Resources Research* 45, W00A12.

Roger-Estrade, J., Richard, G., Caneill, J., Boizard, H., Coquet, Y., Defossez, P. and Manichon, H. 2004. Morphological characterisation of soil structure in tilled fields: from a diagnosis method to the modelling of structural changes over time. *Soil and Tillage Research* 79(1), 33–49.

Sasal, M. C., Boizard, H., Andriulo, A. E., Wilson, M. G. and Léonard, J. 2017. Platy structure development under no-tillage in the northern humid Pampas of Argentina and its impact on runoff. *Soil and Tillage Research* 173, 33–41.

Shepherd, T. G. 2000. *Visual Soil Assessment, Vol. 1: Field Guide for Cropping and Pastoral Grazing on Flat to Rolling Country*. Horizons. Mw/Landcare Research, Palmerston North, 84 pp.

Shepherd, T. G. 2009. *Visual Soil Assessment, Vol. 1: Field Guide for Pastoral Grazing and Cropping on Flat to Rolling Country*. Horizons Regional Council, Palmerston North.

Sillmann, J., Kharin, V. V., Zwiers, F. W., Zhang, X. and Bronaugh, D. 2013. Climate extremes indices in the CMIP5 multimodel ensemble: Part 2. Future climate projections. *Journal of Geophysical Research: Atmospheres* 118(6), 2473–2493.

Sobelius, J. 1995. *Lär känna Din jord!* Division of Ecological Agriculture, Swedish University of Agricultural Sciences, Uppsala.

Sonneveld, M. P. W., Heuvelink, G. B. M. and Moolenaar, S. W. 2014. Application of a visual soil examination and evaluation technique at site and farm level. *Soil Use and Management* 30(2), 263-271.

Stroosnijder, L. 2009. Modifying land management in order to improve efficiency of rainwater use in the African highlands. *Soil and Tillage Research* 103(2), 247-256.

Suhr, K., Dissing, A., Schmidt, K. G., Eriksen, S. and Hansen, L. F. 1995. *Økologisk Landbrug* (1st edn). Landbrugets Informationskontor, Århus.

Sun, F., Bakr, N., Dang, T., Pham, V., Weindorf, D. C., Jiang, Z., Li, H. and Wang, Q.-B. 2020. Enhanced soil profile visualisation using portable X-ray fluorescence (PXRF) spectrometry. *Geoderma* 358, 113997.

Sys, C., Van Ranst, E., Debaveye, J. and Beernaert, F. 1993. *Land Evaluation. Part III: Crop Requirements*. Agricultural Publications No. 7. GADC, Brussels, 191 pp.

Tavares Filho, J., Ralisch, R., Guimarães, M. F., Medina, C. C., Balbino, L. C. and Neves, C. S. V. J. 1999. Método do perfil cultural para avaliação do estado físico de solos em condições tropicais. *Revista Brasileira de Ciência do Solo* 23(2), 393-399.

Tomis, V., Duparque, A. and Boizard, H. 2019. Development of the "mini 3D soil profile" - a visual method derived from the "profil cultural". *Soil and Tillage Research* 194, 104285.

van Leeuwen, M. M. W. J., Heuvelink, G. B. M., Wallinga, J., de Boer, I. J. M., van Dam, J. C., van Essen, E. A., Moolenaar, S. W., Verhoeven, F. P. M., Stoorvogel, J. J. and Stoof, C. R. 2018. Visual soil evaluation: reproducibility and correlation with standard measurements. *Soil and Tillage Research* 178, 167-178.

Vereecken, H., Weynants, M., Javaux, M., Pachepsky, Y., Schaap, M. G. and Genuchten, M. T. 2010. Using pedotransfer functions to estimate the van Genuchten-Mualem soil hydraulic properties: a review. *Vadose Zone Journal* 9(4), 795-820.

Wahlhütter, S., Vogl, C. R. and Eberhart, H. 2016. Soil as a key criteria in the construction of farmers' identities: the example of farming in the Austrian province of Burgenland. *Geoderma* 269, 39-53.

Wani, S. P., Sreedevi, T., Rockström, J. and Ramakrishna, Y. 2009. Rainfed agriculture - past trends and future prospects. In: Wani, S. P., Rockström, J. and Oweis, T. (Eds), *Rainfed Agriculture: Unlocking the Potential*. CABI, Wallingford, pp. 1-35.

Werner, D. and Thaemert, W. 1989. Zur Diagnose des physikalischen Bodenzustandes auf Produktionsflächen. *Archiv für Acker-und Pflanzenbau und Bodenkunde* 33, 729-739.

Wills, S. A., Burras, C. L. and Sandor, J. A. 2007. Prediction of soil organic carbon content using field and laboratory measurements of soil color. *Soil Science Society of America Journal* 71(2), 380-388.

WRB 2015. *World Reference Base for Soil Resources 2014 (Update 2015). International Soil Classification System for Naming Soils and Creating Legends for Soil Maps*. World Soil Resources Reports. Food and Agriculture Organization, Rome.

Young, I. M., Crawford, J. W. and Rappoldt, C. 2001. New methods and models for characterising structural heterogeneity of soil. *Soil and Tillage Research* 61(1-2), 33-45.

Chapter 5

Imaging soil structure to measure soil functions and soil health with X-ray computed micro-tomography

Alexandra Kravchenko and Andrey Guber, Michigan State University, USA

1 Introduction

Soil structure is a well-recognized key characteristic defining soil functioning. It is the framework within which all physical, chemical, and biological soil processes take place. However, even though it is generally clear and agreed-on by both researchers and practitioners on what a good soil structure looks like, it has never been easy to express that understanding quantitatively. Within the research community there are two schools of thought for defining soil structure, which emerged from two contrasting approaches to measure and quantify it. The classical definition describes soil structure as an assemblage of structural units, where structural units typically refer to soil aggregates. For example, the definition from the encyclopedia of soil science states 'the overall architecture of a soil defined by the assemblage of structural units, which it contains. The way the soil is organized in terms of the physically definable parts within it' (Chesworth, 2008). This definition is driven by the tools historically available for soil structure analyses. Such tools

http://dx.doi.org/10.19103/AS.2020.0079.11

involved destructive soil sampling to procure 'the physically definable parts', commonly referred to as aggregates, from either wet or dry sieving of the soil samples, followed by enumerating/weighing aggregates of different size classes or calculating mean aggregate diameters either for dry or for water-stable aggregates.

However, the development of novel techniques for analyzing non-disturbed intact soil, notably X-ray computed tomography (X-ray CT) (Mooney et al., 2012, Pohlmeier et al., 2018, Young et al., 2001), promoted an alternative view on the definition of soil structure. The alternative definition shifts the focus from physical parts themselves to their spatial arrangement and to the properties of the void space surrounding the structural units. From that perspective the soil structure is defined as 'spatial heterogeneity of the different components or properties of soil' (Dexter, 1988) or more specifically as 'the spatial arrangement of solids and voids across different scales' (Rabot et al., 2018). The need in the new definition was initially driven by the operational capabilities of the imaging tools. Indeed, identifying 'structural units' or 'physically definable parts' in three-dimensional images of intact soil samples is typically an impractical effort. On the contrary, the 'spatial arrangement of solids and voids' in intact samples can be easily visualized and quantified at a given spatial resolution. However, besides its practicality for quantifying soil structure from images, the alternative void-focused definition has advantages over the traditional solid-focused one since it better reflects the flow and transport processes crucial for soil functioning (Koestel and Schluter, 2019, Lucas et al., 2019, Schluter et al., 2018).

At present, the basis for the proposed physical indicators of soil health was formed by the traditional soil structure definition, that is, by viewing soil structure as a collection of aggregates that can be procured and enumerated by sieving techniques. For example, the USDA Natural Resources Conservation Service (NRCS) recently recommended wet aggregate stability as the main soil structure indicator for soil health assessment (USDA NRCS Soil Health Technical Note No. 450-03, 2019a). Likewise, wet aggregate stability is one of the two physical indicators of the soil health within Cornell Soil Health Test/ Comprehensive Assessment of Soil Health framework (Van Es and Karlen, 2019).

At the same time, there is a growing realization of the importance of the non-destructive view of soil structure for understanding processes governing soil health (Wander et al., 2019). Characterization of soil structure of intact soil samples using non-invasive techniques can provide extra benefits in assessing soil health and soil functioning, supplementing the information from the traditional destructive soil structure analysis (Wander et al., 2019). For example, even though procured soil structural units, aka aggregates, can be used to gain valuable insights into their interior features, typically, they provide only limited information about the processes that took place outside of them when they were located in the originally intact soil. The reason is that the information about their position within the original soil matrix is irreversibly lost upon their procurement (Kravchenko et al., 2019b). Moreover, the structural units represent the solid part of the soil matrix, while the processes relevant to soil functioning and soil health take place within the void part, that is, within soil pores, including those that form the aggregate boundaries. The processes taking place within the pores are water and gas fluxes, chemical transport, activity and movement of soil biota. Thus, we believe that soil health quantification will benefit from (i) operating within the alternative soil structure definition, that is, soil as an assemblage of solids and voids, and (ii) describing the influence of soil structure on soil processes from the perspective of characteristics and properties of the voids, that is, soil pores, in addition to those of solids, that is, soil aggregates. The imaging techniques have been well developed for this task and the X-ray CT scanners are becoming more and more common for soil analyses and more affordable. The characteristics of the pore space obtained through images provide new knowledge about soil functioning and especially about the characteristics of soil microbial habitats (Nunan et al., 2006, Rabot et al., 2018). It is obvious that soil solids and voids are intrinsically linked and, thus, transferring to the void-focused definition of the soil structure does not undermine the gains in understanding achieved from the solid-focused definition. In fact, X-ray CT imaging can provide information directly relevant to several of the indicators suggested for soil health assessment by USDA NRCS Soil Health (2019b), including infiltration, soil macropores, water holding capacity, retention and transport of water and nutrients, habitat for soil microbes, compaction, water movement, and porosity. With X-ray CT scanning techniques becoming more and more available they grow into popular tools in assessing a variety of soil management impacts on soil ultimately related to soil health.

2 X-ray computed micro-tomography scanning

X-ray CT is currently the most popular tool for non-destructive analyses of intact soil samples. The general principle behind the method can be described as follows: X-rays penetrate the sample and the magnitude of their penetration depends on the properties of the material they encounter. Specifically, for a given X-ray energy, it depends on the atomic number of the elements constituting the material and on the material's density. Among the materials commonly present within the soil the metal-rich particles have the greatest attenuation of the X-rays, followed by quartz sands, clay minerals, organic matter inclusions, and finally by the air, that is, soil pores. Which materials are present within each particular voxel within the soil image and what is the density of the materials within the specific voxel define the value of the voxel's attenuation coefficient. The attenuation coefficients then translate into an image, where grayscale values reflect the internal properties, that is, composition and density, of the sample. An illustration of an image in a typical format used in soil analyses is shown on Fig. 1 where iron-rich stones are of light gray color, followed by clay/silt materials forming soil aggregates, followed by yet darker plant roots and particulate organic matter (POM) particles, and then by nearly black soil pores. The ultimate product of the scanning, that is, the three-dimensional image of a soil sample, is obtained from multiple projections taken from the sample at different angles via CT approach. Detailed descriptions of the technique can be found in many sources, with Cnudde and Boone's (2013) and Wildenschild and Sheppard's (2013) reviews being of particular relevance for applications in soil science. The X-ray CT scanning at relatively high scanning resolutions, for example, from a few to <100 μm, is commonly referred to as computed micro-tomography (μCT), as opposed to the CT that produces > 100 μm results. Some studies can benefit from CT scanning, especially those focusing on preferential flow, earthworm activities, and whole-profile soil fluxes. However, μCT images are preferable to CT when the goal is the understanding of many other aspects

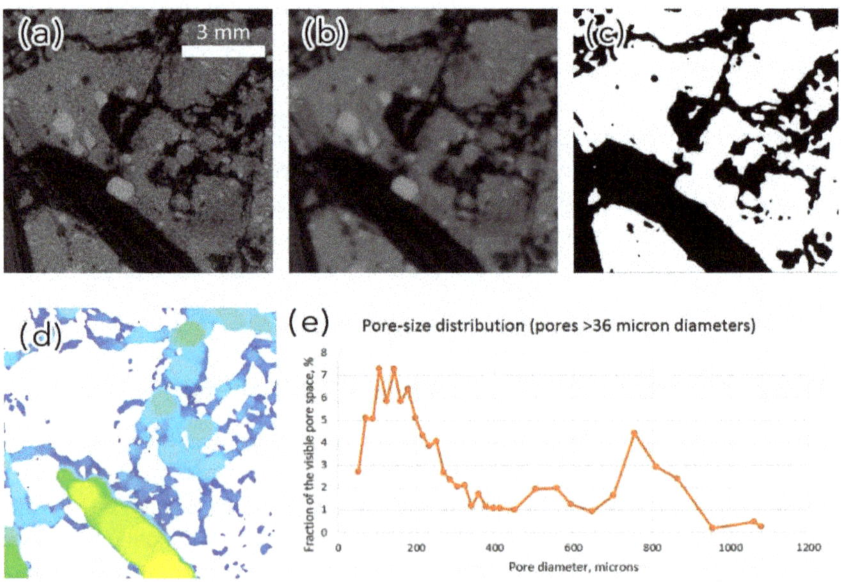

Figure 1 An illustration of image processing and analysis steps: (a) a region of interest from an original X-ray μCT image of a soil sample, (b) the image after application of the three-dimensional Gaussian filter with 2×2×2 filter size, (c) the image segmented into solid/pore voxels using Otzu thresholding procedure, (d) the image of the pore-size distribution obtained using the continuous inscribed spheres algorithm from XLIB, ImageJ, and (e) pore-size distribution.

of soil functioning and soil health characteristics, since μCT enables a much more detailed visualization of the soil matrix.

The two types of facilities/instruments where X-ray μCT for soil samples can be conducted are synchrotron facilities and benchtop μCT scanners. The key differences between the two types of facilities are the properties of their X-ray beams, monochromatic at synchrotron, and polychromatic in benchtop instruments. Prior to encountering the sample, synchrotron radiation passes through a monochromator, resulting in a beam ray with a very narrow range of energies. An X-ray source in benchtop scanners produces a cone-shaped beam with a relatively wide range of energies present. μCT scanning at synchrotron facilities typically produces images of higher resolution and better quality with fewer artifacts than what can be achieved at benchtop scanners. However, it should be noted that the quality of commercially available benchtop scanners continuously improves, while the costs continue to decrease, thus the number of research groups and universities owing such equipment keeps growing and X-ray μCT scanning becomes more and more accessible to soil scientists. Likewise, while originally the synchrotron facilities provided much better resolutions and faster scanning times than benchtop scanners, the recent improvements greatly reduced the differences.

Note that regardless of the μCT scanning equipment, the size of the scanned sample is inversely related to the achievable scanning resolution. At present, in most facilities the scanning resolutions of a few microns can be achieved only for the samples that are no bigger than 1–2 cm, while resolutions of a few tens of microns can be expected for the samples that are 5–10 cm in size, that is, the size of commonly taken intact soil cores.

3 Soil health-related structure characteristics that can be obtained via X-ray computed micro-tomography

The key question in using X-ray μCT images in soil health assessments is which features, from a multitude of those that can be technically obtained from the images, are indicative of the state of the soil structure and are relevant to soil health. Several potential criteria can be suggested; however, it should be noted that the use of X-ray μCT for soil health characterization is at its infancy (Bunemann et al., 2018). No research assessments or discussions on pros and cons of different potential criteria have taken place yet and no standardized methods for obtaining soil structure characteristics from the images have been agreed on. However, even with the traditional approach to soil structure characterization, there is a lot of current debate on which soil characteristics should be used as indicators of soil quality and health (Bunemann et al., 2018), how useful they can be in assessing soil functions (Rabot et al., 2018), and what are the optimal measurement procedures to follow for most of them.

Since at present there is no consensus on the suitable and reliable indicators even for purely physical measures of soil structure (Bunemann et al., 2018, Johannes et al., 2019), it is easy to imagine the difficulties involved in finding such indicators for soil health, a concept crossing over physics, chemistry, and biology. The task of finding appropriate indicators there is indeed daunting.

Suitable indicators of soil health must meet the following criteria: (i) they should represent key soil processes, (ii) their values should be sensitive to the status of the soil health and be interpretable; and (iii) they should be informative for making management decisions (Cornell Assessment of Soil Health, 2016). It is highly promising for the X-ray μCT use in soil health assessments that the majority of the currently suggested physical indicators are related to soil pores (Reynolds et al., 2009, Topp et al., 1997). Soil porosity and its components, that is, macroporosity, mesoporosity, and microporosity, are the most obvious of such indicators. Yet, a number of other soil health indicators are also related to porosity and pore-size distributions, including (i) soil tilth, a qualitative feature that includes soil structuring; (ii) soil water characteristics, including hydraulic conductivity, infiltration rate, water holding capacity; and (iii) soil aggregation and aggregate stability (Soil Health Institute, 2019). X-ray μCT is an excellent tool to visualize soil pores, and the visualized pores can then be subjected to different morphological analyses depicting various aspects of pore structure and architecture (Helliwell et al., 2013). Most of the above-listed proposed physical indicators of soil health are quantitative features, and they are ultimately related to soil pores and are quantifiable from a perspective of a pore size distribution and X-ray μCT pore information. Most promising is that there is a large body of evidence that the information collected from X-ray μCT agrees well with traditional soil physical and hydro-physical measurements (Hansson et al., 2018, Koestel, 2018, Mooney, 2002, Munkholm et al., 2012, Pagenkemper et al., 2014, Tollner et al., 1998, Tracy et al., 2015).

It should be noted that the final identification, comparison, and assessment of the indicators of soil health, which can be obtained using X-ray μCT, are beyond the scope of the current review. Rather we would like to focus on several soil characteristics commonly determined using X-ray μCT, which hold a particular promise for their future use as indicators. In Table 1 we have listed these suggested characteristics along with their values determined from three-dimensional images of the soils under two contrasting land-use and management treatments: a bare soil that was maintained without vegetation for >5 years and a diverse herbaceous perennial plant community that has been in place for the same amount of time. The representative images of the samples and selected characteristics are shown in Fig. 2. We will review these characteristics, discuss their relevance to soil functioning, and briefly introduce common methods for their measurements from X-ray μCT images.

Table 1 X-ray μCT-based soil characteristics suggested as potential indicators of soil health and their values for the soils from two management practices with contrasting soil health implications, that is, bare soil and native prairie vegetation

Criterion	Soil management		ImageJ/Fiji tools used to obtain the measurements
	Bare	Native prairie	
Macro-porosity, % (image-based porosity)	6.8	17.0	Otsu thresholding
Pore connectivity			
Largest connected pore volume, % pore space	83.8	96.4	From Particle Analyzer, BoneJ outputs
Percolation porosity, % pore space	5.7	16.4	From Particle Analyzer, BoneJ outputs
Pore shape			
Surface/volume ratio	1.76	1.89	From Particle Analyzer, BoneJ outputs
Median soil-to-pore distance	3.32	2.23	Distance transform, Beat

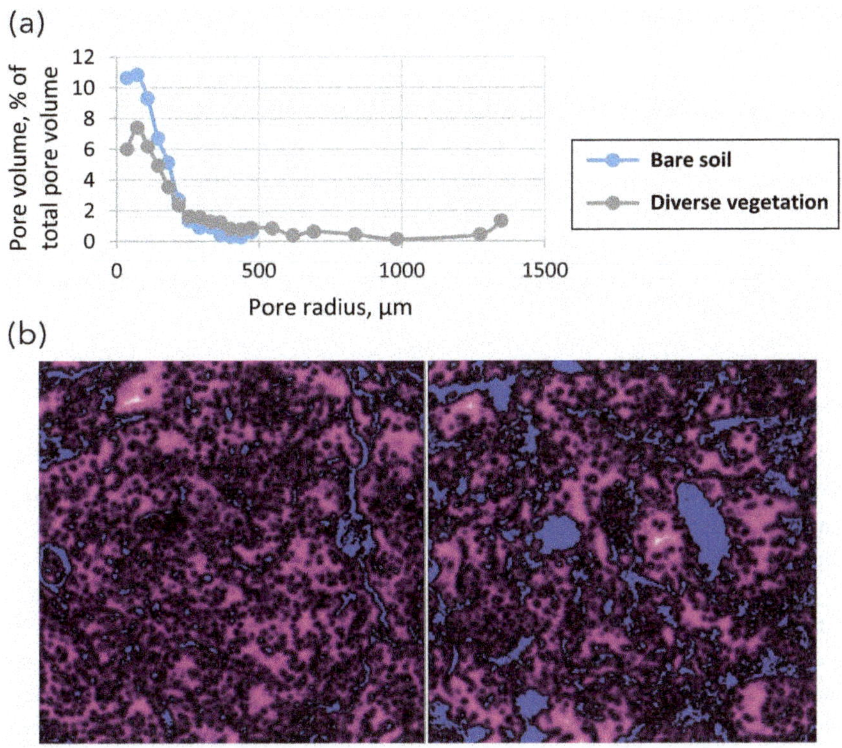

Figure 2 Soil pore size distributuion (a) and examples of soil-to-pore distance maps for samples from a bare and a diverse plant vegetation soils. (b) Pores are shown in blue. The gradient of distances is from dark purple (the lowest) to white (the highest).

4 Image analysis software

The procedures involved in an actual X-ray scanning are beyond the scope of this chapter. We suggest and assume that the readers interested in μCT will seek an access to the scanning facilities where trained personnel can assist them with scanning and image preprocessing.

After the users receive the scans, they will need to extract from the images relevant information about scanned objects, a task which typically must be preceded by an additional image processing for noise reduction and artifact removal. For viewing convenience and data analyses the three-dimensional images are typically presented as a collection of two-dimensional slices. An example of one such two-dimensional slice from a scanned soil sample is shown in Fig. 1a.

A number of software packages can be used for processing X-ray μCT images of soil samples. ImageJ/Fiji is one of the most common tools, as it is a powerful and versatile public domain software (Rasband, 1997–2015, Schindelin et al., 2012). Among the popular commercial software are AVISO, VGStudioMAX, 3DMA-Rock, and MAVI. However, their costs can be very substantial. MATLAB also has a powerful toolbox of image processing techniques. Special software packages focusing on the data analysis demands specific to soil studies also have been developed. One latest example is SoilJ plugin for Fiji (Koestel, 2018) that offers multiple tools to address tasks often faced by soil scientists, including adjusting the position and orientation of the samples during image preprocessing; segmentation, including identification of POM and roots; and tools for soil pore analyses. Another example is ImageToolsWinR100 (Houston et al., 2013) with automatic segmentation option and a variety of pore structure statistical tools.

5 Image processing

It should be noted that the outcomes of the X-ray CT image analyses for soil samples are notoriously dependent on the image analyses methods (Baveye et al., 2010, Houston et al., 2013, Smet et al., 2018). It also does not help that the settings used during image acquisition will affect not only the properties of the produced images but also will determine which specific procedures/procedure settings will be needed to process a given batch of imaged soil samples. The same soil sample scanned at two different settings might require two very different approaches to its image processing. Arguably, the difficulties faced while processing soil images are considerably bigger than those encountered during image analyses of living organisms and/or rocks/mineral deposits. Soil samples lack the organized structure of a living organism's anatomy, and they greatly exceed rocks and most other porous media materials in variability.

Yet, even though the specific details needed for each set of samples might vary substantially, usually, preprocessing involves only three key tasks: (1) selecting the region of interest, (2) reducing the noise, and (3) removing artifacts from the region of interest. While the first two tasks almost always need to be undertaken, fortunately, high-quality scanning might reduce or even completely eliminate the necessity to deal with the third task.

Selecting the region of interest from within the sample's image is a generally recommended and highly commendable practice. Typically, such a region excludes a portion of the sample in an immediate vicinity of its boundaries, the portion that can be affected not only by artifacts generated close to the image borders during image reconstruction, but also by soil deformation during sampling (Hansson et al., 2018, Pohlitz et al., 2018, Schaffer et al., 2007, Vogel et al., 2010). Besides preventing the unreliable portion of the soil sample from influencing the results, focusing only on the selected region of interest, instead of the entire soil sample, reduces the size of the image, and thus speeds up further analyses.

It should be noted that the issue of the sample being representative of the studied soil profile/horizon is more important for intact soil samples than for the homogenized composite soil samples, and, unfortunately, there are no easy answers to what constitutes a representative sample. The problem is that pore-related soil properties, for example, macro- and micro-soil porosity, saturated and unsaturated hydraulic conductivity, soil water retention, and so on can vary significantly depending on the size of the studied samples (Jury and Horton, 2004). It is desirable that the size of the sample used to determine the soil property was representative of the studied soil, otherwise the measurements will lead to unreliable and highly uncertain results. However, the size of what can be considered a representative sample can vary among the soils as well as among the studied soil properties (e.g. Baveye et al., 2002, Al-Raoush and Papadopoulos, 2010). A size range of 10–17 cm^3 has been reported as representative for characterizing soil-pore systems in some soils under diverse tillage regimes (Borges et al., 2018); however, other studies showed that caution must be exercised even with the sizes >100 cm^3 (Baveye et al., 2002), even though they are typically used for X-ray μCT as well as classical soil physics measurements.

Regardless of the quality and resolution of the scanning, a certain amount of noise will always be present in the images, thus reducing/removing the noise is a very important step in the image analysis (Peth et al., 2008). One must be aware, though, that a certain amount of fine resolution features from the sample will also be lost during noise removal, thus care should be taken to strike a balance between obtaining a perfectly smooth image and loosing valuable information. The type of the noise-removing filtering can substantially influence the resultant pore characteristics (Smet et al., 2018). Therefore, the procedure must be tailored for each specific set of soil samples depending on quality of the scans and on the goals of data analyses. Among many different noise-removing filtering approaches suggested and implemented by soil scientists (Schluter et al., 2014), median filtering is by far the most often used, followed by Gaussian blur. An illustration of the effect of Gaussian three-dimensional blur filter is shown in Fig. 1b. Please note that when processing a three-dimensional image it is best to use three-dimensional procedures, instead of their two-dimensional counterparts, which operate only on individual two-dimensional slices. While more time-consuming to implement, the three-dimensional procedures are more suitable for representing the three-dimensional features and nature of the soil samples. After filtering, one might consider the use of additional processing tools, which can help with further analyses, for example, enhancing the image contrast in order to increase the visibility of certain image features, for example, pores or roots.

6 Thresholding

A process of solid/pore segmentation converts a grayscale image into a binary format where voxels identified as solid material or pores are assigned only one of the two colors, say, white and black, respectively. Segmentation is rarely a trivial task, as even though majority of the image voxels are typically easily assigned either solid or pore status, there is always a certain number of voxels with intermediate grayscale values. Such voxels can either belong to materials with attenuation coefficients intermediate of those of solids and air/water, such as organic matter. Or, more often, they result from partial volume effect, which is produced by the image voxels being occupied by more than one phase, for example, being partly solid and partly porous. Such voxels are always present on the borders of solids and pores and during segmentation they are placed into either the solid or pore class. This results in an unavoidable oversimplification, but it is important that the thresholding decisions do not lead to a consistent overestimation or underestimation of the pore space.

A number of segmentation procedures have been developed for solid/pore classifications of porous media, including soils. A 15-year-old study (Sezgin and Sankur, 2004) compared an impressive total of 40 thresholding algorithms, while many more have been added since (e.g. Hapca et al., 2013, Houston et al., 2013). Overall, the segmentation approaches are broadly divided into two groups: global and local. Global thresholding procedures operate with a single thresholding grayscale value or a set of thresholding grayscale values that are applied to the entire image. The voxels with grayscale values above the threshold are classified as solids and those below the level are classified as pores. Local thresholding procedures identify the thresholding level(s) that are specific for each local subregion of the image. How the threshold(s) is(are) decided on is specific to each of the very many available segmentation procedures. The most straightforward approach is to segment based on a single global threshold selected from the histogram of the image grayscale values (e.g. Koestel et al., 2018), as the point that is in the middle between the histogram peaks corresponding to solids and pores. However, such selection is not always possible and then the implementation of other segmentation techniques becomes necessary. Segmentation using the Otsu (Otsu, 1979) method is often popular in soil studies (e.g. Dal Ferro et al., 2014), while other more soil-specific methods also have been proposed (Hapca et al., 2013, Houston et al., 2013). Several reviews and comparisons of the thresholding algorithms were conducted (e.g. Beckers et al., 2014, Iassonov et al., 2009, Wang et al., 2011) and can be used as a starting point in identifying the thresholding method optimal for the study. An illustration of a solid/pore segmentation obtained for the image using the Otsu method is shown in Fig. 1c.

After segmentation it is often advisable to eliminate single voxels classified as pores, as the true classification of such voxels is uncertain (Pires et al., 2017, Schaffer et al., 2007). That can be achieved by different methods, for example, component labeling (Schaffer et al., 2007) that selects the objects of certain size, or, more commonly, by a sequence of erosion/dilation steps.

7 Potential indicators of soil health that can be derived from X-ray computed micro-tomography

7.1 Macroporosity

What should be remembered about pore characterization using X-ray CT, as well as other imaging tools, is that it could only be performed for the pores that are visible at/above the studied scanning resolution. If the total porosity of the sample can be calculated from its bulk density measurement, then the total volume of the pores below the image resolution can be obtained as the difference between the total porosity and the visible porosity. Yet, it is the only reliable quantity that can be generated from the image for the below-resolution pores. For the pores with the diameters at/above image resolution the respective volumes as well as other characteristics can be obtained from the images. However, due to the partial volume effect the characterization of the pores with diameters equal or only somewhat above the scanning resolution should be done with caution. Because of the partial volume effect, a lot of pores with diameters approaching image resolution will cross more than one voxel and upon segmentation many of the voxels including them are likely to be classified as a solid. The outcome is a substantial underestimation of the volume of pores with sizes approaching the image resolution. While no single set of guidelines exist, it has been recommended that fully reliable inferences about soil features from the images, for example, pore volumes, can be obtained only for the features that are at least 2–5 times greater than the image resolution (Koestel et al., 2018, Vogel et al., 2010).

The studies using μCT scanning to look at soil pore structure are typically conducted using sizeable soil samples, those that arguably constitute volume representing all pore group sizes in the soil (e.g. >5 × 5 × 5 cm^3). The scanning resolutions for such samples are typically relatively high, that is, tens of microns. Pores visible at such resolutions are often referred to as mesopores and macropores (Brewer, 1976) and the total volume of such visible pores as mesoporosity and macroporosity. Note that there is not a single size/diameter that is universally accepted as the macropore size. The size of what is called macropores in X-ray μCT studies varies and is commonly a function of the scanning resolution. For example, Dal Ferro et al. (2014) called macropores as pores with 50 μm diameters; several other studies (Borges et al., 2019, Drewry et al., 2008, Munkholm et al., 2012, Pohlitz et al., 2018, 2019) used 60 μm diameter, Schaffer et al. (Schaffer et al., 2007) 90 μm, and Naveed et al. (Naveed et al., 2013) >1 mm. In that respect, the X-ray CT scanning studies are not different from the works where macroporosity is measured using laboratory methods. The separation between micropores and macropores is notoriously arbitrary (Hillel, 1998), and macropore size in laboratory studies ranges anywhere from 30 μm (Drewry, 2006), to 50 μm (Carter, 1988), to 300 μm (Reynolds et al., 2009), to 75 μm (Jongerius, 1957, Brewer, 1976), and to > 1 mm (Colombi et al., 2017). Thus, here we loosely define macroporosity as the percent of soil volume occupied by pores with diameters greater than a few tens of microns. Future standardization of the sample sizes and scanning resolutions would be highly advisable for increasing comparability among the X-ray μCT studies and μCT use in soil health assessments.

Macroporosity is regarded as an important physical indicator of soil health (Miller and Lin, 2018, Slater, 2018). Volume of macropores is indicative of the biological activities in the soil, as macropores are either produced or populated by plant roots and by the soil biota, often constituting hot spots of microbial activities (Banfield et al., 2017, 2018). Macropores commonly serve as the routes for preferential water flow (Jarvis, 2007), which transfer contaminants through the root zone with a bypass flow. Drainage of the excess water is faster in soils with larger macropore presence, leading to reduced waterlogging and less anaerobic conditions within the soil matrix. Macropores play an important role in soil aeration (Gliński and Stepniewski, 1985). They are avenues for gas transport and their presence/absence can influence O_2 supply and soil greenhouse gas emissions (Ben-Noah and Friedman, 2018, Kuncoro et al., 2014).

Macroporosity determined from X-ray μCT scanning has been widely used to characterize effects of land-use and management practices on soil structure. There is a large body of recent literature that explored the effect of tillage (e.g.

Borges et al., 2019, Dal Ferro et al., 2014, Drewry et al., 2008, Lucas et al., 2019, Pires et al., 2017, Pohlitz et al., 2018), agricultural and forest machinery traffic (Hansson et al., 2018, Schaffer et al., 2007), cover crops (Cercioglu et al., 2018), and diversity of plant communities (Kravchenko et al., 2018) on macroporosity. It appears that the use of macroporosity as an indicator of soil structure can be more advantageous than the use of total porosity, since macropore volume is more sensitive to soil compaction (Dal Ferro et al., 2014, Pohlitz et al., 2018). There is also a significant body of research that employs macroporosity in studying soil processes related to soil health. Some examples include root growth (Colombi et al., 2017), temporal dynamics of soil structure (Koestel and Schluter, 2019), and spatial patterns in distributions of bacteria (Juyal et al., 2018, 2019) and fungi (Falconer et al., 2012) within the soil matrix.

How to determine macroporosity from an X-ray μCT image? The specific procedure(s) can vary depending on the specific circumstances. If the image resolution is rather low and the entire image-visible porosity can be regarded as macroporosity, then the determination of the macroporosity will require only the solid-pore thresholding. The volume of the obtained image-visible porosity is then regarded as macroporosity. It can be reported either in reference to the volume of the entire soil sample, or as a fraction of the total porosity of the sample, if the latter has been determined. However, if the image resolution is high, that is, voxel sizes are low, then after the thresholding, it will be necessary to assess pore-size distributions, that is, to obtain information on volumes of pores of different size ranges. Then, the macroporosity can be calculated as the sum of volumes for the pores greater than the chosen macropore size. An illustration of the pore-space classification into pores of different sizes and a plot of the resultant pore-size distribution data are shown in Fig. 1d–e. There are several techniques developed for conducting pore-size distribution analysis, please see Houston et al. (Houston et al., 2017) for a comprehensive review. Among the most commonly used techniques are medial axis identification by morphological thinning, implemented in 3DMA-Rock software (Lindquist et al., 1996, 2000), and maximum inscribed sphere method, implemented in the ImageJ plugins BoneJ (Doube et al., 2010) and Xlib/Beat (Munch and Holzer, 2008, Munch, 2017). New methods better suited for soils, and especially biopores, are being continuously developed (Bauer et al., 2017).

7.2 Pore connectivity

The total volume of pores and/or macropores may not by itself be universally important for soil functioning. It is not the total volume, but the arrangement, and, connections among the pores what matters. Currently, pore connectivity is not mentioned among the physical indicators of soil health because it is not possible to accurately measure it using traditional measurement tools. However, given that it can be assessed from X-ray μCT images, the pore connectivity has a great potential as an indicator of soil health. Pore connectivity is especially important for liquid and gas fluxes in soil. Greater macropore connectivity translates into greater chemical transport and liquid flows as well as in greater microbial transport and fungi and root growth (Colombi et al., 2017, Jarvis et al., 2016). Pore connectivity can be particularly important for providing gas exchange with the atmosphere, thus influencing decomposition of soil POM (Kravchenko et al., 2014a) and affecting N_2O emissions (Kravchenko et al., 2018).

While pore connectivity, as all other pore characteristics, is generally related to bulk density and total porosity, the relationships are by no means unequivocal and can vary depending on several other factors, including tillage and soil moisture (Pohlitz et al., 2019). For example, upon soil compaction, pore connectivity can be initially insensitive to porosity decreases until a certain threshold in compaction is reached followed by a rapid reduction in pore connectivity (Pohlitz et al., 2019, Schluter and Vogel, 2016). Moreover, an absolute value of pore connectivity might not be an unequivocal indicator of the soil status, for example, pore connectivity will be high in freshly conventionally tilled soil, but lower in no-till (Kravchenko et al., 2011, Pires et al., 2019). Thus, just one characteristic of the soil pore space, for example, just macroporosity or just macropore connectivity, might not be sufficient to quantify various aspects of land-use and management effects on soil health. A holistic assessment will need to be developed.

A number of studies already used pore connectivity in evaluating the effects of land use and management on soil structure, including the work of Deurer et al. (2009) comparing porosities of chemically and organically managed orchards; Pagenkemper et al. (2014) studying multiple years of cultivation for different forage species; and Lucas et al. (2019) exploring the effect of tillage, management duration, soil depth, and crop on pore connectivity characteristics.

How to determine pore connectivity from an X-ray μCT image? The first step is the solid-pore thresholding. Then, the connectivity can be calculated for the entire body of the image-visible pores, or one can identify the pores of a certain size(s) and determine the connectivity for such a size subset(s). Several measures of pore connectivity have been proposed and used in soil science. The ones that are more relevant to soil functioning, that is, flows, transports, biota movement, fungal and root growth, are those that are based on calculations of the connected pore volumes. The simplest are the enumeration of connected pore volumes, expressed as the percent of the total pore volume (Kravchenko et al., 2011, Pagenkemper et al., 2014, Zuniga et al., 2019) or as a probability of connection (Koestel and Schluter, 2019, Pohlmeier et al., 2018, Renard and Allard, 2013). A recently proposed connectivity measure is percolation porosity suggested by Koestel and Schluter (2019) and determined as the total volume of the pores connected to both top and bottom of the studied soil domain. Other commonly reported measures of connectivity are the Euler-Poincare and Euler numbers (Toriwaki and Yonekura, 2002, Vogel, 2002). A relative disadvantage of these measures is that they do not reflect the size of connected pore space and do not supply information on how likely it is for the two points within the pore space to be connected (Renard and Allard, 2013). However, when applied in conjunction with other measures of connectivity, for example, to specific connected pore volumes, they provide useful information on pore network complexity (Zuniga et al., 2019).

Pore connectivity measures can be generated by several image analyses tools, including ImageJ's BoneJ plugin (Particle Analyzer) and 3DMA-Rock. After solid/pore thresholding, the pore space can be divided into separate pore clusters, that is, connected pore volumes, using any particle separation or particle labeling tool (e.g. labeling 2D/3D of Xlib/Beat), followed by their volume determination. Euler numbers can be obtained in several ImageJ plugins, including Particle Analyzer of BoneJ and MorphoLibJ (Legland et al., 2016).

7.3 Pore shape

Pore shape can be an informative criterion reflecting prevailing forces of pore formation and providing information about stability of the pore system. Several other image characteristics have comparable equivalents in traditional soil physics methods, for example, soil total porosity can be directly measured, approximate pore size distributions can be deduced from water retention or obtained via mercury porosimetry. On the contrary, images of intact samples are the only way to see and quantify the shape of the pores.

The partitioning of pore types based on their shape that is most important from soil functioning and health perspective is the partitioning into biopores and cracks. Biopores, originated from plant roots and earthworm activities, are of round shape, while cracks formed via physical processes of freezing/thawing, drying/wetting, and mechanical stress actions are typically of elongated shape and may create large interconnected networks. Even when equal in diameters and volumes, biopores and cracks can differ in their contributions to soil functioning, with biopores playing a particularly significant role (Kautz, 2015). Biopores contribute more to soil drainage and solute transport than pores of non-biological origin, they can serve as hot spots of microbial activities (Kuzyakov and Blagodatskaya, 2015) and function as preferred pathways for root growth (Colombi et al., 2017, Hirth et al., 2005) and earthworm activities (Pagenkemper et al., 2015). Biopores can be more stable under soil compaction than pores of non-biological origin (Schaffer et al., 2007). For example, during soil compaction via heavy trafficking it is the non-biological pores that disappeared first, leading to increased proportion of biopores in the soil matrix (Schaffer et al., 2007). Once a system of biopores develops within the soil matrix it can remain surprisingly stable, being continuously reutilized by new plant roots and soil biota (Koestel and Schluter, 2019, Lucas et al., 2019). Soil crack networks are more dynamic as compared to the biopores. They can disappear due to soil swelling, and form again in the same locations during soil draughts due to soil swelling. These pore networks are very important during heavy precipitation events after prolonged draughts, when dry soil surface has very low water conductivity and most precipitation turn to runoff. Particularly for these pores, X-ray CT is a unique tool allowing the study of the dynamics of crack formation in undisturbed soils.

Several measures of pore shape have been developed for separating pores of biological and non-biological origin as well as for the overall pore shape characterization. Most of them are derivatives of pore volumes and surfaces, such as a pore surface-to-volume ratio (e.g. Leue et al., 2019) or structure-model-index (Hildebrand and Ruegsegger, 1997, Schaffer et al., 2007).

Another characteristic of soil pore shape is potentially important for assessing soil heath and structure in pore anisotropy (Zuniga et al., 2019). Soil compaction under tillage and grazing can result in formation of platy soil structure reflected in a pore space with primarily horizontally oriented pores (Dorner and Horn, 2006). However, how sensitive the measures of anisotropy obtained from X-ray μCT images can be to such influences is not clear yet (Pohlitz et al., 2019).

How to calculate pore shape? Pore surfaces and volumes can be obtained in Particle Analyzer (BoneJ) and then used to calculate the ratios and structure-model index values. Anisotropy can be calculated using BoneJ-Anizotropy option.

Given the importance of biopores for soil functioning and a lack of traditional measurement tools for biopore identification, it is great to know that X-ray μCT can be used to separate biopores from pores of non-biological origin, aka cracks, and then to quantify the volumes and connectivities of different pore types separately. However, such separation is not trivial and requires several image processing steps. Unfortunately, at present, there are no packaged image analyses tools available for such separation. Multiple efforts for identification and analysis of the biopores are primarily based on in-house developed algorithms (Bauer et al., 2017, Leue et al., 2019, Lucas et al., 2019, Zhang et al., 2018).

7.4 Solid-to-pore distance

A new promising criterion recently proposed for soil structure characterization is a solid-to-pore distance (Schluter and Vogel, 2016), a measure that can be of particular relevance to soil functioning and soil health. This criterion is based on determining the distances from points within the solid soil matrix to the nearest pore. It might be viewed as an indicator of the travel distance that physical, chemical, and biological inputs must concur before they arrive at a destination within the soil matrix, that is, the volume of the soil matrix experiencing gas and nutrient exchanges. Soil pores are the main routes by which such inputs can travel through the soil and by which the exchanges can take place. The farther the location within the solid soil matrix is from a near-by pore, the less likely it is for it to receive new inputs or to have gas/liquid/chemical exchange with the rest of the soil profile. Of particular importance is the influx of O_2 into the soil matrix which plays a major role in driving soil organic decomposition (Keiluweit et al., 2017, 2018). The anaerobic conditions generated in the areas within the soil matrix at large distances from the air transferring pores can slow down the decomposition of the organic matter (Keiluweit et al., 2017, 2018, Kravchenko et al., 2015), while enhancing denitrification and facilitating N_2O production (Kravchenko et al., 2018). Pores are also the locations of the plant roots and the root exudates, the major source of C inputs into the soil. How far the plant inputs can be spread within the soil matrix depends on the spatial distribution of the soil pores, hence, on the soil-to-pore distances. Large soil-to-pore distances can translate into poor nutrient supply of soil microbial communities and, hence, into overall lower microbial activity and lower impact of soil microorganisms on a variety of soil functions, including soil organic matter accrual (Kravchenko et al., 2019a).

Even though the concept of soil-to-pore distance was relatively recently introduced, there are good examples of how useful it can be in separating the land-use and management influences on soil structure. Figure 2 provides an illustration of the concept for the soils experiencing two contrasting plant diversity levels: a bare soil and a mixture of 6 native grasses and 24 forbs. The latter has much greater spatial coverage of the soil matrix with pores, translating into overall shorter soil-to-pore distances. While the long-term bare soil has sizeable volumes that are far from any near-by pores. Quantitatively, the distribution of solid-to-pore distances from the entire image can be aggregated into a single criterion, that is, mean or median, or the distribution of distances can be presented in its entirety as a frequency distribution (Schluter and Vogel, 2016). The average solid-to-pore distance was a good indicator of the critical change in the soil pore structure due to intensive compaction, that is, when soil bulk density was reduced from 1.1 g/cm^3 to 1.5 g/cm^3 (Schluter and Vogel, 2016). Variations in solid-to-pore distances reflected well the pore structure dynamic in an intact soil column during multiple observation seasons (Koestel and Schluter, 2019).

It should be noted that as any image-based measure, the solid-to-pore distances can only be interpreted in the context of the studied image resolution, and the comparisons among different studies are only possible if their resolutions are comparable. Of particular importance for solid-to-pore distance implementation as the soil health and functioning criterion is a decision of what pore size(s) to use in its calculations. Pores of different sizes are involved in flow and transport processes driven by different forces and occurred at different velocities. For example, water is driven mostly by gravity in large pores, while capillary flow dominates in fine pores. Pores with 30–150 μm diameters might be of particular relevance for soil microbial functioning (Negassa et al., 2015). The presence of pores of this size range marked greater organic matter decomposition (Negassa et al., 2015) and greater C inputs from growing plants (Quigley et al., 2018). Likely, it is because pores of this size are accessible by the smallest plant roots, which play an important role in generating root exudates and supplying C to soil microbes. However, further analyses of the solid-to-pore distances in soils of different textures and mineralogy are needed to identify the meaningful pore-size ranges to be used for calculating this measure. While solid-to-pore distance is positively correlated with porosity, it does not carry the same information. It can be regarded as more meaningful for soil function assessment than porosity or pore-size distributions, because it reflects not just the volumes of the pore space but also their spatial distribution patterns within the soil matrix.

How to calculate the soil-to-pore distance? After solid/pore thresholding the distance transform tool of ImageJ/Fiji can be used to generate a map of the soil-to-pore distances. An efficient version of the distance transformation is provided in the plugin XLIB (Beat) (Meijster et al., 2000, Saito and Toriwaki, 1994). Once a map of soil-to-pore distances is generated, histograms or means and medians of the obtained distances can be generated using the ImageJ/Fiji measuring tools. Note that similar information to soil-to-pore distance can be expressed as the mean pore separation (Schaffer et al., 2007), that is, the mean difference between the pores.

7.5 Soil particulate organic matter

Particulate organic matter (POM) is one of the soil characteristics commonly considered among the indicators of soil health (Obrycki et al., 2018, Wegner et al., 2018). POM consists of the organic particles ranging in size from 0.053 to 2 mm (Cambardella and Elliott, 1992), it relatively quickly responds to changes in soil management and land use (Christensen, 2001, Six et al., 2002), and is relatively easy to measure in the lab. However, laboratory determination of POM is known to generate highly variable results due to significant variability of POM presence in individual soil samples (Ladoni et al., 2015). Thus, POM determination from the X-ray μCT images of intact soil samples can be a promising alternative to traditional laboratory measurements (Kravchenko et al., 2014b). POM can be determined from the soil images simultaneously with the previously discussed pore-related characteristics.

How to determine POM from X-ray μCT images? POM particles are identified based on their grayscale values, and a segmentation similar to that used for solid-pore thresholding is commonly the first step in POM identification. But POM identification can present additional challenges, since the gray values representative of POM are intermediate of those typical to solid and pore fractions. Thus, the contrasting differences are harder to obtain, the influence of the partial volume effects can be substantial, and a substantial number of artifacts can be generated. Indeed, a considerable number of voxels with the same range of grayscale values as POM are often located on the boundaries of the pores and solid materials, and thus are more difficult to separate than just solids from pores. Segmentation of the sample images in Fig. 1 illustrates the problem with POM identification: POM from plant root residues visible within the large pores in the bottom left part of the image is thresholded as pore space when using a default thresholding via Otsu method.

It also does not help that the range of grayscale values of different POM particles can vary substantially within the same soil image depending on the POM's origin and nature. A few examples: (i) on an X-ray μCT image fresh plant residues will be represented by somewhat darker colors than the residues at an advanced stage of decomposition; (ii) POM originated from plants with higher Si content will tend to have lighter grayscale values; and (iii) biochar particles will be yet lighter than the fresh plant-residue-originated POM. While a number of studies have been devoted to segmentation of live plant roots (e.g., Maenhout et al., 2019, Mooney et al., 2012, Perret et al., 2007), less attention has been paid to POM determination (Kravchenko et al., 2014b, Maenhout et al., 2019), and no tools specifically devoted to POM determination currently exist. However, the ImageJ procedures such as Particle Analyzer from BoneJ along with 3D Erosion and Dilation steps can be combined in order to facilitate POM identification.

8 Where to look for further information

Listed below are the reviews and comparative studies where further information on the selected topics is presented in detail and depth:

- Soil structure paradigms, different methods of soil structure characterization, and background on conflicting views on soil structure definitions (Dexter, 1988, Rabot et al., 2018, Young et al., 2001).
- Overviews of X-ray μCT scanning and its applications in geosciences, including soil science (Cnudde and Boone, 2013, Landis and Keane, 2010, Wildenschild and Sheppard, 2013).

- Soil quality and soil health index selection approaches (Bunemann et al., 2018).
- Solid-pore segmentation methods with X-ray μCT (Iassonov et al., 2009, Sezgin and Sankur, 2004).
- Pore size distribution determinations using X-ray μCT images (Houston et al., 2017).
- An in-depth comparison of different approaches to measure connectivity (Renard and Allard, 2013). Even though this work is not focused in X-ray μCT images, but on hydrological data sets, the information it provides is very relevant for connectivity measures related to water and gas flows in soil.
- Root segmentation methods with X-ray μCT (Mooney et al., 2012).

9 References

Al-Raoush, R. and Papadopoulos, A. (2010). Representative elementary volume analysis of porous media using x-ray computed tomography. *Powder Technology* 200(1–2), 69–77.

Banfield, C. C., Dippold, M. A., Pausch, J., Hoang, D. T. T. and Kuzyakov, Y. (2017). Biopore history determines the microbial community composition in subsoil hotspots. *Biology and Fertility of Soils* 53(5), 573–588.

Banfield, C. C., Pausch, J., Kuzyakov, Y. and Dippold, M. A. (2018). Microbial processing of plant residues in the subsoil – the role of biopores. *Soil Biology and Biochemistry* 125, 309–318.

Bauer, B., Cai, X. H., Peth, S., Schladitz, K. and Steidl, G. (2017). Variational-based segmentation of bio-pores in tomographic images. *Computers and Geosciences* 98, 1–8.

Baveye, P., Rogasik, H., Wendroth, O., Onasch, I. and Crawford, J. W. (2002). Effect of sampling volume on the measurement of soil physical properties: simulation with x-ray tomography data. *Measurement Science and Technology* 13(5), 775–784.

Baveye, P. C., Laba, M., Otten, W., Bouckaert, L., Dello Sterpaio, P., Goswami, R. R., Grinev, D., Houston, A., Hu, Y., Liu, J., Mooney, S., Pajor, R., Sleutel, S., Tarquis, A., Wang, W., Wei, Q. and Sezgin, M. (2010). Observer-dependent variability of the thresholding step in the quantitative analysis of soil images and X-ray microtomography data. *Geoderma* 157(1–2), 51–63.

Beckers, E., Plougonven, E., Roisin, C., Hapca, S., Leonard, A. and Degre, A. (2014). X-ray microtomography: a porosity-based thresholding method to improve soil pore network characterization? *Geoderma* 219, 145–154.

Ben-Noah, I. and Friedman, S. P. (2018). Review and evaluation of root respiration and of natural and agricultural processes of soil aeration. *Vadose Zone Journal* 17(1), 1–47.

Borges, J. R., Pires, L. F., Cassaro, F. M., Auler, A. C., Rosa, J. A., Heck, R. J. and Roque, W. L. (2019). X-ray computed tomography for assessing the effect of tillage systems on topsoil morphological attributes. *Soil and Tillage Research* 189, 25–35.

Borges, J. R., Pires, L. F., Cassaro, F. M., Roque, W. L., Heck, R. J., Rosa, J. A. and Wolf, F. G. (2018). X-ray microtomography analysis of representative elementary volume (REV) of soil morphological and geometrical properties. *Soil and Tillage Research* 182, 112–122.

Brewer, R. (1976). *Fabric and Mineral Analysis of Soils*. New York, NY: Krieger Publishing, 482 p.

Bunemann, E. K., Bongiorno, G., Bai, Z. G., Creamer, R. E., De Deyn, G., de Goede, R., Fleskens, L., Geissen, V., Kuyper, T. W., Mäder, P., Pulleman, M., Sukkel, W., van Groenigen, J. W. and Brussaard, L. (2018). Soil quality - a critical review. *Soil Biology and Biochemistry* 120, 105-125.

Cambardella, C. A. and Elliott, E. T. (1992). Particulate soil organic-matter changes across a grassland cultivation sequence. *Soil Science Society of America Journal* 56(3), 777-783.

Carter, M. R. (1988). Temporal variability of soil macroporosity in a fine sandy loam under moldboard plowing and direct drilling. *Soil and Tillage Research* 12(1), 37-51.

Cercioglu, M., Anderson, S. H., Udawatta, R. P. and Haruna, S. I. (2018). Effects of cover crop and biofuel crop management on computed tomography-measured pore parameters. *Geoderma* 319, 80-88.

Chesworth, W. (2008). Structure. In: *Encyclopedia of Soil Science* Chesworth, W. (Ed.). Dordrecht, Netherlands: Springer.

Christensen, B. T. (2001). Physical fractionation of soil and structural and functional complexity in organic matter turnover. *European Journal of Soil Science* 52(3), 345-353.

Cnudde, V. and Boone, M. N. (2013). High-resolution X-ray computed tomography in geosciences: a review of the current technology and applications. *Earth-Science Reviews* 123, 1-17.

Colombi, T., Braun, S., Keller, T. and Walter, A. (2017). Artificial macropores attract crop roots and enhance plant productivity on compacted soils. *Science of the Total Environment* 574, 1283-1293.

Dal Ferro, N., Sartori, L., Simonetti, G., Berti, A. and Morari, F. (2014). Soil macro- and microstructure as affected by different tillage systems and their effects on maize root growth. *Soil and Tillage Research* 140, 55-65.

Deurer, M., Grinev, D., Young, I., Clothier, B. E. and Muller, K. (2009). The impact of soil carbon management on soil macropore structure: a comparison of two apple orchard systems in New Zealand. *European Journal of Soil Science* 60(6), 945-955.

Dexter, A. R. (1988). Advances in characterization of soil structure. *Soil and Tillage Research* 11(3-4), 199-238.

Dorner, J. and Horn, R. (2006). Anisotropy of pore functions in structured Stagnic luvisols in the Weichselian moraine region in N Germany. *Journal of Plant Nutrition and Soil Science* 169(2), 213-220.

Doube, M., Klosowski, M. M., Arganda-Carreras, I., Cordelières, F. P., Dougherty, R. P., Jackson, J. S., Schmid, B., Hutchinson, J. R. and Shefelbine, S. J. (2010). BoneJ Free and extensible bone image analysis in ImageJ. *Bone* 47(6), 1076-1079.

Drewry, J. J. (2006). Natural recovery of soil physical properties from treading damage of pastoral soils in New Zealand and Australia: a review. *Agriculture, Ecosystems and Environment* 114(2-4), 159-169.

Drewry, J. J., Cameron, K. C. and Buchan, G. D. (2008). Pasture yield and soil physical property responses to soil compaction from treading and grazing - a review. *Australian Journal of Soil Research* 46(3), 237-256.

Falconer, R. E., Houston, A. N., Otten, W. and Baveye, P. C. (2012). Emergent behavior of soil fungal dynamics: influence of soil architecture and water distribution. *Soil Science* 177(2), 111-119.

Gliński, J. and Stepniewski, W. (1985). *Soil Aeration and Its Role for Plants*. Boca Raton: CRC Press.

Hansson, L. J., Koestel, J., Ring, E. and Gardenas, A. I. (2018). Impacts of off-road traffic on soil physical properties of forest clear-cuts: X-ray and laboratory analysis. *Scandinavian Journal of Forest Research* 33(2), 166–177.

Hapca, S. M., Houston, A. N., Otten, W. and Baveye, P. C. (2013). New local thresholding method for soil images by minimizing grayscale intra-class variance. *Vadose Zone Journal* 12(3), 1–13.

Helliwell, J. R., Sturrock, C. J., Grayling, K. M., Tracy, S. R., Flavel, R. J., Young, I. M., Whalley, W. R. and Mooney, S. J. (2013). Applications of X-ray computed tomography for examining biophysical interactions and structural development in soil systems: a review. *European Journal of Soil Science* 64(3), 279–297.

Hildebrand, T. and Ruegsegger, P. (1997). A new method for the model-independent assessment of thickness in three-dimensional images. *Journal of Microscopy* 185(1), 67–75.

Hillel, D. (1998). *Environmental Soil Physics*. San Diego, CA: Academic Press.

Hirth, J. R., McKenzie, B. M. and Tisdall, J. M. (2005). Ability of seedling roots of *Lolium perenne* L. to penetrate soil from artificial biopores is modified by soil bulk density, biopore angle and biopore relief. *Plant and Soil* 272(1–2), 327–336.

Houston, A. N., Otten, W., Baveye, P. C. and Hapca, S. (2013). Adaptive-window indicator kriging: a thresholding method for computed tomography images of porous media. *Computers and Geosciences* 54, 239–248.

Houston, A. N., Otten, W., Falconer, R., Monga, O., Baveye, P. C. and Hapca, S. M. (2017). Quantification of the pore size distribution of soils: assessment of existing software using tomographic and synthetic 3D images. *Geoderma* 299, 73–82.

Iassonov, P., Gebrenegus, T. and Tuller, M. (2009). Segmentation of X-ray computed tomography images of porous materials: a crucial step for characterization and quantitative analysis of pore structures. *Water Resources Research* 45(9), 1–12.

Jarvis, N., Koestel, J. and Larsbo, M. (2016). Understanding preferential flow in the vadose zone: recent advances and future prospects. *Vadose Zone Journal* 15(12), 1–11.

Jarvis, N. J. (2007). A review of non-equilibrium water flow and solute transport in soil macropores: principles, controlling factors and consequences for water quality. *European Journal of Soil Science* 58(3), 523–546.

Johannes, A., Weisskopf, P., Schulin, R. and Boivin, P. (2019). Soil structure quality indicators and their limit values. *Ecological Indicators* 104, 686–694.

Jongerius, A. (1957). Morfologische Onderzoekingen over de bodemstructuur. Diss. Wageningen. Versl. Landbouwk. Onderz. *Bodemk. Stud.* 63, 12.

Jury, W. A. and Horton, R. (2004). *Soil Physics*. Hoboken, NJ: John Wiley & Sons.

Juyal, A., Eickhorst, T., Falconer, R., Baveye, P. C., Spiers, A. and Otten, W. (2018). Control of pore geometry in soil microcosms and its effect on the growth and spread of Pseudomonas and Bacillus sp. *Frontiers in Environmental Science* 6(73), 1–12.

Juyal, A., Otten, W., Falconer, R., Hapca, S., Schmidt, H., Baveye, P. C. and Eickhorst, T. (2019). Combination of techniques to quantify the distribution of bacteria in their soil microhabitats at different spatial scales. *Geoderma* 334, 165–174.

Kautz, T. (2015). Research on subsoil biopores and their functions in organically managed soils: a review. *Renewable Agriculture and Food Systems* 30(4), 318–327.

Keiluweit, M., Gee, K., Denney, A. and Fendorf, S. (2018). Anoxic microsites in upland soils dominantly controlled by clay content. *Soil Biology and Biochemistry* 118, 42–50.

Keiluweit, M., Wanzek, T., Kleber, M., Nico, P. and Fendorf, S. (2017). Anaerobic microsites have an unaccounted role in soil carbon stabilization. *Nature Communications* 8(1), 1771.

Koestel, J. (2018). SoilJ: an ImageJ plugin for the semiautomatic processing of three-dimensional X-ray images of soils. *Vadose Zone Journal* 17(1), 1-7.

Koestel, J., Dathe, A., Skaggs, T. H., Klakegg, O., Ahmad, M. A., Babko, M., Giménez, D., Farkas, C., Nemes, A. and Jarvis, N. (2018). Estimating the permeability of naturally structured soil from percolation theory and pore space characteristics imaged by X-ray. *Water Resources Research* 54(11), 9255–9263.

Koestel, J. and Schluter, S. (2019). Quantification of the structure evolution in a garden soil over the course of two years. *Geoderma* 338, 597–609.

Kravchenko, A. N., Guber, A. K., Quigley, M. Y., Koestel, J., Gandhi, H. and Ostrom, N. E. (2018). X-ray computed tomography to predict soil N2O production via bacterial denitrification and N2O emission from soils in contrasting bioenergy cropping systems. *Global Change Biology Bioenergy* 10(11), 894–909.

Kravchenko, A. N., Guber, A. K., Razavi, B. S., Koestel, J., Quigley, M. Y., Robertson, G. P. and Kuzyakov, Y. (2019a). Microbial spatial footprint as a driver of soil carbon stabilization. *Nature Communications* 10, 3121.

Kravchenko, A., Otten, W., Garnier, P., Pot, V. and Baveye, P. C. (2019b). Soil aggregates as biogeochemical reactors: not a way forward in the research on soil-atmosphere exchange of greenhouse gases. *Global Change Biology* 25(7), 2205–2208.

Kravchenko, A. N., Negassa, W. C., Guber, A. K., Hildebrandt, B., Marsh, T. L. and Rivers, M. L. (2014a). Intra-aggregate pore structure influences phylogenetic composition of bacterial community in macroaggregates. *Soil Science Society of America Journal* 78(6), 1924-1939.

Kravchenko, A. N., Negassa, W., Guber, A. K. and Schmidt, S. (2014b). New approach to measure soil particulate organic matter in intact samples using X-ray computed microtomography. *Soil Science Society of America Journal* 78(4), 1177–1185.

Kravchenko, A. N., Negassa, W. C., Guber, A. K. and Rivers, M. L. (2015). Protection of soil carbon within macro-aggregates depends on intra-aggregate pore characteristics. *Scientific Reports* 5, 16261.

Kravchenko, A. N., Wang, A. N. W., Smucker, A. J. M. and Rivers, M. L. (2011). Long-term differences in tillage and land use affect intra-aggregate pore heterogeneity. *Soil Science Society of America Journal* 75(5), 1658–1666.

Kuncoro, P. H., Koga, K., Satta, N. and Muto, Y. (2014). A study on the effect of compaction on transport properties of soil gas and water I: Relative gas diffusivity, air permeability, and saturated hydraulic conductivity. *Soil and Tillage Research* 143, 172–179.

Kuzyakov, Y. and Blagodatskaya, E. (2015). Microbial hotspots and hot moments in soil: concept & review. *Soil Biology and Biochemistry* 83, 184–199.

Ladoni, M., Basir, A. and Kravchenko, A. (2015). Which soil carbon fraction is the best for assessing management differences? A statistical power perspective. *Soil Science Society of America Journal* 79(3), 848–857.

Landis, E. N. and Keane, D. T. (2010). X-ray microtomography. *Materials Characterization* 61(12), 1305–1316.

Legland, D., Arganda-Carreras, I. and Andrey, P. (2016). MorphoLibJ: integrated library and plugins for mathematical morphology with ImageJ. *Bioinformatics* 32(22), 3532–3534.

Leue, M., Uteau-Puschmann, D., Peth, S., Nellesen, J., Kodesova, R. and Gerke, H. H. (2019). Separation of soil macropore types in three-dimensional X-ray computed tomography images based on pore geometry characteristics. *Vadose Zone Journal* 18(1), 1-13.

Lindquist, W. B., Lee, S. M., Coker, D. A., Jones, K. W. and Spanne, P. (1996). Medial axis analysis of void structure in three-dimensional tomographic images of porous media. *Journal of Geophysical Research: Solid Earth* 101(B4), 8297-8310.

Lindquist, W. B., Venkatarangan, A., Dunsmuir, J. and Wong, T. F. (2000). Pore and throat size distributions measured from synchrotron X-ray tomographic images of Fontainebleau sandstones. *Journal of Geophysical Research: Solid Earth* 105(B9), 21509-21527.

Lucas, M., Schluter, S., Vogel, H. J. and Vetterlein, D. (2019). Soil structure formation along an agricultural chronosequence. *Geoderma* 350, 61-72.

Maenhout, P., Sleutel, S., Xu, H., Van Hoorebeke, L., Cnudde, V. and De Neve, S. (2019). Semi-automated segmentation and visualization of complex undisturbed root systems with X-ray mu CT. *Soil and Tillage Research* 192, 59-65.

Meijster, A., Roerdink, J. B. T. M. and Hesselink, W. H. (2000). A general algorithm for computing distance transforms in linear time. *Mathematical Morphology and Its Applications to Image and Signal Processing* 18, 331-340.

Miller, M. and Lin, H. (2018). The role of soil hydrology in soil health. *Managing Soil Health for Sustainable Agriculture, Vol 1: Fundamentals*, 48, 245-263.

Moebius-Clune, B. N., Moebius-Clune, D. J., Gugino, B. K., Idowu, O. J., Schindelbeck, R. R., Ristow, A. J., van Es, H. M., Thies, J. E., Shayler, H. A., McBride, M. B., Kurtz, K. S. M., Wolfe, D. W., and Abawi, G. S. (2016). *Comprehensive Assessment of Soil Health - The Cornell Framework*. Cornell University. Available at: https://soilhealth.cals.cornell.edu/ (Accessed 09/02/2020).

Mooney, S. J. (2002). Three-dimensional visualization and quantification of soil macroporosity and water flow patterns using computed tomography. *Soil Use and Management* 18(2), 142-151.

Mooney, S. J., Pridmore, T. P., Helliwell, J. and Bennett, M. J. (2012). Developing X-ray computed tomography to non-invasively image 3-D root systems architecture in soil. *Plant and Soil* 352(1-2), 1-22.

Munch, B. and Holzer, L. (2008). Contradicting geometrical concepts in pore size analysis attained with electron microscopy and mercury intrusion. *Journal of the American Ceramic Society* 91(12), 4059-4067.

Munch, E. A. (2017). User's guide to topological data analysis. *Journal of Learning Analytics* 4(2), 47-61.

Munkholm, L. J., Heck, R. J. and Deen, B. (2012). Soil pore characteristics assessed from X-ray micro-Cf derived images and correlations to soil friability. *Geoderma* 181, 22-29.

Naveed, M., Moldrup, P., Arthur, E., Wildenschild, D., Eden, M., Lamandé, M., Vogel, H. and de Jonge, L. W. (2013). Revealing soil structure and functional macroporosity along a clay gradient using X-ray computed tomography. *Soil Science Society of America Journal* 77(2), 403-411.

Negassa, W. C., Guber, A. K., Kravchenko, A. N., Marsh, T. L., Hildebrandt, B. and Rivers, M. L. (2015). Properties of soil pore space regulate pathways of plant residue decomposition and community structure of associated bacteria. *PLoS ONE* 10(4), e0123999.

Nunan, N., Ritz, K., Rivers, M., Feeney, D. S. and Young, I. M. (2006). Investigating microbial micro-habitat structure using X-ray computed tomography. *Geoderma* 133(3-4), 398-407.

Obrycki, J. F., Karlen, D. L., Cambardella, C. A., Kovar, J. L. and Birrell, S. J. (2018). Corn stover harvest, tillage, and cover crop effects on soil health indicators. *Soil Science Society of America Journal* 82(4), 910-918.

Otsu, N. (1979). A threshold selection method from gray-level histograms. *IEEE Transactions on Systems, Man, and Cybernetics* 9(1), 62-66.

Pagenkemper, S. K., Athmann, M., Uteau, D., Kautz, T., Peth, S. and Horn, R. (2015). The effect of earthworm activity on soil bioporosity - investigated with X-ray computed tomography and endoscopy. *Soil and Tillage Research* 146, 79-88.

Pagenkemper, S. K., Puschmann, D. U., Peth, S. and Horn, R. (2014). Investigation of time dependent development of soil structure and formation of macropore networks as affected by various precrop species. *International Soil and Water Conservation Research* 2(2), 51-66.

Perret, J. S., Al-Belushi, M. E. and Deadman, M. (2007). Non-destructive visualization and quantification of roots using computed tomography. *Soil Biology and Biochemistry* 39(2), 391-399.

Peth, S., Horn, R., Beckmann, F., Donath, T., Fischer, J. and Smucker, A. J. M. (2008). Three-dimensional quantification of intra-aggregate pore-space features using synchrotron-radiation-based microtomography. *Soil Science Society of America Journal* 72(4), 897-907.

Pires, L. F., Borges, J. A. R., Rosa, J. A., Cooper, M., Heck, R. J., Passoni, S. and Roque, W. L. (2017). Soil structure changes induced by tillage systems. *Soil and Tillage Research* 165, 66-79.

Pires, L. F., Roque, W. L., Rosa, J. A. and Mooney, S. J. (2019). 3D analysis of the soil porous architecture under long term contrasting management systems by X-ray computed tomography. *Soil and Tillage Research* 191, 197-206.

Pohlitz, J., Rucknagel, J., Koblenz, B., Schluter, S., Vogel, H. J. and Christen, O. (2018). Computed tomography and soil physical measurements of compaction behaviour under strip tillage, mulch tillage and no tillage. *Soil and Tillage Research* 175, 205-216.

Pohlitz, J., Rucknagel, J., Schluter, S., Vogel, H. J. and Christen, O. (2019). Computed tomography as an extension of classical methods in the analysis of soil compaction, exemplified on samples from two tillage treatments and at two moisture tensions. *Geoderma* 346, 52-62.

Pohlmeier, A., Garre, S. and Roose, T. (2018). Noninvasive imaging of processes in natural porous media: From pore to field scale. *Vadose Zone Journal* 17(1).

Quigley, M. Y., Negassa, W. C., Guber, A. K., Rivers, M. L. and Kravchenko, A. N. (2018). Influence of pore characteristics on the fate and distribution of newly added carbon. *Frontiers in Environmental Science*. doi:10.3389/fenvs.2018.00051.

Rabot, E., Wiesmeier, M., Schluter, S. and Vogel, H. -J. (2018). Soil structure as an indicator of soil functions: a review. *Geoderma* 314, 122-137.

Rasband, W. S. (1997-2015). *ImageJ*. Bethesda, MD: U.S. National Institutes of Health.

Renard, P. and Allard, D. (2013). Connectivity metrics for subsurface flow and transport. *Advances in Water Resources* 51, 168-196.

Reynolds, W. D., Drury, C. F., Tan, C. S., Fox, C. A. and Yang, X. M. (2009). Use of indicators and pore volume-function characteristics to quantify soil physical quality. *Geoderma* 152(3-4), 252-263.

Saito, T. and Toriwaki, J. I. (1994). New algorithms for Euclidean distance transformation of an N-dimensional digitized picture with applications. *Pattern Recognition* 27(11), 1551–1565.

Schaffer, B., Stauber, M., Muller, R. and Schulin, R. (2007). Changes in the macro-pore structure of restored soil caused by compaction beneath heavy agricultural machinery: a morphometric study. *European Journal of Soil Science* 58(5), 1062–1073.

Schindelin, J., Arganda-Carreras, I., Frise, E., Kaynig, V., Longair, M., Pietzsch, T., Preibisch, S., Rueden, C., Saalfeld, S., Schmid, B., Tinevez, J. Y., White, D. J., Hartenstein, V., Eliceiri, K., Tomancak, P. and Cardona, A. (2012). Fiji: an open-source platform for biological-image analysis. *Nature Methods* 9(7), 676–682.

Schluter, S., Blaser, S. R. G. A., Weber, M., Schmidt, V. and Vetterlein, D. (2018). Quantification of root growth patterns from the soil perspective via root distance models. *Frontiers in Plant Science* 9, 1084.

Schluter, S., Sheppard, A., Brown, K. and Wildenschild, D. (2014). Image processing of multiphase images obtained via X-ray microtomography: a review. *Water Resources Research* 50(4), 3615–3639.

Schluter, S. and Vogel, H. J. (2016). Analysis of soil structure turnover with garnet particles and X-ray microtomography. *PLoS ONE* 11(7), e0159948.

Sezgin, M. and Sankur, B. (2004). Survey over image thresholding techniques and quantitative performance evaluation. *Journal of Electronic Imaging* 13(1), 146–168.

Six, J., Conant, R. T., Paul, E. A. and Paustian, K. (2002). Stabilization mechanisms of soil organic matter: implications for C-saturation of soils. *Plant and Soil* 241(2), 155–176.

Slater, B. K. (2018). Soil health assessment and inventory: indices and databases. *Managing Soil Health for Sustainable Agriculture, Vol 2: Monitoring and Management*, 49, 3–21.

Smet, S., Plougonven, E., Leonard, A., Degre, A. and Beckers, E. (2018). X-ray micro-CT: how soil pore space description can be altered by image processing. *Vadose Zone Journal* 17(1), 1–14.

Soil Health Institute (2019). North American project to evaluate soil health measurements. Available at: https://soilhealthinstitute.org/north-american-project-to-evaluate-soil-health-measurements/ (Checked 11/18/2019).

Tollner, E. W., Melear, N. D., Rodriguez, L. A. and Wright, M. E. (1998). Soil aggregate size distributions using X-ray images. *Transactions of the ASAE* 41, 1207–1215.

Topp, G. C., Reynolds, W. D., Cook, F. J., Kirby, J. M. and Carter, M. R. (1997). Physical attributes of soil quality. In: *Soil Quality for Crop Production and Ecosystem Health* Gregorich, E. G. and Carter, M. R. (Eds). New York, NY: Elsevier.

Toriwaki, J. and Yonekura, T. (2002). Euler number and conectivity indexes of a three dimentional digital picture. *Forma* 17, 183–209.

Tracy, S. R., Daly, K. R., Sturrock, C. J., Crout, N. M. J., Mooney, S. J. and Roose, T. (2015). Three-dimensional quantification of soil hydraulic properties using X-ray computed tomography and image-based modeling. *Water Resources Research* 51(2), 1006–1022.

USDA NRCS (2019a). Soil health technical note no. 450-03. Available at: https://directives.sc.egov.usda.gov/OpenNonWebContent.aspx?content=43754.wba (Checked 11/18/2019).

USDA NRCS (2019b). Soil health. Available at: https://www.nrcs.usda.gov/wps/portal/nrcs/main/soils/health/assessment/ (Checked 11/18/2019).

Van Es, H. M. and Karlen, D. L. (2019). Reanalysis validates soil health indicator sensitivity and correlation with long-term crop yields. *Soil Science Society of America Journal* 83(3), 721-732.

Vogel, H. J. (2002). Topological characterization of porous media. *Morphology of Condensed Matter: Physics and Geometry of Spatially Complex Systems* 600, 75-92.

Vogel, H.-J., Weller, U. and Schluter, S. (2010). Quantification of soil structure based on Minkowski functions. *Computers and Geosciences* 36(10), 1236-1245.

Wander, M. M., Cihacek, L. J., Coyne, M., Drijber, R. A., Grossman, J. M., Gutknecht, J. L. M., Horwath, W. R., Jagadamma, S., Olk, D. C., Ruark, M., Snapp, S. S., Tiemann, L. K., Weil, R. and Turco, R. F. (2019). Developments in agricultural soil quality and health: reflections by the Research Committee on Soil Organic Matter Management. *Frontiers in Environmental Science* 7(109), 1-9.

Wang, W., Kravchenko, A. N., Smucker, A. J. M. and Rivers, M. L. (2011). Comparison of image segmentation methods in simulated 2D and 3D microtomographic images of soil aggregates. *Geoderma* 162(3-4), 231-241.

Wegner, B. R., Osborne, S. L., Lehman, R. M. and Kumar, S. (2018). Seven-year impact of cover crops on soil health when corn residue is removed. *Bioenergy Research* 11(2), 239-248.

Wildenschild, D. and Sheppard, A. P. (2013). X-ray imaging and analysis techniques for quantifying pore-scale structure and processes in subsurface porous medium systems. *Advances in Water Resources* 51, 217-246.

Young, I. M., Crawford, J. W. and Rappoldt, C. (2001). New methods and models for characterising structural heterogeneity of soil. *Soil and Tillage Research* 61(1-2), 33-45.

Zhang, Z. B., Liu, K. L., Zhou, H., Lin, H., Li, D. M. and Peng, X. H. (2018). Three dimensional characteristics of biopores and non-biopores in the subsoil respond differently to land use and fertilization. *Plant and Soil* 428(1-2), 453-467.

Zuniga, F., Horn, R., Rostek, J., Peth, S., Uteau, D. and Dorner, J. (2019). Anisotropy of intensity-capacity parameters on Aquands with contrasting swelling-shrinkage cycles. *Soil and Tillage Research* 193, 101-113.

Chapter 6

Geophysical methods to assess soil characteristics

Ho-Chul Shin, Rothamsted Research, UK; Guillaume Blanchy, Lancaster University, UK; Ian Shield, Peter Fruen, Timothy Barraclough and Christopher W. Watts, Rothamsted Research, UK; Andrew Binley, Lancaster University, UK; and William R. Whalley, Rothamsted Research, UK

1 Introduction

In this chapter, we explore the use of geophysical methods to assess soil physical characteristics. Several geophysical methods can be used for this purpose, such as electrical resistivity, induced polarisation, electromagnetic induction methods, ground-penetrating radar, nuclear magnetic resonance as well as methods based on seismic wave analysis (Table 1). Most geophysical methods do not directly measure soil characteristics but rather physical properties that may be related to properties or states of interest. Relationships to convert the physical property measured to the soil characteristics of interest play an essential role in using geophysical tools for soil science. It is also important to note that different soil characteristics can affect the same physical property, which complicates the interpretation of the geophysical results. For instance, soil moisture and salinity can both increase the soil electrical conductivity. Different methodological strategies can be used to separate these multiple effects.

With the exception of ground-penetrating radar, geophysical methods used in soil studies originate from mineral and oil exploration. The scalability

http://dx.doi.org/10.19103/AS.2020.0079.12

Table 1 A summary of the different geophysical methods for soil probing and their associated geophysical properties and soil characteristics that can be derived

Method	Basic physical property affecting measurement	Soil characteristics
Electrical resistivity	Electrical resistivity	Saturation, porosity, clay content, ionic concentration of pore water, salinity
Induced polarisation	Chargeability	Clay content, grain surface area
Electromagnetic induction	Electrical conductivity	Saturation, porosity, clay content, ionic concentration of pore water, salinity
Ground-penetrating radar	Electrical permittivity	Soil horizon boundaries, saturation
Nuclear Magnetic Resonance	Resonant properties of hydrogen nuclei	Saturation and pore size (in saturated soils)
Seismic method	Elastic properties	Soil strength and internal stress

of the methods has allowed the application to the shallow investigation of the subsurface. Growth in their use in hydrology expanded in the 1990s (e.g. Binley et al., 2015) and, more recently, has been widely integrated with many agronomy investigations (e.g. Allred et al., 2008). Static surveys (i.e. conducted at one point in time) can be effective for mapping variation of soil properties; the adoption of time-lapse surveys can help reveal changes to states (e.g. soil moisture) or properties (e.g. density) over time (e.g. Blanchy et al., 2020a).

Electrical resistivity methods measure the subsurface electrical resistivity distribution using an array of electrodes, typically on the ground surface. Such methods can be used for mapping spatial patterns of electrical resistivity using mobile arrays but are commonly used with a fixed array of electrodes to determine a 2D (vertical section) or 3D image of the subsurface (Binley, 2015; Binley and Slater, 2020). The term electrical resistivity tomography (ERT) is commonly used in reference to imaging with a fixed array of electrodes. Samouëlian et al. (2005) provide an introduction to ERT for soil studies, while the recent review of Cimpoiaşu et al. (2020) describe the application of geoelectrical methods for root studies. Electromagnetic induction (EMI) relies on the measurement of the electromagnetic field from induced eddy currents to infer the subsurface electrical conductivity (the conductivity is the inverse of the resistivity). As a contactless method, it is widely used for mapping large areas. Doolittle and Brevik (2014), Sudduth et al. (2005) and Corwin et al. (2006), amongst others, demonstrate the use of EMI for soil studies.

Electrical resistivity (or conductivity) methods can sometimes result in ambiguous interpretation because of the number of factors that can change the resistivity of soil. Induced polarisation (Binley and Slater, 2020) is an extension of the resistivity method that is sensitive to the storage of electrical charge, and consequently senses electrical characteristics of the soil grain surface, making it effective at differentiating variation in resistivity due to both pore water ionic content and clay content (Slater and Lesmes, 2002). The direct link between induced polarisation and grain surface characteristics has prompted a number of studies using induced polarisation to infer permeability of earth materials (e.g. Weller et al., 2015).

Ground-penetrating radar (GPR) sends radar pulses into the ground and records their multiple 'echos' to build an image of the subsurface revealing material with contrasting electrical permittivity. This can be extremely effective at mapping lithological boundaries or soil horizons. The speed of the radar wave is directly related to soil water content (as is exploited in the commonly used dielectric methods for soil moisture) and thus knowledge of the radar velocity allows the assessment of soil moisture. Analysis of radar velocity (and hence soil moisture) can be carried out at shallow depths using reflected signals or the direct wave that travels along the ground surface (e.g. Algeo et al., 2018). Like GPR, nuclear magnetic resonance (NMR) is also sensitive to water content in soils (Paetzold et al., 1985). In saturated soils, NMR signals can be used to infer a pore size distribution and permeability. The method is based on the interaction of hydrogen nuclei moments and an externally applied magnetic field. Most applications of NMR have focussed on relatively deep investigations (up to hundreds of metres), although recent advances in instruments are permitting very shallow depths of investigation, opening up opportunities for soil studies.

Seismic waves methods, with which acoustic methods are related, rely on the analysis of the 'echos' of elastic waves through the soil. Seismic methods share some similarity with GPR which relies on an electromagnetic pulse. While geophysical methods were initially used to characterise lithological features, often to depths of hundreds of metres such as for mineral exploration, their use in agriculture is focussed on shallow layers, usually to the depth of crop roots (a few metres). Their minimal invasiveness and high throughput offer great potential for the non-destructive study of the soil characteristics for agriculture.

It is not possible to cover all these methods and approaches in a single chapter, instead, we focus on three methods, which have specific values for soil investigations. For a more exhaustive overview of the different geophysical methods for agriculture, we redirect the reader to Allred et al. (2008). The first two methods we cover are electromagnetic induction and electrical resistivity, which are well established in agriculture. The third method is an emerging method based on acoustic to seismic coupling. The first two methods have been used in agriculture and soil science for both mapping soil characteristics, such as salinity and clay content, and for more detailed studies of soil processes, often related to water and solute movement (Beff et al., 2013; Binley et al., 1996; Blanchy et al., 2020b; Dahlin et al., 2014; Koestel et al., 2008; Reyes, et al., 2018; Shanahan et al., 2015; Whalley et al., 2017). Methods based on the analysis of seismic waves are sensitive to the elastic properties of soil but also depend on soil water status and soil compaction (Whalley et al., 2011, 2012).

2 Geophysical properties of soil

In practice, geoelectrical methods such as EMI and ERT are sensitive to a number of properties that are reasonably constant (clay content, organic matter content) as well as ephemeral states such as soil temperature, pore water electrical conductivity, bulk density and volumetric soil water content. The soil structure affects electrical conductivity by virtue of pore connectivity and porosity.

Archie's law (Archie, 1942), originally developed for oil reservoir investigations, is a commonly used empirically derived model that relates the soil condition to the bulk electrical conductivity. In terms of resistivity, Archie's law may be expressed, for fluid-saturated porous media, by $\rho_o = \rho_w \Phi^{-m}$ with the resistivity of saturated media (ρ_o), the resistivity of pore fluid (ρ_w), the porosity (Φ) and the exponent (m). The resistivity of partly-saturated media (ρ) can be expressed as $\rho = \rho_0 S^{-n}$ with the saturation (S) and exponent (n). Waxman and Smits (1968) extended Archie's law by accounting for the effect of clay minerals (forming surface electrical conductivity). Several other approaches have been developed specifically for soils (e.g. Rhoades et al., 1976). Laloy et al. (2011) compared a range of models for soil electrical conductivity, adopting the term "pedo-electrical" model to differentiate this from classical petrophysical approaches.

Electrical resistivity (respectively, conductivity) is a measure of how difficult (respectively, easy) an electrical current can flow through and is a property of the material independent of its dimension. The resistivity and conductivity are reciprocally related. Their MKS units are ohm metre (Ωm) and siemens per metre (S/m), respectively. A textbook introduction to electrical resistivity and conductivity often refers to a direct current flow through a finite-length (L) rod with a uniform cross-section (A). Then, resistivity (ρ) and conductivity (σ) are defined as $\sigma = RA/L$ and $\sigma = GL/A$ where $R = V/I = 1/G$; then, R, G, V and I are the resistance, conductance, voltage difference and current, respectively. When an alternating current with varying frequency is employed instead of a direct current, the electrical property can also be a function of frequency. However, for materials with conductivity in the range (1 - 1000 mS/m) usually measured by most electromagnetic induction devices, the electrical properties are largely independent of frequency. Therefore, the electrical conductivity measured by an electromagnetic induction equipment operating in low-frequency and low induction number is likely the same as the inverse of the electrical resistivity measured by a direct-current resistivity method (McNeill, 1980a).

Soils are the result of various weathering processes acting on parent materials whose physical and chemical properties are altered consequently. Structurally, soils are made up of mineral matrix and interwoven voids (or pores). Most soil minerals are effectively electrical insulators with negligible electrical conductivity, except some rare exceptions. Therefore, soil electrical conductivity is usually associated with the pores having the moisture in various stages of saturation. Soil electrical conductivity is electrolytic in nature and hence is affected by the number and mobility of ions. The ease of movement of the ions is largely determined by the fluid viscosity. The viscosity of fluid depends on the temperature of the fluid. Therefore, the electrical conductivity is also affected by the temperature (McNeill, 1980a).

The relation between electrical conductivity and temperature is generally nonlinear. However, over normal ambient temperatures, the degree of nonlinearity is relatively small, a linear equation known as the ratio model is commonly used (McNeill, 1980a):

$$\sigma(T)=\sigma(25)\left[1+a(T-25)\right],$$

where $\sigma(T)$ is the electrical conductivity measured at a temperature T (°C) and $\sigma(25)$ is a reference electrical conductivity at a common temperature of 25°C. The level of change is determined by a temperature compensation factor a. Several different choices have been used for a. McNeill (1980a) used a from a sodium chloride solution ($a = 0.022°C^{-1}$), that is 2.2% change per degree Celsius. Other values are $a = 0.0191°C^{-1}$ based on 0.01 M KCL solution; $a = 0.025°C^{-1}$ commonly used by geophysicists; $a = 0.02°C^{-1}$ for '2% increase of electrical conductivity per 1°C increase of temperature' frequently cited (Hayashi, 2004). Ma et al. (2011) offer a comparison of the different temperature corrections available.

Temperature effects can be accounted for given local vertical soil temperature profiles, which can be assumed to have limited spatial variability within a site, although effects of daily or seasonal variation in temperature do need to be considered. The electrical conductivity of the pore water contributes significantly to the bulk conductivity of the soil. In temperate climates, the variation of the pore water electrical conductivity should be minimal in rain-fed settings, but it may be greatly affected by irrigation depending on the source of the irrigation water (e.g. groundwater sourced). In semi-arid environments, pore water conductivity effects may be significant due to enhanced salinity arising from high evaporative fluxes (Corwin and Lesch, 2005).

Soils can also be examined by using elastic waves which require a medium (or a matter) to propagate through: sounds (or acoustic waves) are also part of elastic waves. In this regard, soils are the media whose particles are disturbed while the elastic waves are propagated. The elasticity inherent in soils acts as a restoring force.

The soil as a wave-propagating medium, for elastic waves, can be approximated as either purely solid or porous media. For solid elastic media without constituent pores, there are two types of elastic waves which are well-known primary (P)-wave and secondary (S)-wave, also known as a shear wave. The P-wave is a compressional (or longitudinal) wave in that the movement of particles is in the direction of wave propagation. The S-wave is a transverse (or shear) wave in that the disturbed particles oscillate perpendicular to the direction of waves. As the names suggest, P-wave propagates faster than S-wave. These are body waves which means that the medium of interest is unbounded. When a boundary or surface is considered, in addition to body waves, so-called surface waves are encountered such as Rayleigh waves, Love waves, etc.

When the wave-propagating medium is porous, more waves can be present. If the pore network of a poroelastic (porous and elastic) medium is filled with a single fluid, for example, either air-saturated or water-saturated, then we have three waves in total: two compressional waves and a shear wave. The phenomenon can be modelled by a two-phase poroelasticity theory pioneered by Biot (1956a,b). If the pores are occupied by two different fluids such as unsaturated soils, four distinct waves are encountered: three compressional waves and a shear wave, when there is a wave path through water in addition to those through the solid and air (Arora and Tomar, 2008; Lo and Sposito, 2013). Theoretically, the aforementioned poroelasticity models assume the solid phase represented by a single material. In practice, the solid part is likely to be composed of different materials. However, the wavelength considered in these models is much larger than the inhomogeneity; therefore, different solid constituents can be treated as a single 'effective' material that represents the overall properties of various solid elements.

In the case of both geoelectrical and seismic probing of soil, the first challenge is to determine the basic physical property of soil. The second challenge is to interpret that basic physical property in terms of some soil property of interest (see Table 1). For example, Reyes et al. (2018) employed geostatistical interpolation such as ordinary kriging and cokriging with the electrical conductivity data to identify the spatial variability of soil clay content at the field scale. A laboratory work by Gao et al. (2013) demonstrated a relationship to calibrate the soil penetration resistance (i.e. soil strength) and the shear modulus (i.e. elastic property); their relationship was later adopted by Shin et al. (2017) to predict the penetration resistance from the shear modulus deduced by the acoustic to seismic method.

3 Electromagnetic induction

The electromagnetic induction (EMI) method is a non-invasive, high-throughput conductivity survey method that can be done over a large area with minimal setup (McNeill, 1980b). Modern equipment can also be mounted on a mobile platform such as a non-conducting sledge (Corwin and Lesch, 2005), therefore enabling a more efficient survey of large areas. Also in soil with higher clay content, cracks can appear around the electrodes for the resistivity method (discussed in Section 4), resulting in poor galvanic contact; EMI does not suffer from this limitation (Sheets and Hendrickx, 1995). However, as EMI instruments are sensitive to electrical conductors, the method can be limited in very resistive environments (e.g. very dry clay-free soils). Note that what we refer to as EMI is more precisely frequency-domain electromagnetic induction (FDEM). Time-domain electromagnetic methods are out of the scope of this chapter as they are much less frequently used for agricultural applications.

EMI instruments are composed of at least two coils. One transmitter coil (Tx) and one receiver coil (Rx). The time-varying magnetic field arising from the alternating current in the transmitting coil at a certain frequency (H_p) induces very small currents (eddy currents) in the ground. These currents generate a secondary magnetic field H_s which is sensed, together with the primary field H_p, by the receiving coil (McNeill, 1980b). Stronger eddy currents will be induced in more conductive soil resulting in a larger secondary magnetic field. From the ratio between the primary and the secondary magnetic field, a value of the ground electrical conductivity can be obtained. This value is called 'apparent' electrical conductivity (often abbreviated ECa). Its apparent nature is derived from the fact that it represents a weighted average of the depth-specific electrical conductivity. The ECa represents the electrical conductivity of a homogeneous ground that would yield the signal measured by the EMI instrument.

Different coil configurations of the Tx and Rx coils are used. When the plane of both Tx and Rx coils is vertical to the surface of the ground, the direction of dipoles is horizontal: this configuration is known as the horizontal dipole mode (H) or the vertical co-planar mode. The other is the vertical dipole mode (V), also called the horizontal co-planar mode, where the plane of the coils is horizontal. Both configurations can be achieved by a single instrument by simply rotating it 90°. Additionally, the plane of the Tx and Rx coil can be perpendicular to each other: this is called the perpendicular orientation (PRP). In this section, we focus on the vertical and the horizontal dipole modes only, because these are more commonly available commercially and both modes can be exploited by a single instrument.

The orientation and the distance between the Tx and Rx coil define the shape of the electromagnetic fields and hence the depth and the volume investigated by the EMI device. Larger coil separation between Tx and Rx results in a greater volume investigated. While original instruments had only one Tx and one Rx coil (such as the Geonics EM38 and EM35), more recent instruments (e.g. the GF Instruments CMD Mini-Explorer shown in Fig. 1) are

Figure 1 Operator using a CMD Mini-Explorer (GF Instruments, Brno, Czech Republic) to survey a wheat field.

multi-coil, meaning they incorporate multiple Rx coils and one Tx coil. This enables the simultaneous measurement from multiple Tx-Rx coil separations and hence different depths of investigation at one time.

Analytical expression of the ratios of the secondary field to the primary magnetic field for a homogeneous subsurface can be derived from Maxwell's equation and are given by (McNeill, 1980b):

$$\left(\frac{H_s}{H_p}\right)_V = \frac{2}{(\gamma s)^2}\left[9 - \left\{9 + 9\gamma s + 4(\gamma s)^2 + (\gamma s)^3\right\} e^{-\gamma s}\right],$$

for a vertical dipole mode, and

$$\left(\frac{H_s}{H_p}\right)_H = 2\left[1 - \frac{3}{(\gamma s)^2} + \left\{3 + 3\gamma s + (\gamma s)^2\right\} \frac{e^{-\gamma s}}{(\gamma s)^2}\right],$$

for a horizontal dipole mode. $\gamma = \sqrt{i\omega\mu_0\sigma}$ with μ_0 the free-space magnetic permeability, ω the angular frequency and i the imaginary unit. Use of μ_0 instead of μ (the magnetic permeability of the soil) means that we are largely concerned with the non-magnetic environment.

The variable γ can be represented by the electrical skin depth δ with the relationship $\gamma\delta = \sqrt{2i}$. The skin depth is defined as the distance of travel at which an electromagnetic wave is attenuated by $1/e \cong 0.368$. The ratio of the coil spacing s (i.e. transmitter-receiver separation) to the skin depth δ is known as the induction number B, that is, $B = s/\delta$. Then, the combined variable γs can be expressed by $\gamma s = B\sqrt{2i}$. Therefore, a low induction number ($B \ll 1$), which is a common requirement for most commercial electromagnetic induction devices, is a condition where the coil spacing is much shorter than the skin depth and hence an important indicator of wave attenuation. A very conductive environment will increase the value of γ, decrease the skin depth δ and increase the induction number B, potentially breaking the low induction number condition.

Under the low induction number condition, the imaginary part (i.e. quadrature) of the ratios of H_s to H_p can be simplified to

$$\left(\frac{H_s}{H_p}\right)_V \cong \left(\frac{H_s}{H_p}\right)_H \cong \frac{(\gamma s)^2}{4} = \frac{iB^2}{2} = \frac{i\omega\mu_0\sigma s^2}{4},$$

for both dipole orientations, by the Taylor series expansion of $e^{-\gamma s}$ with at least five terms. Therefore, at low values of induction number, the secondary magnetic field is a very simple function of ω, σ and s. When the operating frequency and coil spacings are arranged to ensure the low induction number for the range of conductivity of interest, the apparent electric conductivity σ_a (also denoted by ECa) can be determined by using the imaginary part of the ratio of H_s to H_p:

$$\sigma_a = \frac{4}{\omega\mu_o s^2}\mathrm{Im}\left(\frac{H_s}{H_p}\right).$$

The condition of a low induction number ensures not only the aforementioned simple relationship to deduce the apparent conductivity, but also negligible magnetic coupling between the current loops. In addition, regardless of the value of the induction number, all current flows are assumed to be horizontal. In a horizontally layered soil (which is often assumed theoretically), the effect of the refraction of the current flows is assumed negligible, rendering further theoretical development more accessible. Therefore, the secondary magnetic field measured at the surface of a horizontally layered ground can be considered as the sum of all the contributions from each layer.

Under the condition of a low induction number ($B \ll 1$), the relative sensitivity of the vertical (S_V) and horizontal (S_H) dipole configurations to layers at various depths are known as (McNeill, 1980b)

$$S_V(z) = 4z(4z^2 + 1)^{-3/2}$$

$$S_H(z) = 2 - 4z(4z^2 + 1)^{-1/2}$$

where z is a normalised depth, that is, a depth divided by the coil spacing s. Therefore, s and the coil orientation are the sole factor affecting the relative sensitivity distribution with depth. Note that $S_H(z)$ peaks at the surface ($z=0$), that is, $S_H(0)=2$, then gradually decreases (Fig. 2a). By contrast, $S_V(z)$ vanishes at the surface ($S_V(0)=0$), peaks at $z=1/\sqrt{8}$, that is, $S_V\left(1/\sqrt{8}\right)=4\sqrt{3}/9$, then gradually decreases (Fig. 2a).

More useful may be its cumulative sensitivity function (CS_V, CS_H) which accumulates the relative contribution to the secondary magnetic field from all layers below a given depth z:

$$CS_V(z)=\int_z^{\infty} S_V(z)dz=\left(4z^2+1\right)^{-1/2}$$

$$CS_H(z)=\int_z^{\infty} S_H(z)dz=\left(4z^2+1\right)^{1/2}-2z$$

which means all the layers below a depth of z-times coil spacings produce a relative contribution of $CS_V(z)$ or $CS_H(z)$ to the indicated meter reading of the apparent conductivity. At the surface, the complete contribution is made from all the depths; therefore, $CS_V(0)=CS_H(0)=1$ (Fig. 2b). For cases when the EMI instrument is not operated on the ground, a rescaled version of the CS function can be used (Andrade and Fischer, 2018).

The apparent conductivity (σ_a) from an electromagnetic induction meter on a horizontally layered earth model can be simply calculated by using the

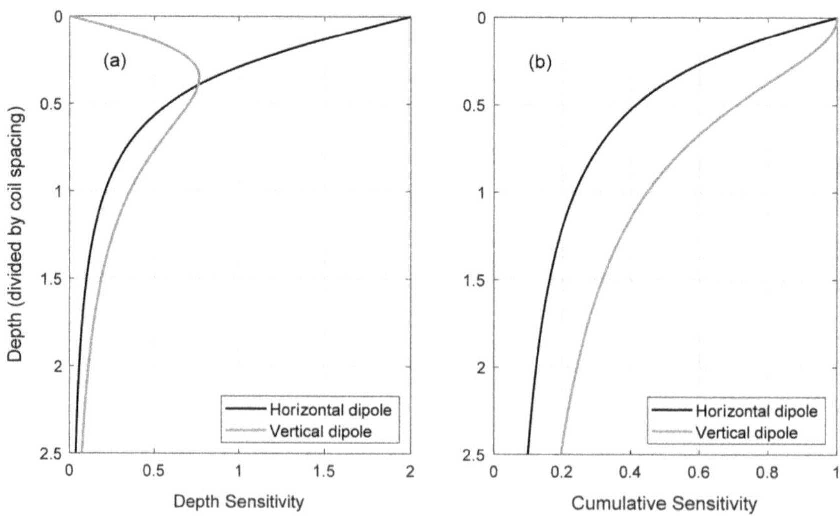

Figure 2 (a) Relative sensitivity and (b) cumulative sensitivity of horizontal dipole and vertical dipole.

cumulative sensitivity functions, that is, by adding up the contribution from each layer independently (e.g. JafarGandomi and Binley, 2013):

$$\sigma_a = \sigma_1\left(1 - CS\left(z_1\right)\right) + \sum_{j=2}^{N-1} \sigma_j\left(CS\left(z_{j-1}\right) - CS\left(z_j\right)\right)$$
$$+ \sigma_N CS\left(z_{N-1}\right), \quad \text{for } N \geq 2;$$

$$\sigma_a = \sigma_1\left(1 - CS\left(\infty\right)\right) = \sigma_1, \text{ for a homogeneous half-space } (N = 1),$$

where CS could be either CS_V or CS_H depending on the orientation of dipoles. The total number of layers is N. The layer numbering is such that the one nearest to the surface is identified by $j = 1$. The normalised depth z_j is measured from the surface to the lower boundary of a layer j.

It is useful to know how much secondary-field signals contribute from a body of soils from the surface down to a certain depth. It is a complementary concept to the CS function. The normalised depth of soils with p $(0 \leq p \leq 1)$ contribution is

$$z = \frac{\sqrt{p(2-p)}}{2(1-p)} \quad \text{for a vertical dipole mode,}$$

$$z = \frac{p(2-p)}{4(1-p)} \quad \text{for a horizontal dipole mode.}$$

The effective depth of investigation is commonly determined by setting $p = 0.7$ which amounts to about 1.59 and 0.76 times the inter-coil spacing for the vertical and horizontal dipole modes, respectively.

The cumulative sensitivity function (based on the low-induction number assumption) is still widely used but can be limited mainly in a very conductive environment (Callegary et al., 2007, 2012; Andrade et al., 2016). Indeed, the shape of the cumulative sensitivity function does not depend on the electrical conductivity of the layers. An alternative approach consists of using Maxwell's equation that Wait (1982) derived for a 1D layered earth. This strategy has been successfully used in a number of studies (e.g. Lavoué et al., 2010; von Hebel et al., 2014; von Hebel et al., 2019; Hanssens et al., 2019).

As discussed above, EMI instruments usually output apparent electrical conductivity values for each unique combinations of coil orientations and separations. Given multiple apparent values for one position, it is possible, using the shape of the CS function, for instance, to rebuild the profile of depth-specific electrical conductivity. This process is called inversion and is widely employed in other geophysical methods such as the electrical resistivity method discussed in Section 4. Inversion of EMI data can be challenging given the smaller number of measurements at each location, but when successful provides a vertical profile of electrical conductivity. While a full description of EMI inversion is out of the scope of this chapter, useful information is provided by von Hebel et al. (2019) and Monteiro Santos (2004). An example of the inversion is shown in Fig. 3(a). Commercial software such as EM4Soil (Monteiro Santos, 2004) or Aarhus workbench (Auken et al., 2015) or open-source ones such as EMagPy (McLachlan et al., 2021), SimPEG (Heagy et al., 2017) and pyGIMLi (Rücker et al., 2017) are examples of software able to perform EMI inversion.

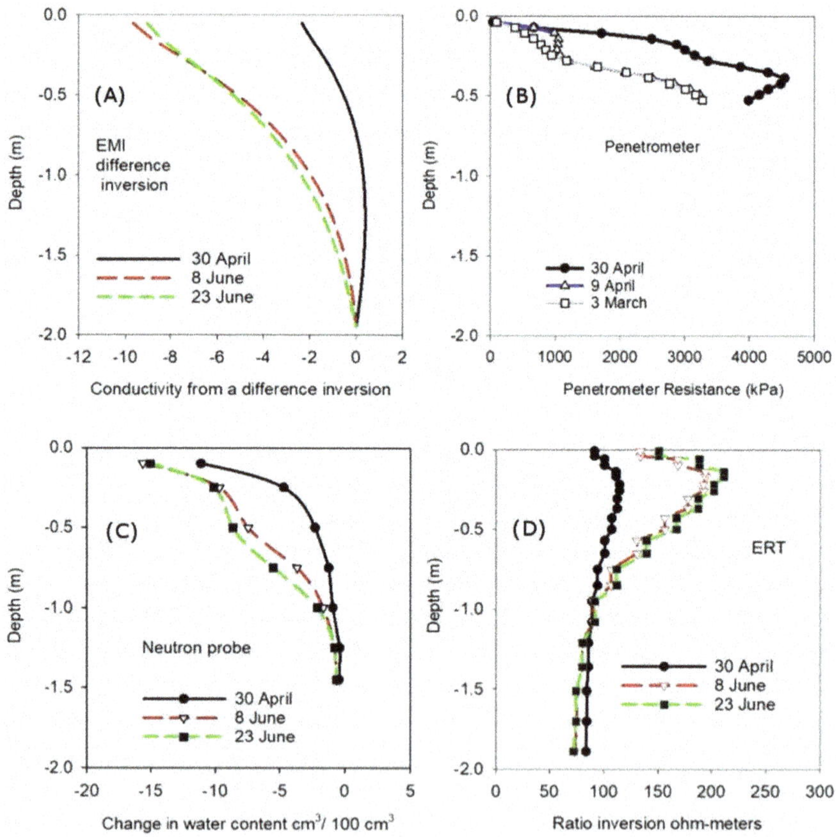

Figure 3 Comparison of EMI and other data from a 2015 survey of a wheat field: These data were collected on Warren field, Woburn, UK. The effect of temporal patterns in soil drying with depth on data determined from the inversion of ERT (d) and EMI (a) data, penetrometer measurements (b) and neutron probe (c) measurements is shown. EMI and ERT data are obtained from the inversion routine that provides a continuous distribution over depth (From Whalley et al., 2017).

3.1 Applications of electromagnetic induction

The measurement of electrical conductivity from EMI instruments has been extensively used for mapping different properties and states such as soil salinity (Corwin, 2008), water content (Corwin and Rhoades, 1984), soil texture (Triantafilis and Lesch, 2005) or soil organic matter (Huang et al., 2017). EMI mapping of agricultural fields can serve to delineate management zones (King et al., 2005; Rossi et al., 2018; Moral and Serrano, 2019). Hence enabling spatially tailored crop management for precision agriculture.

Often, a key issue remains to understand how the electrical conductivity is related to a soil property of interest. For instance, Korsaeth et al. (2008) showed that maps of electrical conductivity obtained from EMI measurements could be used to map soil organic matter. This was because soil organic matter was proportional to the soil water content by virtue of the relationship between soil organic matter and porosity. Identically linking apparent electrical conductivity to soil moisture content is challenging and often results in site-specific relationships. Calamita et al. (2015) present a review of the different soil moisture content – electrical conductivity relationships from multiple studies. In field studies where small variations between treatments are measured, accounting for the spatial heterogeneity of soil moisture – electrical conductivity relationships are useful to improve the discriminatory power of geophysical data (Blanchy et al., 2020c).

A way forward is the use of time-lapse surveys that enables the isolation of the dynamic component of the electrical conductivity which can be related, under some circumstances, to the change in soil moisture (Blanchy et al., 2020a; Robinson et al., 2012; Martini et al., 2017). The time-lapse approach consists of subtracting a reference survey from other surveys, hence allowing the removal of the static contribution of electrical conductivity from soil texture, for example.

To allow quantitative application, EMI measurements often need to be calibrated against other soil conductivity data, for example, from electrical resistivity surveys (Lavoué et al., 2010; Shanahan et al., 2015; von Hebel et al., 2019). Figure 3 shows the progressive patterns of soil drying over time by wheat estimated with EMI, ERT, neutron probe and penetrometer measurements. The penetrometer could not be manually operated later than 30 April because the soil became too strong as it dried, while the other three modalities were successfully applied afterwards. However, the penetrometer data clearly demonstrated soil drying with depth and over time earlier in the season. The other three methods returned considerable change of metrics to reflect soil drying from April to June, but registered insignificant variations during June suggesting that the soils were already relatively dry by then. The temporal patterns of soil drying recorded by EMI compare well with those obtained with a neutron probe.

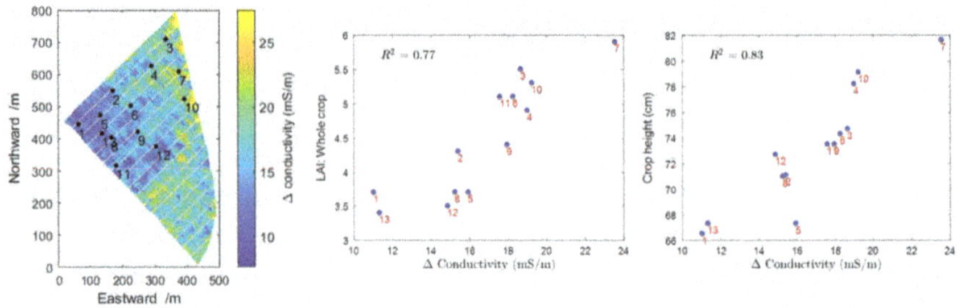

Figure 4 A map of the change in apparent conductivity between February and July 2019 obtained with a GF Instruments Mini-Explorer EMI instrument in vertical dipole mode on a field near Cheddington, UK. At various locations in the field, the height and leaf area index of wheat were measured and these are plotted against the change in apparent conductivity. The numbers from 1 to 13 identify the locations of crop data sampling. The white curves in the image show the survey path of EMI acquisition.

Figure 4 shows how changes in apparent electrical conductivity can be correlated with plant height and leaf area in wheat. This relationship is predicated on the assumption that a change in apparent conductivity is proportional to a change in water content (Whalley et al., 2017), or root water uptake, and hence crop growth. The electrical conductivity in Fig. 4 was measured by an operator, guided by the tramlines spaced 36 m, carrying an EMI instrument and a GPS receiver unit; for the field shown in Fig. 4, the EMI survey can be done in a half-day. The sampling locations of crop data were designed to have statistical significance.

Information collected from EMI surveys can also be used to improve simulation of crop productivity (Brogi et al., 2019; Brogi et al., 2020). Zhu et al. (2013) used EMI maps to help explain variation in crop yield in terms of soil moisture and clay content. They developed an approach for using EMI data to help develop a functional map of soil variation in time and space to explain crop yield.

4 Electrical resistivity

The resistivity method requires galvanic contact with the soil (Binley and Kemna, 2005) and hence is more invasive than EMI. The method is based on injecting current between a pair of electrodes (A and B) and measuring the voltage difference between another electrode pair (M and N). Figure 5 illustrates three quadrupole (A, B, M, N) configurations. The measurement provides an apparent resistivity (the equivalent electrical resistivity of the ground if it were homogenous). Increasing the electrode spacing results in a deeper depth of investigation. Measurements are typically made with an array of several tens of electrodes installed along a transect, providing a 2D vertical section of resistivity (3D images can be derived from data using multiple transects). For such applications, the term ERT is commonly used. It requires the application

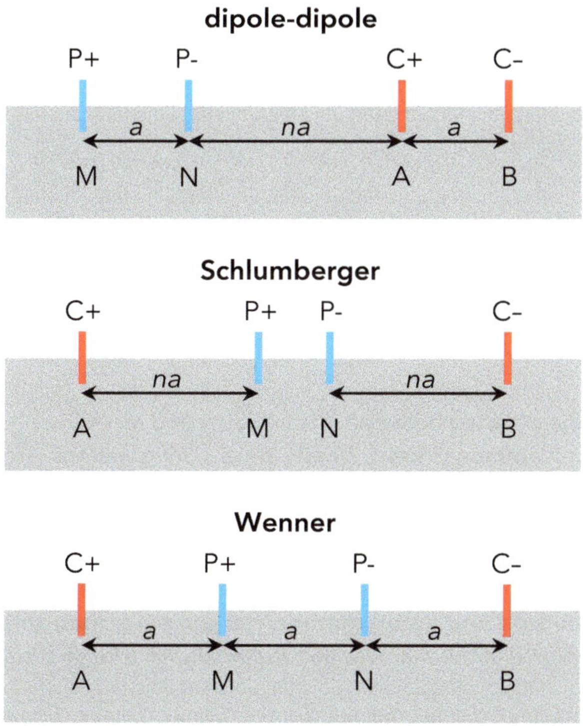

Figure 5 Schematic diagrams of dipole-dipole, Schlumberger and Wenner arrays. The current pair is denoted by C+ and C− (A and B), while the voltage pair is by P+ and P− (M and N). The design separation is indicated by *a* with an integer *n*.

of inverse methods (as discussed below) to determine the spatial distribution of resistivity. Mobile methods have also been developed in which the ground contact is achieved using heavy electrodes (e.g. Sørensen, 1996) or coulter-based electrodes (Corwin and Lesch, 2005; Panissod et al., 1998; Reyes et al., 2018). Such approaches typically involve making measurements on a few voltage electrode pairs and thus can be used, like EMI, to provide an estimate of resistivity at different depths. Mobile systems are limited to relatively flat sites with limited vegetation (due to the disturbance inevitable from the towing vehicle).

The spatial distribution of resistivity $\rho(x)$ is defined in Poisson's equation (Binley, 2015; Binley and Slater, 2020):

$$\nabla \cdot \left(\frac{1}{\rho(x)} \nabla V(x) \right) = -I\delta(x_0), \tag{1}$$

in which $V(x)$ represents the spatial distribution of the electric voltage or potential subject to a point current source with strength I located at x_0. The

Dirac delta is denoted by δ and ∇ is the gradient operator. For an idealised subsurface of uniform resistivity ρ with current electrodes on the flat surface of the soil, the half-space fundamental solution of Eq. (1) is

$$V(r) = \frac{I\rho}{2\pi r},$$

where r is the distance from the current source when the current sink is located at infinity. The voltage difference can be calculated by

$$\Delta V = \frac{I\rho}{2\pi}\left(\frac{1}{AM} - \frac{1}{BM} - \frac{1}{AN} + \frac{1}{BN}\right),$$

where AM is the distance between electrodes A and M, etc.

The most commonly used quadrupole configurations are the dipole-dipole, Schlumberger and Wenner configurations (Fig. 5). The receiver (voltage) electrodes of the Schlumberger and Wenner arrays are placed within their transmitter (current) pair; this arrangement can provide relatively strong voltage signals at the receiver pair. By contrast, the dipole-dipole array has its voltage pair outside and away from the current electrodes; this can cause a weak voltage signal at the receiver pair especially for a large distance between the two pairs.

Among these arrays, the dipole-dipole configuration is realised by setting the same spacing for current and voltage dipoles (AB = MN = a) on the surface and the separation na between the nearest electrodes of these two dipole sets (with an integer n). Then, the apparent resistivity ρ_a of the dipole-dipole array may be calculated by (Binley, 2015)

$$\rho_a = \pi a n(n+1)(n+2)\Delta V / I, \quad (2)$$

where I is the injected current and ΔV is the measured voltage difference.

The resistivity in Eq. (2) is an apparent resistivity from the measurement and can be a true resistivity of the subsurface only if the subsurface is homogeneous and its surface is flat, which is an unlikely scenario in real applications. Therefore, inverse methods have to be applied to determine the resistivity structure of the subsurface (Binley, 2015; Binley and Slater, 2020). The depth of investigation, defined as a layer contributing the maximum amount to the total measured signal, of a dipole-dipole array is 0.18L for homogeneous subsurface (Roy and Apparao, 1971) which is deeper than the Schlumberger (0.125L) and Wenner (0.11L) arrays; here, L being the distance between the two extreme active electrodes. For example, ERT measurements shown in Fig. 3(d) (and later in Figs. 7 and 8) were carried out by arrays of the dipole-dipole configuration having 96 electrodes with 32 cm spacing between adjacent electrodes. The vertical resolution of a dipole-dipole array can be poorer than the other two array configurations, but its lateral resolution is better. The Schlumberger array has a weak sensitivity to the lateral variability which is however advantageous when it comes to vertical exploration; hence, it is commonly used for vertical electrical sounding (Binley, 2015; Binley and Slater, 2020).

Due to advances in multi-channel switching systems in data acquisition, modern-day ERT measurement systems can measure on a number and configuration of electrodes. A resulting 2D image was traditionally presented in a pseudosection in which the apparent resistivities from individual four-electrode units are discretely marked in vertical and lateral positions according to the geometric configuration of the four electrodes. However, the pseudosection is merely a way of graphically presenting the apparent resistivity; hence, the subsurface structure of resistivity is not easily gleaned from the pseudosection. Nowadays, inverse methods are routinely used to determine an image of the subsurface resistivity that is consistent with the apparent resistivity measurements.

Forward models essentially return a synthetic (or simulated) apparent resistivity (or equivalent parameters) for a given set of underground structure and distribution of true resistivity by solving Eq. (1). The problem is usually handled numerically for more general geometries of 2D and 3D by implementing the finite difference method or the finite element method (Binley, 2015). The spatial region of interest is first discretised to approximate the area by either structured or unstructured meshes; then, local resistivities are assigned to (or solved at) in cells (or elements) of the mesh which represent the degree of freedom for the problem.

The goal of inverse methods is to find a set of model parameters that reduce the difference between the measured data and the simulated data from forward models by means of numerical optimisation. Least-squares fitting is commonly used for this purpose by minimising the sum of the squares of the difference between the observed data and the forward model outputs. The goodness of fit for this purpose may be expressed by (Binley and Kemna, 2005)

$$\Phi_d = \left\| W_d\left(d - F(m)\right) \right\|^2 = \left(d - F(m)\right)^T W_d^T W_d \left(d - F(m)\right), \tag{3}$$

where d is the column vector composed of the observed data, $F(m)$ is the output of the forward model with model-parameter vector m. A matrix of data weighting is W_d; for the case of uncorrelated errors, it is a diagonal matrix with the inverse of data errors. The superscript T represents the transpose of a vector or matrix.

There are many numerical optimisation algorithms to minimise Eq. (3). In geophysical inversions, minimisation of Eq. (3) may lead to non-unique or unstable inversions. Therefore, some level of constraint is needed to guide the solution.

One of the most successful and widely adopted approaches in this regard is to introduce the smoothness constraint by spatial regularisation. Regularisation in this context can be implemented by considering a model penalty term:

$$\Phi_m = \|W_m m\|^2 = m^T W_m{}^T W_m m = m^T Rm.$$

Here, the spatial smoothness of the model parameters is constrained by a roughness matrix R which is usually a tridiagonal matrix and composed of a model-weighting matrix W_m.

The data and model misfit terms, Φ_d and Φ_m, respectively, can be combined by adding the two with an extra proportionality coefficient:

$$\Phi = \Phi_d + \alpha \Phi_m, \tag{4}$$

which is minimised instead of Φ_d alone. The scalar regularisation parameter α can be fixed throughout the optimisation or adjusted by iterative reduction starting from a large value.

Minimisation of Eq. (4) can be carried out by using the Gauss-Newton approach (Binley and Kemna, 2005; Binley and Slater, 2020):

$$\left(J^T W_d^T W_d J + \alpha R\right)\Delta m = J^T W_d^T W_d \left(d - F(m_i)\right) - \alpha R m_i;$$

$$m_{i+1} = m_i + \Delta m,$$

where J is the Jacobian matrix; m_i is a set of model parameters at the ith iteration. The Δm represents the updates in model parameters.

The approach used above is commonly adopted for the analysis of ERT data. For example, they are implemented in an ERT freeware R2 (http://www.es.lancs.ac.uk/people/amb/Freeware/R2/R2.htm); its GUI environment is also available in a Python-based open-source software ResIPy (Blanchy et al., 2020d). Binley (2015) describes a range of modifications to the approach (including the analysis of time-lapse data) and more information on the inversion process.

4.1 Applications of electrical resistivity

ERT is commonly used to image variations in resistivity associated with soil texture. In a few cases, it has been used, in time-lapse mode, to assess variations of spatial changes of resistivity due to soil water, for example, the uptake by roots (Whalley et al., 2017). The photograph (Fig. 6) below shows an ERT array connected to an instrument capable of making the measurements shown schematically in Fig. 5.

In this application, the ERT arrays were installed after experimental plots were sown, with the purpose of monitoring soil drying. The plots were

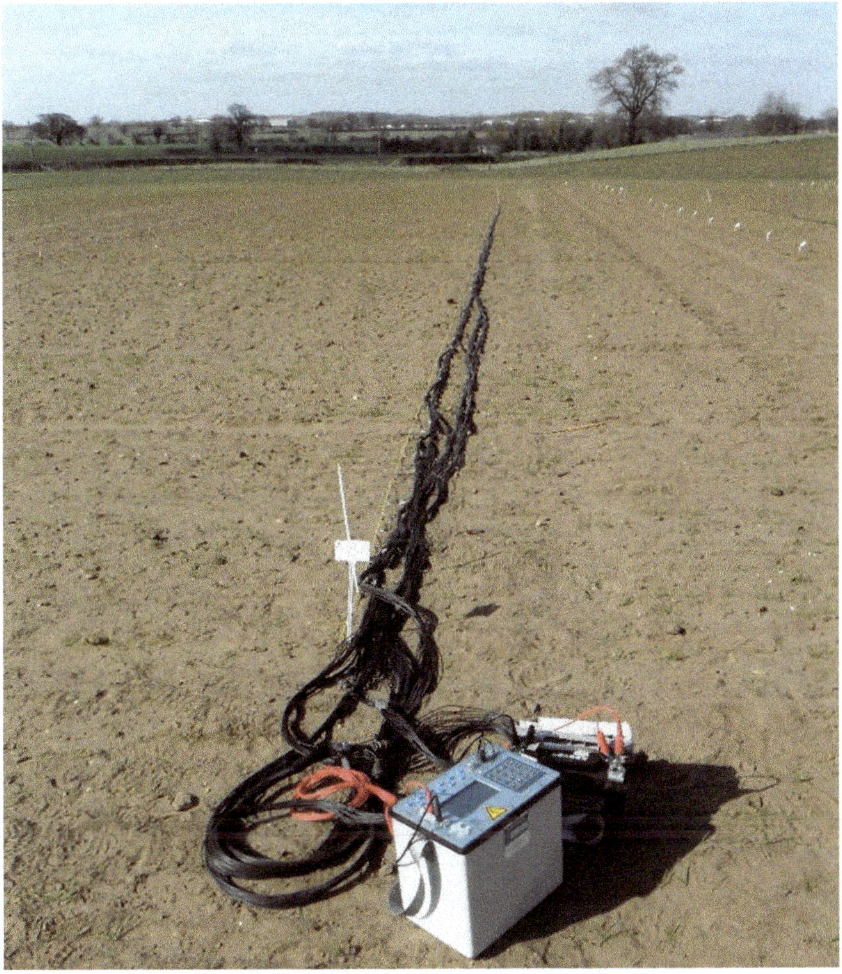

Figure 6 An ERT instrument connected to an ERT array set in Rothamsted's Experimental Farm at Woburn, Bedfordshire, UK.

approximately 2 m wide and 10 m long and the arrays were laid out to give a 2 m ERT image of consecutive plots. Figure 7 shows resistivity changes with time. It indicates that by July resistivity increases can be observed at depth of nearly 1 m and by inference soil dry to that depth. The names of the different wheat lines are indicated in Fig. 7 along with a fallow plot with no wheat, and hence no drying, used for the purpose of reference.

At the study site of Whalley et al. (2017), there are two types of soil: one sandy with a cemented hard layer at a depth of approximately 30 cm and one with a deep clay-rich profile. Figure 8 compares spatial resistivity maps on these

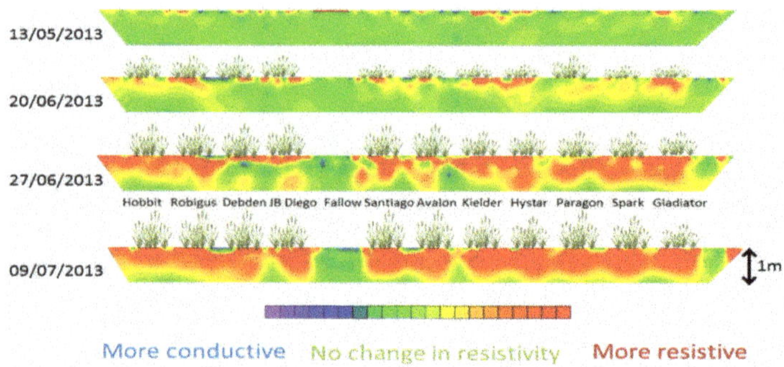

Figure 7 A time-lapse ERT image showing changes in resistivity (from 23 April 2013) because of soil drying by different winter wheat varieties (indicated). A fallow plot is also shown. The drier the soil the more resistive it becomes. The plots are approximately 2 m in width.

different soils in July following soil drying. It can be seen clearly that on the sandy soil (Butt Close) that only the very surface layers are dried by roots. In contrast, on the clay-rich profile with no cemented layer (Warren Field), there is soil drying to depth.

While ERT methods may not give accurate soil water content data compared with a specialised soil water content sensor, they do give a wealth of spatial information. The spatial resolution depends on the electrode spacing and resolution, but in a field application, the greatest resolution is likely to be in the order of several centimetres.

ERT was also employed to monitor the three-dimensional distribution of soil water content in a maize field (Beff et al., 2013), in which the effect of plants and rainfall event was investigated during the growing season. They deduced

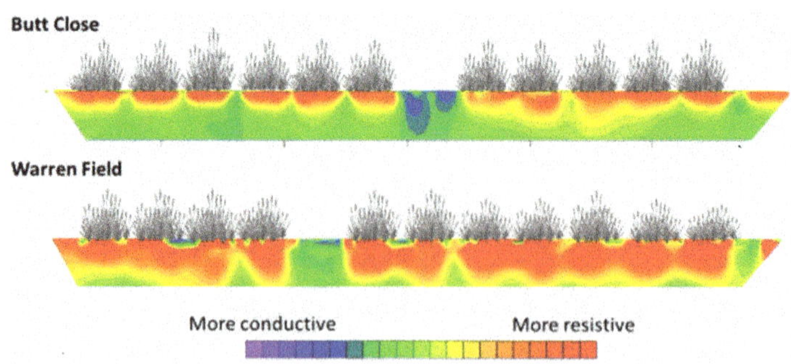

Figure 8 A comparison of ERT difference images (change from April to July 2013) following soil drying by wheat. The top image is from Butt Close near Woburn, UK which is a sandy site with a cemented layer at a depth of approximately 30 cm. At this site, soil drying is limited to the shallow layers (less than 30 cm depth). The bottom image is from a neighbouring Warren Field with a deep clay-rich profile. This site shows soil drying to depth. Each plot is approximately 2 m in width.

the water contents from the inversion of ERT data which were validated by TDR measurements. The high resolution of the ERT-based water content distribution allowed the identification of the drying pattern around maize root profiles.

A system of permanently installed ERT arrays was used to measure the temporal and spatial variation of the resistivities in an irrigation experiment over a three-year duration (Dahlin et al., 2014). They also inverted the ERT data to construct the 2D and 3D images of resistivity patterns and observed the resistivity changes linked to the irrigation quantities and the growth of willow. Their result also confirmed the effectiveness of their groundwater sampling protocol carried out in pipes installed in depth. Other examples of the use of ERT in crop studies include Michot et al. (2003) and Garré et al. (2011). Macleod et al. (2013) used ERT to help identify grass varieties that enhance soil structure and potentially minimise the risk of flooding. Blanchy et al. (2020b) use hourly ERT measurements to couple below-ground processes with the evolution of above-ground traits under a high-throughput automated phenotyping platform.

ERT arrays have also been used in laboratory conditions. Rings of electrodes can be installed around the circumference of soil cores. Such configuration can be used to determine the 3D distribution of resistivity. Using this setup, Binley et al. (1996) demonstrated the effectiveness of ERT in solute tracing experiment in soils, using the ERT images to identify preferential flow pathways in the core. In a subsequent study, Koestel et al. (2008) used a combination of ERT and TDR measurements to develop solute transport properties of a soil core.

5 Acoustic-to-seismic coupling

When the soil is assumed as homogeneous and isotropic elastic solid without voids, then there are two waves that propagate separately: P-wave and S-wave. The phase velocity of P-wave is well established as

$$V_p = \sqrt{\frac{K + 4\mu / 3}{\rho}}. \tag{5}$$

For S-wave, we have

$$V_s = \sqrt{\mu / \rho}, \tag{6}$$

where the mass density of the solid is ρ; two elastic moduli are K for bulk modulus and μ for shear modulus (also known as Lamé's second parameter).

For poroelastic media, models for two-phase poroelastic materials comprised of solid and a single fluid (either gas or liquid) have been widely studied since the seminal works by Biot (1956a,b) and have been expanded by many researchers since then. To some extent, the theory behind two-phase modelling can be judged almost matured while interesting applications are still being reported. By contrast, three-phase modelling is still being actively studied despite the first of the kind was reported some 55 years ago by Brutsaert (1964).

Just as simple as their phase velocities (Eqs. 5 and 6), wave numbers of isotropic elastic solids are also succinctly expressed: $\omega\sqrt{\rho/(K+4\mu/3)}$ for a compressional wave and $\omega\sqrt{\rho/\mu}$ for a shear wave. However, for fluid-saturated two-phase porous materials, the wavenumbers (l_1 and l_2) of two compressional waves and the wavenumber (l_3) for a shear wave can be obtained by solving the following equations (Shin et al., 2013, 2017; Stoll, 1974)

$$\left(\rho_f\omega^2 - Cl_{1,2}^2\right)^2 = \left(\rho\omega^2 - Hl_{1,2}^2\right)\left(\rho_f q^2\omega^2/\Omega - Ml_{1,2}^2 + i\omega F\eta/k\right),$$

$$\left(\rho_f\omega^2\right)^2 = \left(\rho\omega^2 - \mu l_3^2\right)\left(\rho_f q^2\omega^2/\Omega + i\omega F\eta/k\right)$$

with fluid density (ρ_f), bulk density of soil (ρ), tortuosity (q^2), fluid permeability (k), porosity (Ω), fluid viscosity (η), imaginary unit (i): the time-harmonic convention of $\exp(-i\omega t)$ is used with time t. The elastic constants H, C, M and D are further defined as (Stoll, 1974)

$$H = \frac{\left(K_r - K_b\right)^2}{D - K_b} + K_b + \frac{4}{3}\mu,$$

$$C = \frac{K_r\left(K_r - K_b\right)}{D - K_b},$$

$$M = \frac{K_r^2}{D - K_b},$$

$$D = K_r\left\{1 + \Omega\left(K_r/K_f - 1\right)\right\}$$

with a bulk modulus of the solid particle (K_r), bulk modulus of the porous medium (K_b), bulk modulus of fluid (K_f), and shear modulus of the porous medium (μ). In reality, the soil matrix is unlikely to be perfectly elastic. To this effect, the imaginary part accounting for damping can be introduced to the bulk moduli (K_b and μ) to incorporate the damping effect for elastic waves (Stoll, 1974). The details of the viscosity correction factor (F) depend on a model of a pore structure idealised for given soils (Biot, 1956b).

5.1 Model for acoustic-to-seismic coupling

When waves propagating in poroelastic materials encounter layer boundaries, the amplitudes of the reflected and transmitted waves can be determined by solving a boundary value problem, as long as the wave numbers of the three-component waves have been calculated for each poroelastic layer.

To solve the boundary value problem, equations of motions for each layer must be established. With a point source having axial symmetry, the wave fields can be represented by only two components of wave motions in the radial (r) and vertical (z) coordinates. When the height of an airborne acoustic source is z_s, the Green's function (Ψ_0) for an airborne layer and those ($\Psi_{1,n}$, $\Psi_{2,n}$, $\Psi_{3,n}$) for poroelastic waves in soils are (Shin et al., 2013; Tooms et al., 1993)

$$\psi_0(\xi,z)=\frac{1}{\beta_0}e^{i|z-z_s|\beta_0}+A_0^{\uparrow}e^{-iz\beta_0},$$

$$\psi_{1,n}(\xi,z)=A_{1,n}^{\downarrow}e^{i(z-d_n^+)\beta_{1,n}}+A_{1,n}^{\uparrow}e^{i(z-d_n^-)\beta_{1,n}} + A_{2,n}^{\downarrow}e^{i(z-d_n^+)\beta_{2,n}}+A_{2,n}^{\uparrow}e^{i(z-d_n^-)\beta_{2,n}},$$

$$\psi_{2,n}(\xi,z)=M_{1,n}\left[A_{1,n}^{\downarrow}e^{i(z-d_n^+)\beta_{1,n}}+A_{1,n}^{\uparrow}e^{i(z-d_n^-)\beta_{1,n}}\right] + M_{2,n}\left[A_{2,n}^{\downarrow}e^{i(z-d_n^+)\beta_{2,n}}+A_{2,n}^{\uparrow}e^{i(z-d_n^-)\beta_{2,n}}\right],$$

$$\psi_{3,n}(\xi,z)=A_{3,n}^{\downarrow}e^{i(z-d_n^+)\beta_{3,n}}+A_{3,n}^{\uparrow}e^{i(z-d_n^-)\beta_{3,n}},$$

where $M_{1,n}$ and $M_{2,n}$ indicate the ratio of the fluid-borne wave to the soil frame-borne wave; β with subscripts are the vertical wave numbers; ξ is the horizontal wave number. The vertical coordinates of upper and lower soil boundary of a layer n are denoted by d_n^+ and d_n^-. The upward ($A_{m,n}^{\uparrow}$) and downward ($A_{m,n}^{\downarrow}$) amplitudes of Green's functions can be calculated by applying appropriate boundary conditions at layer interfaces. The systems of equations for the boundary conditions of each layer interface are then mapped together with those from others to form a linear system of a global matrix (Tooms et al., 1993). Then, the amplitudes of propagating wave components in layers are calculated all together through standard methods for solving a system of linear equations. Once these amplitudes ($A_{m,n}^{\uparrow}$ and $A_{m,n}^{\downarrow}$) are determined for each layer and wave types, the acoustic pressure in the fluid (air) layer and particle velocities on the ground can be calculated; for a detailed procedure, see Shin et al. (2013) and Tooms et al. (1993).

These simulated acoustic pressure and particle velocities on the soil surface were compared to the measured counterparts. The acoustic pressure was measured by a pair of microphones; the soil particle velocity was recorded by a laser Doppler vibrometer (LDV). An acoustic-source system was composed of a compression driver extended by a metal pipe (Fig. 9). Instead of one-to-one comparisons, the measured and modelled signals were compared in terms of ratios: that is, transfer functions. Then, the numerical optimisation is carried out to minimise the differences between the transfer functions from the measurement and the simulation. For details about the optimisation procedure, see Shin et al. (2013).

The minimum set of properties required by a numerical optimisation is composed of air permeability, porosity, complex shear modulus, Poisson's ratio and the layer thickness for each layer: these are updated during the optimisation. The other properties in the poroelasticity model are well documented. In place of complex shear modulus, we adopted the loss factor and S-wave speed which were first converted to the shear modulus, then to the penetration resistance.

5.2 Applications of acoustic-to-seismic coupling model

Field measurements were carried out on the Woburn Experimental Farm operated by Rothamsted Research, UK. The acoustic-to-seismic coupling measurements were made in plots of loamy sand with different winter wheat varieties of Paragon and Hystar hybrid. A fallow plot was also included for comparison.

Figure 10 compares the soil strength profile deduced from the acoustic-to-seismic coupling (on 26 and 27 June) with that measured by a penetrometer (on 28 June). These are the extracts from Shin et al. (2017); full results can be found therein. The blue line inside the grey strip shows the arithmetic mean of five

Figure 9 Schematic diagram and photograph of measurement configuration showing acoustic source, two microphones and a laser Doppler vibrometer (LDV). The left-hand side is adapted from Shin et al. (2013).

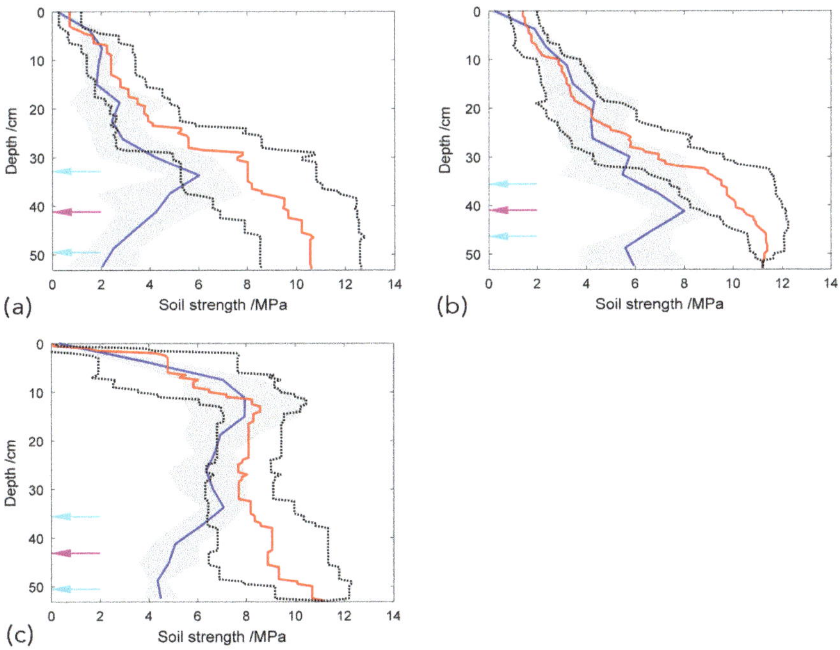

Figure 10 Plots of soil strength measured by a penetrometer (28 June) and deduced from measurements of the acoustic-to-seismic coupling (a and b on 26 June; c on 27 June) for the experimental plots: (a) fallow (b) Paragon variety and (c) Hystar hybrid variety. The mean and standard deviation of (i) penetrometer readings, (ii) the acoustic-to-seismic soil strength and (iii) estimated hardpan depth are illustrated by (i) blue lines and grey boundaries, (ii) red lines and black dotted lines, and (iii) magenta arrows and cyan arrows, respectively. Adapted from Shin et al. (2017).

penetrometer readings at a given depth and the boundaries of the grey region indicate one standard deviation. In some cases, the soil was too strong for penetrometer operation in depth; this may have resulted in an underestimated penetration resistance and a smaller variance.

The red line in Fig. 10 indicates the average soil strength deduced from the acoustic-to-seismic coupling. The number of optimised results taken for the average is 12, 28 and 9, respectively, for (a), (b) and (c) of Fig. 10. The two dotted black lines on either side of the red lines are the estimates of a standard deviation at a given depth. There are three arrows: the magenta one represents the average of the estimated depths of a hardpan and the two cyan arrows indicate a standard deviation. The deduced depths of hardpans are similar among the three plots.

The comparison in Fig. 10 of soil penetration strength between the fallow plot and the other wheat plots suggests that as the crop grows it tends to extract more water from the soil than supplied by rain, leading to increased soil strength. The similarity between the penetrometer readings and deduced soil strength is also good, especially for the upper layers. The agreement is less encouraging in the lower layers, partly because some of the penetrometer operations were not successful at those depths and partly because of lack of high-fidelity low-frequency signals for the acoustic-seismic measurements.

It is emphasised that Fig. 10 is the outcome of three independent efforts. First, acoustic-to-seismic coupling (measurements and numerical optimisation) was carried out to estimate the shear modulus (μ). Second, a penetrometer was operated to measure the penetration resistance (Q). Third, there was an independently obtained relation between Q and μ. For Fig. 10, we have adopted an empirical relation proposed by Gao et al. (2013):

$$Q \approx (0.0286 \pm 0.0013)\mu + (-53.0 \pm 12.9\,\text{kPa}), \tag{7}$$

for the arable loamy sand in the area for the acoustic-to-seismic measurements. The shear modulus from the optimisation is about 35 times larger than the penetration resistance. Therefore, without the help of Eq. (7), it is not possible to observe good agreement of soil strengths from both acoustic-to-seismic coupling and penetrometer in the shallow layers in Fig. 10.

5.3 Deducing the soil water contents by wave-speed measurements

Fluid-saturated two-phase models may describe the wave propagation in air-saturated sound-absorbing materials or water-saturated sediments on sea floors. However, two-phase models can be too ideal to represent most of (arable) soils with partial water saturation, except low-level water contents as demonstrated in the previous section. In this regard, Brutsaert (1964) is widely credited to have initiated the theory of elastic wave propagation through unconsolidated unsaturated granular materials having three distinct phases of a solid and two immiscible fluids. Due to the introduction of the second fluid component, three compressional waves and one shear wave were predicted to propagate when the medium is unbounded. One noticeable shortcoming of his model was the lack of the inertial coupling between phase components as also pointed by others (Arora and Tomar, 2008; Lo and Sposito, 2013).

In his follow-up work (Brutsaert and Luthin, 1964), Brutsaert investigated the relationship between wave speed and soil moisture content. Among four waves, he focussed on the speed of the fastest wave which is the one closest to the P-wave in characteristics and hence often referred to as P1-wave (Lo and Sposito, 2013). Due perhaps to the lack of inertial coupling between different phases, Brutsaert contemplated the low frequency by proposing its velocity as:

$$c = \sqrt{\frac{0.306 a p_e^{1/3} Z}{\rho_m \Omega b^{2/3}}}$$

with $p_e = p_t - p_g - Sp_c$ where p_e, p_t, p_g and p_c are effective pressure, total pressure, equilibrium pressure in the gas, and capillary pressure in the liquid, respectively. The degree of saturation with liquid is S. The equation has the mass density $\rho_m = \rho_s + \rho_g + \rho_l$ with the masses of solid (ρ_s), gas (ρ_g) and liquid (ρ_l) per unit bulk volume. The porosity is denoted by Ω. The factors a and b are empirical. The remaining term Z is a nonlinear function of b, p_e and effective bulk modulus (K_{eff}). Brutsaert and Luthin (1964) adopted $K_{eff} = (1-S)K_g + SK_l$ with K_g and K_l for the bulk moduli of gas and liquid, respectively. Later, the choice of this specific formula of K_{eff} was considered as a typing error by others (Flammer et al., 2001) who suggested a correct form might have been $K_{eff}^{-1} = (1-S)K_g^{-1} + SK_l^{-1}$.

Despite the uncertainty on the use of K_{eff}, Brutsaert's model has been revisited with the aim of establishing the relationship between the wave speeds and the soil water contents (Adamo et al., 2009, 2010; Flammer et al., 2001; Sharma and Gupta, 2010). Their works were essentially conducted and reported as forward approaches (Shin et al., 2016); however, the rationales behind some of these efforts were investigating the feasibility of inverse tools. Here, by the forward method, it studies 'the velocity of sound in soils ... by function of the moisture content' which is actually part of the article title of Brutsaert and Luthin (1964). By contrast, the inverse model aims to solve 'the moisture content ... by function of the velocity of sound in soils'. Logically, the success of the former proposition (i.e. forward model) does not necessarily guarantee the success of the latter (i.e. inverse approach). More in-depth discussion of Brutsaert's model and its applications by others (Adamo et al., 2009, 2010; Flammer et al., 2001; Sharma and Gupta, 2010) can be found in Shin et al. (2016).

There is also a question of whether the water content is the right parameter to investigate from the measurement of the elastic wave speeds in unsaturated soils (Shin et al., 2016). In Brutsaert's model, it is related to the occurrence of the water content and effective pressure (or water potential) as a product of Sp_c in which whether S is an appropriate parameter is also an interesting topic (Bishop and Blight 1963; Gray and Schrefler, 2001; Jennings and Burland, 1962; Khalili and Khabbaz, 1998; Nuth and Laloui, 2008); more discussion can be found in Shin et al. (2016). Lu and Sabatier (2009) and Lu (2014) demonstrated that the sound speed in the unsaturated soils is closely related to the water potential rather than the moisture content and temperature. In the case of shear waves, Whalley et al. (2012) reported that, at matric potentials higher than the air entry potential, V_s in sands increased even at constant water content while the matric potential decreased. Therefore, in the context of possible inverse problems, the water potential could be the one to pursue rather than the water content.

Whether it is the water content or the water potential, its estimation based on the measurement of a single parameter, the elastic wave speed in this case, can be challenging even in the laboratory setting (Adamo et al., 2009, 2010;

Sharma and Gupta, 2010), let alone in the field of arable soils. Additional information is required. As shown briefly in a previous section, Shin et al. (2013, 2017) demonstrated the feasibility of using the acoustic-seismic coupling method to deduce various soil properties non-invasively. The acoustic-seismic coupling means that the method has the potential to investigate the pore-related and structure-related properties at the same time by a single set of related measurements. However, one of the theoretical bases behind the method is a fluid-saturated poroelastic model. Therefore, the method is currently limited to either dry or low-moisture soils. There is certainly room for improving the method by incorporating the three-phase poroelastic models to account for the more general status of unsaturated soils. In this implementation, the hysteretic nature (Gallipoli et al., 2003; Gregory et al., 2010; Lu and Sabatier, 2009) of soil moisture-suction relationships may need to be considered.

6 Conclusion

There are several geophysical methods that can deduce the characteristics of agricultural soils by measuring physical properties such as electric resistivity (or conductivity), elasticity constants, etc. These are the electrical resistivity method, electromagnetic induction method, induced polarisation, ground-penetrating radar, nuclear magnetic resonance, and acoustic-seismic methods. Among these, we have focussed on three: electromagnetic induction, electrical resistivity, and acoustic-to-seismic. For each method, a basic theoretical background has been briefly introduced, followed by agricultural applications: we have presented the spatial and temporal mapping of soil water or drying by crops.

In the electromagnetic induction method, a time-varying primary magnetic field induces very small eddy currents in the ground resulting in a secondary magnetic field. From the ratio of the primary to the secondary magnetic field, apparent electrical conductivity can be measured. Knowing the cumulative sensitivity pattern, inversion can be applied to deduce the depth-specific conductivity. The measurement itself can be done swiftly and non-invasively; hence it is widely used for surveying large areas, which can provide the basis for precision agriculture applying spatially customised treatment within the field (Corwin and Lesch, 2003; Lück, et al., 2009). For the electromagnetic induction method, we presented the observation of the temporal pattern of soil drying by wheat, which was also accompanied by the measurement of electrical resistivity, penetration resistance and neutron probe. Another example was the correlation of apparent electrical conductivity with height and leaf area of wheat on the assumption that a change in soil conductivity is related to a change in soil water content associated with root water uptake.

The resistivity method requires galvanic contact with the soil. Basic sensors are often arranged in a quadrupole configuration of a pair of current-injecting electrodes and another pair of voltage-measuring electrodes; nowadays, this configuration is extended in an array of numerous electrodes capable of 2D or 3D investigation due to advances in multi-channel switching systems. Measured apparent resistivity can be further processed with inverse methods to assess the spatial distribution of resistivity. This spatial information, likewise that of EMI, also helps estimate within-field variations required for the successful application of precision farming. By using the ERT, we presented a time-lapse image showing considerable changes in resistivity because of soil drying by different winter wheat varieties; by contrast, a fallow plot experienced the least change. In another example, the presence of a cemented hard layer was identifiable by the analysis of tomography images: only the very surface layers are dried by roots in presence of a hard layer; otherwise, soil drying was observed to depth.

The acoustic-to-seismic method exploits the coupling mechanism between the airborne acoustic source and underground seismic activity. The soil is treated as poroelastic (i.e. porous and elastic) material; hence, the model is sensitive to the soil elastic characteristics and porosity. The poroelasticity theory is incorporated with the wave propagation model in layered media. Several soil physical properties can be estimated through numerical optimisation against measured acoustic-to-seismic transfer functions. As applications, the penetration resistance of wheat-growing soils is compared between the measurement by a penetrometer and the estimation made from shear modulus deduced from the acoustic-to-seismic method. It was shown that as the wheat grew, soil strength increased due to the extraction of water.

7 Where to look for further information

Among many geophysical methods, we have focussed here on three: EMI, ERT and acoustic to seismic method. The first two methods measure the electrical conductivity or resistivity of the soils, while the third estimates the elastic constants. Like many other methods of measuring soil properties, these require instrumentations to acquire the raw data and mathematical modelling to analyse those data. There have been considerable advances in the development of instrumentations such as automated multichannel acquisition of hundreds of electrodes for ERT, simultaneous multicoil reading for EMI and wide-frequency precision microphones and LDV for acoustic to seismic method. However, development is continuously underway when it comes to portability, cost of instruments, speed of measurement, etc. Further development in the mathematical modelling of pedophysical relationships is desired, especially about the mathematical representation of multiphase and multiple-porosity nature of soils. Development of the joint inverse modelling of EMI (and/or ERT) and acoustic to seismic method will be an interesting avenue of future research. This collaboration will ensure the holistic measurement of both electrical and elastic properties of soils.

Lancaster University is a leading institution in the UK for development and application of the inverse modelling of ERT and EMI. The Open University has amassed decades of research in outdoor acoustic and seismic propagation. Rothamsted Research has special interests in applying these methods in agricultural context.

8 Acknowledgements

Rothamsted Research receives grant-aided support from the Biotechnology and Biological Sciences Research Council (BBSRC) through the Designing Future Wheat programme [BB/P016855/1] and the joint NERC/BBSRC ASSIST project (NE/N018117/1). We acknowledge support from BBSRC Crop Improvement Research Club (BB/J019461/1 and BB/J01950X/1). GB was supported by a Lancaster University - Rothamsted Research- CEH Graduate School for Environment PhD studentship.

9 References

Adamo, F., Attivissimo, F., Fabbiano, L. and Giaquinto, N. (2009). Velocity-moisture relationships for sandy soils: experimental results and data analysis, *IEEE Trans. Instrum. Meas.* 58(2), 311-317.

Adamo, F., Attivissimo, F., Fabbiano, L., Giaquinto, N. and Spadavecchia, M. (2010). Soil moisture assessment by means of compressional and shear wave velocities: theoretical analysis and experimental setup, *Measurement* 43(3), 344-352.

Algeo, J., Slater, L., Binley, A., Van Dam, R. L. and Watts, C. (2018). A comparison of ground-penetrating radar early-time signal approaches for mapping changes in shallow soil water content, *Vadose Zone J.* 17(1), 180001.

Allred, B. J., Daniels, J. J. and Ehsani, M. R. (Eds) (2008). *Handbook of Agricultural Geophysics*, Boca Raton, FL: CRC Press.

Andrade, F. C. and Fischer, T. (2018). Generalised relative and cumulative response functions for electromagnetic induction conductivity meters operating at low induction numbers: relative and cumulative response functions, *Geophys. Prospect.* 66(3), 595-602.

Andrade, F. C. M., Fischer, T. and Valenta, J. (2016). Study of errors in conductivity meters using the low induction number approximation and how to overcome them. In: Near Surface Geoscience 2016-22nd European Meeting of Environmental and Engineering Geophysics, Barcelona, Spain: EAGE.

Archie, G. E. (1942). The electrical resistivity log as an aid in determining some reservoir characteristics, *Trans. AIME* 146(1), 54-62.

Arora, A. and Tomar, S. K. (2008). The effect of inertial coupling on seismic reflection amplitudes, *Geophys. Prospect.* 56(5), 643-654.

Auken, E., Christiansen, A. V., Kirkegaard, C., Fiandaca, G., Schamper, C., Behroozmand, A. A., Binley, A., Nielsen, E., Effersø, F., Christensen, N. B., Sørensen, K., Foged, N.

and Vignoli, G. (2015). An overview of a highly versatile forward and stable inverse algorithm for airborne, ground-based and borehole electromagnetic and electric data, *Explor. Geophys.* 46(3), 223–235.

Beff, L., Günther, T., Vandoorne, B., Couvreur, V. and Javaux, M. (2013). Three-dimensional monitoring of soil water content in a maize field using electrical resistivity tomography, *Hydrol. Earth Syst. Sci.* 17(2), 595–609.

Binley, A. (2015). Tools and techniques: electrical methods. In: Schubert, G. (Ed), *Treatise on Geophysics*, Oxford: Elsevier, Chap. 11.08, pp. 233–259.

Binley, A. and Kemna, A. (2005). DC resistivity and induced polarization methods. In: Rubin, Y. and Hubbard, S. S. (Eds), *Hydrogeophysics*, Dordrecht, The Netherlands: Springer, Chap. 5, pp. 129–156.

Binley, A. and Slater, L. (2020). *Resistivity and Induced Polarization: Theory and Applications to the Near-Surface Earth*, Cambridge, UK: Cambridge University Press.

Binley, A., Henry-Poulter, S. and Shaw, B. (1996). Examination of solute transport in an undisturbed soil column using electrical resistance tomography, *Water Resour. Res.* 32(4), 763–769.

Binley, A., Hubbard, S. S., Huisman, J. A., Revil, A., Robinson, D. A., Singha, K. and Slater, L. D. (2015). The emergence of hydrogeophysics for improved understanding of subsurface processes over multiple scales, *Water Resour. Res.* 51(6), 3837–3866.

Biot, M. A. (1956a). Theory of propagation of elastic waves in a fluid-saturated porous solid. I. low-frequency range, *J. Acoust. Soc. Am.* 28(2), 168–178.

Biot, M. A. (1956b). Theory of propagation of elastic waves in a fluid-saturated porous solid. II. higher frequency range, *J. Acoust. Soc. Am.* 28(2), 179–191.

Bishop, A. W. and Blight, G. E. (1963). Some aspects of effective stress in saturated and partly saturated soils, *Géotechnique* 13, 177–197.

Blanchy, G., Saneiyan, S., Boyd, J., McLachlan, P. and Binley, A. (2020a). ResIPy, an intuitive open source software for complex geoelectrical inversion/modeling, *Comput. Geosci.* 137, 104423.

Blanchy, G., Virlet, N., Sadeghi-Tehran, P., Watts, C. W., Hawkesford, M. J., Whalley, W. R. and Binley, A. (2020b). Time-intensive geoelectrical monitoring under winter wheat, *Near Surf. Geophys.* 18(4), 413–425.

Blanchy, G., Watts, C. W., Ashton, R. W., Webster, C. P., Hawkesford, M. J., Whalley, W. R. and Binley, A. (2020c). Accounting for Heterogeneity in the θ–σ Relationship: application to Wheat Phenotyping Using EMI, *Vadose Zone J.* 19(1).

Blanchy, G., Watts, C. W., Richards, J., Bussell, J., Huntenburg, K., Sparkes, D. L., Stalham, M., Hawkesford, M. J., Whalley, W. R. and Binley, A. (2020d). Time-lapse geophysical assessment of agricultural practices on soil moisture dynamics, *Vadose Zone J.* 19(1), e20080.

Brogi, C., Huisman, J. A., Herbst, M., Klosterhalfen, A., Weihermüller, L., Montzka, C., van der Kruk, J. and Vereecken, H. (2019). Geophysics-based soil mapping improves the simulation of crop productivity at the field scale and beyond, *Geophys. Res. Abs.* 21, 7979.

Brogi, C., Huisman, J. A., Herbst, M., Weihermüller, L., Klosterhalfen, A., Montzka, C., Reichenau, T. G. and Vereecken, H. (2020). Simulation of spatial variability in crop leaf area index and yield using agroecosystem modeling and geophysics-based quantitative soil information, *Vadose Zone J.* 19(1), e20009.

Brutsaert, W. (1964). The propagation of elastic waves in unconsolidated unsaturated granular mediums, *J. Geophys. Res.* 69(2), 243–257.

Brutsaert, W. and Luthin, J. N. (1964). The velocity of sound in soils near the surface as a function of the moisture content, *J. Geophys. Res.* 69(4), 643–652.

Calamita, G., Perrone, A., Brocca, L., Onorati, B. and Manfreda, S. (2015). Field test of a multi-frequency electromagnetic induction sensor for soil moisture monitoring in southern Italy test sites, *J. Hydrol.* 529, 316–329.

Callegary, J. B., Ferré, T. P. A. and Groom, R. W. (2007). Vertical spatial sensitivity and exploration depth of low-induction-number electromagnetic-induction instruments, *Vadose Zone J.* 6(1), 158–167.

Callegary, J. B., Ferré, T. P. A. and Groom, R. W. (2012). Three-dimensional sensitivity distribution and sample volume of low-induction-number electromagnetic-induction instruments, *Soil Sci. Soc. Am. J.* 76(1), 85–91.

Cimpoiaşu, M. O., Kuras, O., Pridmore, T. and Mooney, S. J. (2020). Potential of geo-electrical methods to monitor root zone processes and structure: a review, *Geoderma* 365, 114232.

Corwin, D. L. (2008). Past, present, and future trends in soil electrical conductivity measurements using geophysical methods. *Handbook of Agricultural Geophysics*, Boca Raton, FL: CRC Press, 17–44.

Corwin, D. L. and Lesch, S. M. (2003). Application of soil electrical conductivity to precision agriculture: theory, principles, and guidelines, *Agron. J.* 95(3), 455–471.

Corwin, D. L. and Lesch, S. M. (2005). Apparent soil electrical conductivity measurements in agriculture, *Comput. Electron. Agr.* 46(1-3), 11–43.

Corwin, D. L. and Rhoades, J. D. (1984). Measurement of inverted electrical conductivity profiles using electromagnetic induction, *Soil Sci. Soc. Am. J.* 48(2), 288–291.

Corwin, D. L., Lesch, S. M., Ostler, S. R. and Kakkka, S. R. (2006). Monitoring management-induced spatio-temporal changes in soil quality through soil sampling directed by apparent conductivity, *Geoderma* 131, 369–387.

Dahlin, T., Aronsson, P. and Thörnelöf, M. (2014). Soil resistivity monitoring of an irrigation experiment, *Near Surf. Geophys.* 12(1), 35–44.

Doolittle, J. A. and Brevik, E. C. (2014). The use of electromagnetic induction techniques in soils studies, *Geoderma* 223–225, 33–45.

Flammer, I., Blum, A., Leiser, A. and Germann, P. (2001). Acoustic assessment of flow patterns in unsaturated soil, *J. Appl. Geophys.* 46(2), 115–128.

Gallipoli, D., Wheeler, S. J. and Karstunen, M. (2003). Modelling the variation of degree of saturation in a deformable unsaturated soil, *Géotechnique* 53, 105–112.

Gao, W., Watts, C. W., Ren, T., Shin, H. -C., Taherzadeh, S., Attenborough, K., Jenkins, M. and Whalley, W. R. (2013). Estimating penetrometer resistance and matric potential from the velocities of shear and compression waves, *Soil Sci. Soc. Am. J.* 77(3), 721–728.

Garré, S., Javaux, M., Vanderborght, J., Pagès, L. and Vereecken, H. (2011). Three-dimensional electrical resistivity tomography to monitor root zone water dynamics, *Vadose Zone J.* 10(1), 412–424.

Gray, W. G. and Schrefler, B. A. (2001). Thermodynamic approach to effective stress in partially saturated porous media, *Eur. J. Mech. A Solids* 20(4), 521–538.

Gregory, A. S., Bird, N. R. A., Whalley, W. R., Matthews, G. P. and Young, I. M. (2010). Deformation and shrinkage effects on the soil water release characteristic, *Soil Sci. Soc. Am. J.* 74(4), 1104–1112.

Hanssens, D., Delefortrie, S., Bobe, C., Hermans, T. and De Smedt, P. (2019). Improving the reliability of soil EC-mapping: robust apparent electrical conductivity (rECa) estimation in ground-based frequency domain electromagnetics, *Geoderma* 337, 1155–1163.

Hayashi, M. (2004). Temperature-electrical conductivity relation of water for environmental monitoring and geophysical data inversion, *Environ. Monit. Assess.* 96(1–3), 119–128.

Heagy, L. J., Cockett, R., Kang, S., Rosenkjaer, G. K. and Oldenburg, D. W. (2017). A framework for simulation and inversion in electromagnetics, *Comput. Geosci.* 107, 1–19.

Huang, J., Pedrera-Parrilla, A., Vanderlinden, K., Taguas, E. V., Gómez, J. A. and Triantafilis, J. (2017). Potential to map depth-specific soil organic matter content across an olive grove using quasi-2d and quasi-3d inversion of DUALEM-21 data, *CATENA* 152, 207–217.

JafarGandomi, A. and Binley, A. (2013). A Bayesian trans-dimensional approach for the fusion of multiple geophysical datasets, *J. Appl. Geophys.* 96, 38–54.

Jennings, J. E. B. and Burland, J. B. (1962). Limitations to the use of effective stresses in partly saturated soils, *Géotechnique* 12, 125–144.

Khalili, K. and Khabbaz, M. H. (1998). A unique relationship for χ for the determination of the shear strength of unsaturated soils, *Géotechnique* 48, 681–687.

King, J. A., Dampney, P. M. R., Lark, R. M., Wheeler, H. C., Bradley, R. I. and Mayr, T. R. (2005). Mapping potential crop management zones within fields: use of yield-map series and patterns of soil physical properties identified by electromagnetic induction sensing, *Precis. Agric.* 6(2), 167–181.

Koestel, J., Kemna, A., Javaux, M., Binley, A. and Vereecken, H. (2008). Quantitative imaging of solute transport in an unsaturated and undisturbed soil monolith with 3-D ERT and TDR, *Water Resour. Res.* 44(12), W12411.

Korsaeth, A., Riley, H., Kværnø, S. H. and Vestgarden, L. S. (2008). Relations between a commercial soil survey map based on soil apparent electrical conductivity (ECa) and measured soil properties on a morainic soil in Southeast Norway. In: Allred, B. J., Daniels, J. J. and Ehsani, M. R. (Eds), *Handbook of Agricultural Geophysics*, Boca Raton, FL: CRC Press, Chap. 13.

Laloy, E., Javaux, M., Vanclooster, M., Roisin, C. and Bielders, C. L. (2011). Electrical resistivity in a loamy soil Identification of the appropriate pedo-electrical model, *Vadose Zone J.* 10(3), 1023–1033.

Lavoué, F., van der Kruk, J., Rings, J., André, F., Moghadas, D., Huisman, J. A., Lambot, S., Weihermüller, L., Vanderborght, J. and Vereecken, H. (2010). Electromagnetic induction calibration using apparent electrical conductivity modelling based on electrical resistivity tomography, *Near Surf. Geophys.* 8(6), 553–561.

Lo, W.-C. and Sposito, G. (2013). Acoustic waves in unsaturated soils, *Water Resour. Res.* 49(9), 5674–5684.

Lu, Z. (2014). Feasibility of using a seismic surface wave method to study seasonal and weather effects on shallow surface soils, *J. Environ. Eng. Geophys.* 19(2), 71–85.

Lu, Z. and Sabatier, J. M. (2009). Effects of soil water potential and moisture content on sound speed, *Soil Sci. Soc. Am. J.* 73(5), 1614–1625.

Lück, E., Gebbers, R., Ruehlmann, J. and Spangenberg, U. (2009). Electrical conductivity mapping for precision farming, *Near Surf. Geophys.* 7(1), 15–26.

Ma, R., McBratney, A., Whelan, B., Minasny, B. and Short, M. (2011). Comparing temperature correction models for soil electrical conductivity measurement, *Precis. Agric.* 12(1), 55–66.

Macleod, C. K., Humphreys, M. W., Whalley, W. R., Turner, L., Binley, A., Watts, C. W., Skøt, L., Joynes, A., Hawkins, S., King, I. P., O'Donovan, S. and Haygarth, P. M. (2013). A novel grass hybrid to reduce flood generation in temperate regions, *Sci. Rep.* 3, 1683.

Martini, E., Werban, U., Zacharias, S., Pohle, M., Dietrich, P. and Wollschläger, U. (2017). Repeated electromagnetic induction measurements for mapping soil moisture at the field scale: validation with data from a wireless soil moisture monitoring network, *Hydrol. Earth Syst. Sci.* 21(1), 495–513.

McLachlan, P., Blanchy, G. and Binley, A. (2021). EMagPy: open-source standalone software for processing, forward modeling and inversion of electromagnetic induction data, *Comput. Geosci.* 146, 104561.

McNeill, J. D. (1980a). *Electrical Conductivity of Soils and Rocks*, Technical Note TN-5.

McNeill, J. D. (1980b). *Electromagnetic Terrain Conductivity Measurement at Low Induction Numbers*, Technical Note TN-6.

Michot, D., Benderitter, Y., Dorigny, A., Nicoullaud, B., King, D. and Tabbagh, A. (2003). Spatial and temporal monitoring of soil water content with an irrigated corn crop cover using surface electrical resistivity tomography, *Water Resour. Res.* 39(5), 1138.

Monteiro Santos, F. A. (2004). 1-D laterally constrained inversion of EM34 profiling data, *J. Appl. Geophys.* 56(2), 123–134.

Moral, F. J. and Serrano, J. M. (2019). Using low-cost geophysical survey to map soil properties and delineate management zones on grazed permanent pastures, *Precis. Agric.* 20(5), 1000–1014.

Nuth, M. and Laloui, L. (2008). Effective stress concept in unsaturated soils: clarification and validation of a unified framework, *Int. J. Numer. Anal. Met.* 32, 771–801.

Paetzold, R. F., Matzkanin, G. A. and Santos, A. D. L. (1985). Surface soil water content measurement using pulsed nuclear magnetic resonance techniques, *Soil Sci. Soc. Am. J.* 49(3), 537–540.

Panissod, C., Dabas, M., Hesse, A., Jolivet, A., Tabbagh, J. and Tabbagh, A. (1998). Recent developments in shallow-depth electrical and electrostatic prospecting using mobile arrays, *Geophysics* 63(5), 1542–1550.

Reyes, J., Wendroth, O., Matocha, C., Zhu, J., Ren, W. and Karathanasis, A. D. (2018). Reliably mapping clay content Coregionalized with electrical conductivity, *Soil Sci. Soc. Am. J.* 82(3), 578–592.

Rhoades, J. D., Raats, P. A. C. and Prather, R. J. (1976). Effects of liquid-phase electrical conductivity, water content, and surface conductivity on bulk soil electrical conductivity, *Soil Sci. Soc. Am. J.* 40(5), 651–655.

Robinson, D. A., Abdu, H., Lebron, I. and Jones, S. B. (2012). Imaging of hill-slope soil moisture wetting patterns in a semi-arid oak savanna catchment using time-lapse electromagnetic induction, *J. Hydrol.* 416–417, 39–49.

Rossi, R., Pollice, A., Bitella, G., Labella, R., Bochicchio, R. and Amato, M. (2018). Modelling the non-linear relationship between soil resistivity and alfalfa NDVI: a basis for management zone delineation, *J. Appl. Geophys.* 159, 146–156.

Roy, A. and Apparao, A. (1971). Depth of investigation in direct current methods, *Geophysics* 36(5), 943–959.

Rücker, C., Günther, T. and Wagner, F. M. (2017). pyGIMLi: an open-source library for modelling and inversion in geophysics, *Comput. Geosci.* 109, 106–123.

Samouëlian, A., Cousin, I., Tabbagh, A., Bruand, A. and Richard, G. (2005). Electrical resistivity survey in soil science: a review, *Soil Till. Res.* 83(2), 173–193.

Shanahan, P. W., Binley, A., Whalley, W. R. and Watts, C. W. (2015). The use of electromagnetic induction to monitor changes in soil moisture profiles beneath different wheat genotypes, *Soil Sci. Soc. Am. J.* 79(2), 459–466.

Sharma, R. K. and Gupta, A. K. (2010). Continuous wave acoustic method for determination of moisture content in agricultural soil, *Comput. Electron. Agr.* 73(2), 105–111.

Sheets, K. R. and Hendrickx, J. M. H. (1995). Noninvasive soil water content measurement using electromagnetic induction, *Water Resour. Res.* 31(10), 2401–2409.

Shin, H.-C., Taherzadeh, S., Attenborough, K., Whalley, W. R. and Watts, C. W. (2013). Non-invasive characterization of pore-related and elastic properties of soils in linear Biot-Stoll theory using acoustic-to-seismic coupling, *Eur. J. Soil Sci.* 64(3), 308–323.

Shin, H.-C., Whalley, W. R., Attenborough, K. and Taherzadeh, S. (2016). On the theory of Brutsaert about elastic wave speeds in unsaturated soils, *Soil Till. Res.* 156, 155–165.

Shin, H.-C., Watts, C. W., Whalley, W. R., Attenborough, K. and Taherzadeh, S. (2017). Non-invasive estimation of the depth profile of soil strength with acoustic-to-seismic coupling measurement in the presence of crops, *Eur. J. Soil Sci.* 68(5), 758–768.

Slater, L. D. and Lesmes, D. (2002). IP interpretation in environmental investigations, *Geophysics* 67(1), 77–88.

Sørensen, K. (1996). Pulled array continuous electrical profiling, *First Break* 14(3), 85–90.

Stoll, R. D. (1974). Acoustic waves in saturated sediments. In: Hampton, L. (Ed), *Physics of Sound in Marine Sediments*, New York: Plenum Press, pp. 19–39.

Sudduth, K. A., Kitchena, N. R., Wiebold, W. J., Batchelor, W. D., Bollero, G. A., Bullock, D. G., Clay, D. E., Palm, H. L., Pierce, F. J., Schuler, R. T. and Thelen, K. D. (2005). Relating apparent electrical conductivity to soil properties across the north-central USA, *Comput. Electron. Agr.* 46(1–3), 263–283.

Tooms, S., Taherzadeh, S. and Attenborough, K. (1993). Sound propagation in a refracting fluid above a layered fluid-saturated porous elastic material, *J. Acoust. Soc. Am.* 93(1), 173–181.

Triantafilis, J. and Lesch, S. M. (2005). Mapping clay content variation using electromagnetic induction techniques, *Comput. Electron. Agr.* 46(1–3), 203–237.

von Hebel, C., Rudolph, S., Mester, A., Huisman, J. A., Kumbhar, P., Vereecken, H. and van der Kruk, J. (2014). Three-dimensional imaging of subsurface structural patterns using quantitative large-scale multiconfiguration electromagnetic induction data, *Water Resour. Res.* 50(3), 2732–2748.

von Hebel, C., van der Kruk, J., Huisman, J. A., Mester, A., Altdorff, D., Endres, A. L., Zimmermann, E., Garré, S. and Vereecken, H. (2019). Calibration, conversion, and quantitative multi-layer inversion of multi-coil rigid-boom electromagnetic induction data, *Sensors* 19(21), 4753.

Wait, J. R. (1982). *Geo-Electromagnetism*, Academic Press: New York.

Waxman, M. H. and Smits, L. J. M. (1968). Electrical conductivities in oil-bearing shaly sands, *Soc. Petrol. Eng. J.* 8(2), 107–122.

Weller, A., Slater, L., Binley, A., Nordsiek, S. and Xu, S. (2015). Permeability prediction based on induced polarization: insights from measurements on sandstone and unconsolidated samples spanning a wide permeability range, *Geophysics* 80(2), D161–D173.

Whalley, W. R., Jenkins, M. and Attenborough, K. (2011). The velocity of shear waves in saturated soil, *Soil Sci. Soc. Am. J.* 75(5), 1652–1657.

Whalley, W. R., Jenkins, M. and Attenborough, K. (2012). The velocity of shear waves in unsaturated soil, *Soil Till. Res.* 125, 30–37.

Whalley, W. R., Binley, A., Watts, C. W., Shanahan, P., Dodd, I. C., Ober, E. S., Ashton, R. W., Webster, C. P., White, R. P. and Hawkesford, M. J. (2017). Methods to estimate changes in soil water for phenotyping root activity in the field, *Plant Soil* 415(1), 407–422.

Zhu, Q., Lin, H. S. and Doolittle, J. A. (2013). Functional soil mapping for site specific soil moisture and crop yield management, *Geoderma* 200–201, 45–54.

Chapter 7

Advances in techniques to assess soil erodibility

R. J. Rickson, E. Dowdeswell Downey, G. Alegbeleye and S. E. Cooper, Cranfield University, UK

1 Introduction

Soil erodibility is the susceptibility of soil to the erosive forces of rainsplash (raindrop impact on the soil surface), runoff (overland flow) and wind (Bryan et al., 1989). Some definitions of erodibility only consider the detachment of individual soil particles or small aggregates from the soil mass (e.g. Torri, 1987). Others consider detachment and transport of material. Some argue that erodibility is not just about the rate at which soil is detached, but how much of that detached material can be transported by overland flow (Parsons et al., 1994).

This chapter will focus on soil erodibility with regard to erosion by water (rainfall and overland flow/runoff). Wind erodibility has been considered elsewhere (e.g. Chepil et al., 1962; Chepil and Woodruff, 1954; Johannsen et al., 1998; Leys and McTainsh, 2002; de Oro et al., 2019).

2 Factors affecting soil erodibility

Soil erodibility is highly dynamic in time and space (Bryan et al., 1989; Webb and Strong, 2011; Fig. 1). It is influenced by physical, mechanical, hydrologic, chemical, rheological, mineralogical and biological factors, not to mention soil profile characteristics such as soil depth and vegetation (Veihe, 2002). Sanchis et al. (2008) state that erodibility is determined by static and dynamic soil

http://dx.doi.org/10.19103/AS.2020.0079.14

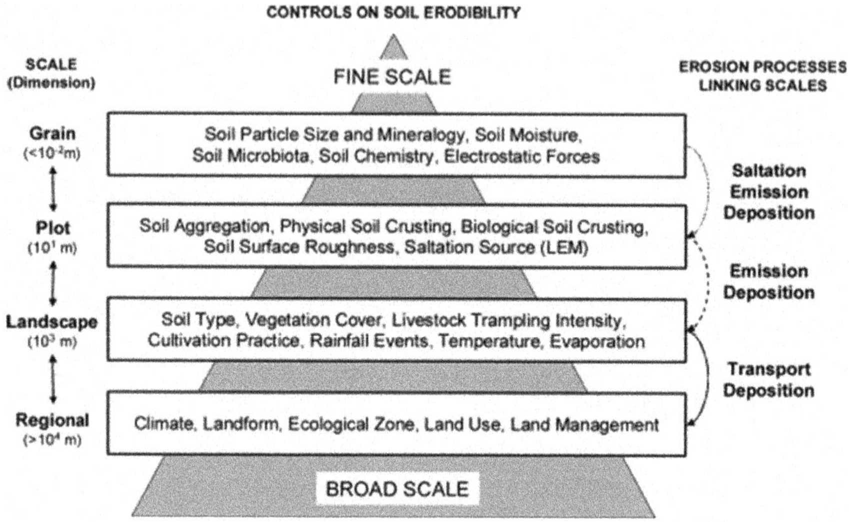

Figure 1 The control on soil erodibility at different spatial scales, including within and between the soil grain ($<10^{-2}$ m), plot (10^1 m), landscape (10^3 m) and regional (10^4 m) scales. Arrows down the right-hand side of the figure show the erosion processes that functionally connect the scale domains (From Webb and Strong (2011)).

properties that control a range of sub-processes affecting soil erosion. In other words, anything that affects soil-water interactions is important (i.e. infiltration rate, permeability, water retention forces, porosity, exchangeable ions, primary particle sizes and mineralogy, aggregate size and stability, soil organic matter) and their interactions, directly and indirectly, with rainfall and surface run-off. Wang et al. (2016) argue that erodibility is not determined by any individual soil property, but is dependent on the interactions among multiple soil properties. A full review of the factors affecting erodibility can be found in Amézketa (1999), but a summary is presented here.

2.1 Soil properties

2.1.1 Soil texture

Soil texture (particle size) is important because different sized fractions (as individual particles or aggregates) require more energy to become detached. For example, clay particles are relatively less erodible due to cohesion. Sand particles, although less cohesive, are also relatively less erodible due to large grain size. These characteristics are demonstrated in the Hjulström curves, where higher energies are needed to detach and transport clay and sand-sized particles (Fig. 2). Clay particles, along with soil organic matter (SOM), also bind and cement soil particles into larger aggregates that may be more resistant

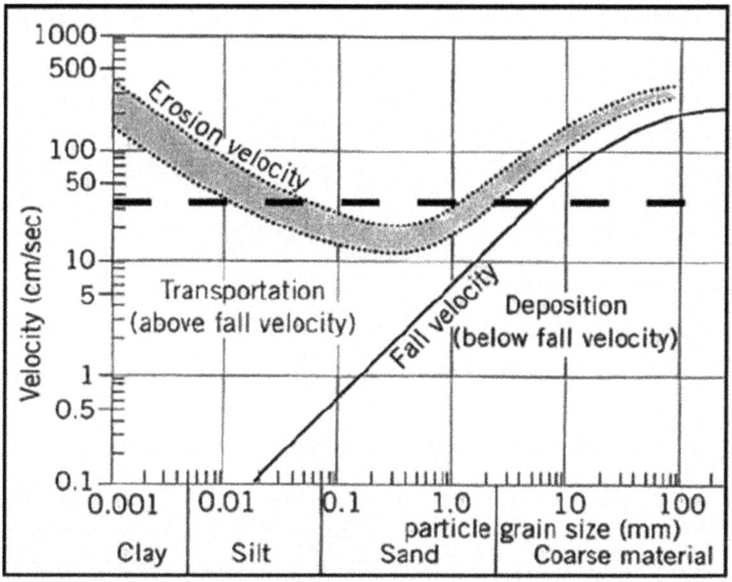

Figure 2 Hjulström curves showing critical flow velocities for detachment, transport and deposition of different soil particle sizes (taken from Puscas et al., 2010).

(less erodible) to erosive forces (Middleton, 1930). These relationships were found on tropical soils tested by Bruce-Okine and Lal (1975). Veihe (2002) found that the sand fraction was the best indicator of erodibility due to its low spatial heterogeneity and its contribution to the stability of soil aggregates.

Clay mineralogy also affects aggregate stability, as soils prone to shrink and swell on drying and wetting (e.g. montmorillonite) are less stable than less expansive soils (e.g. kaolin).

2.1.2 Percentage of rock fragments/stone cover

The protective effect of surface stone cover is usually not considered in soil erodibility estimation, although Poesen et al. (1994) found that surface-level stones can lead to an exponential decrease in soil erosion. Panagos et al. (2014) found this factor resulted in a 15% decrease in soil erodibility on average, particularly in areas with high percentages of surface stone cover. Also, Algayer et al. (2014) found that erodibility values vary depending on whether a surface crust has formed. Stone cover is likely to increase surface roughness, which is associated with lower erodibility. This is because a rougher surface will (a) intercept material splashed by raindrop impact and (b) hinder the generation (and energy) of surface overland flow to transport detached material.

2.1.3 Soil infiltration capacity and rate, and soil permeability

Soil hydraulic properties affect the ability of the soil to absorb rainfall before potentially erosive runoff is generated. In turn, these properties are affected by complex relationships between soil antecedent moisture content, soil water-holding capacity, porosity, rainfall characteristics, soil type, soil depth (e.g. depth to impermeable layer) and bulk density.

2.1.4 Soil shear strength

The resistance of particles or aggregates to the shearing forces of rainfall, wind and flow will affect the ease of detachment (Bones et al., 2017).

2.1.5 Soil organic matter (SOM)/organic carbon content

SOM binds with clay minerals to improve aggregate strength and reduce erodibility (Fig. 3). Threshold levels of SOM have been used to define erodible/non-erodible soil (e.g. 3% SOM/2% soil organic carbon; Greenland et al. (1975)). The presence of SOM promotes aggregate stability by (i) increasing inter-particle cohesion with organo-mineral bonds and (ii) increasing hydrophobicity, though different types of organic matter may have differing effects on cohesion and hydrophobicity (Abiven et al., 2009). Some relationships between SOM and aggregate stability are presented in Loveland and Webb (2003).

2.1.6 Biological factors

While many studies acknowledge the role of soil microbes, roots and soil fauna in aggregation (Amézketa, 1999; Blankinship et al., 2016), relatively few studies consider the specific role of soil biology on soil erodibility (see also Bearden

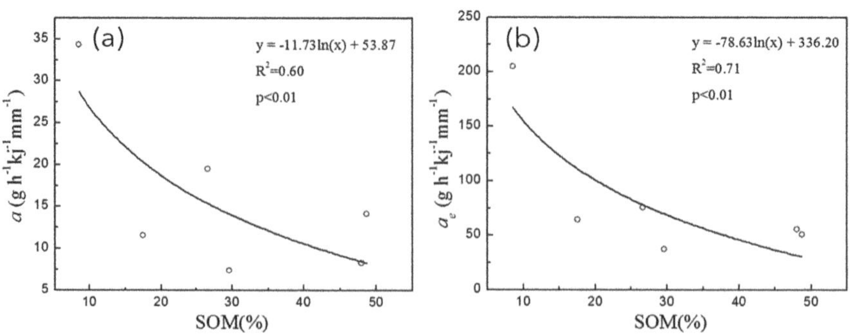

Figure 3 Soil erodibility indicators as a function of SOM: (a) index 'a' as function of SOM; and (b) index 'a_e' as a function of SOM (from Wang et al., 2016).

and Petersen, 2000; Bossuyt et al., 2001; Degens, 1997; Morel et al., 1991; Tisdall et al., 1997).

Deeb et al. (2017) studied erodibility in soils treated with composts, containing earthworms and plants. They found that soil organisms accounted for 50% of soil aggregation variance. The combined effect of earthworms and plants was complex: plants had a dominant effect on the distribution of aggregate sizes by disrupting earthworm casts, but earthworms influenced aggregate stability more than the plants. Amézketa (1999) found the contribution of earthworms to soil aggregation was related to specific species and their diets. Amézketa (1999) proposed the manipulation of soil microbes and their by-products (e.g. microbial extracellular polysaccharides) to promote aggregation in agricultural soils.

Oades (1993) distinguished between biological factors which create aggregates, and those which stabilise or degrade soil aggregates. At large scales, plant roots and associated hyphae enmesh soil particles, acting as a 'sticky string bag'. At the micro-scale, mucilages from roots, hyphae, bacteria and fauna such as earthworms help to stabilise smaller aggregates. Both microflora and fauna cause the degradation of stabilising agents. Fauna may breakdown the roots and hyphae that stabilise larger aggregates. Microorganisms use mucilaginous stabilising agents as an energy source, resulting in a slow breakdown of structural stability.

Gheytasi et al. (2019) considered how biological crusts in dry and semi-arid ecosystems affect erodibility. There was no significant difference in erodibility between physical crusts with lichens ($P \geq 0.05$), but there were significant differences between those with moss and with moss-lichens ($P \leq 0.05$).

However, the role of soil biology in erodibility may be soil-specific: Degens et al. (1994) express doubts about the role of hyphae and fine roots as 'enmeshers' in the macro-aggregate stabilisation in sandy soils.

2.1.7 Soil chemistry

Aluminium, silica, magnesium, $CaCO_3$, Fe and Al oxides, and gypsum have been related to soil erodibility. These chemicals affect bonds between soil particles and binding substances such as SOM. Soil chemistry also affects the absorptivity and infiltration characteristics of soil, which determine the generation of potentially erosive surface run-off. Barbosa et al. (2019) assessed the efficiency of magnetic susceptibility (MS) as a predictor of soil erodibility indices (K for the Universal Soil Loss Equation/USLE model (Wischmeier and Mannering, 1969); K_i and K_r for the Water Erosion Prediction Project/WEPP model (Nearing et al., 1989)) for Oxisols with iron contents ranging from 1 to 20% Fe_2O_3 in northeastern São Paulo State, Brazil. Magnetic susceptibility predicted the erodibility factors, with a precision of up to 60% and an accuracy of up to 85%.

2.1.8 Cation exchange capacity

Cations stimulate soil aggregation with the formation of cationic bridges between organic matter, minerals and clay particles (Bronick and Lal, 2005). The effect of cations on aggregation is dependent on cation composition, concentration and specific cation properties, and influences the structure of cationic bridges, the dispersibility of clay and the resistance to slaking.

2.1.9 Aggregate-size distribution

Comparing the aggregate-size distribution of a soil before and after it is exposed to erosive forces has been used as an indicator of soil erodibility. Also, the greater the percentage of small aggregates, the less energy is required for erosion to occur, according to the Hjulström curves (Fig. 2). However, this is contradicted by Martínez-Mena et al. (1999) who argue that larger aggregates are more likely to be broken down by raindrop impact than smaller sizes.

2.1.10 Aggregate stability

Many authors consider aggregate stability as the most representative indicator of erodibility. Le Bissonnais (1996, 2016) identifies four main mechanisms leading to aggregate breakdown and subsequent production of easily eroded material. These are: slaking; differential swelling; mechanical breakdown by raindrop impact; and physico-chemical dispersion. Aggregate stability is related to clay content (and type of clay mineral); SOM; soil chemistry; environmental conditions at formation (i.e. time for formation (aging), moisture content (Almajmaie et al., 2017), soil density); freeze thaw activity; slaking processes; and microbiology.

Barthes and Roose (2002) investigated aggregate stability and soil erodibility on a wide range of soil types in the field. Wang et al. (2016) correlate erodibility indices with water-stable aggregates (WSA) (Fig. 4).

2.2 Land use and management practices

Soil properties affecting erodibility are all influenced by land use and management practices. Vegetation/crop type (Demenois et al., 2017), rotational context, tillage practices (tillage direction; conservation tillage etc.; Obalum et al., (2019)), trampling by livestock, soil compaction by agricultural vehicles,

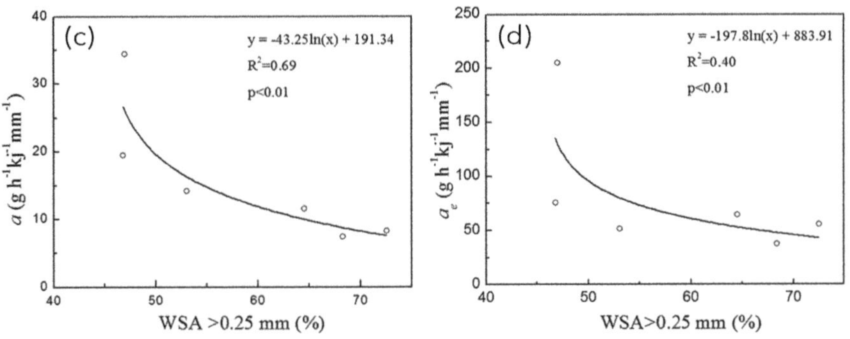

Figure 4 Soil erodibility indicators as a function of WSA: (c) 'a' as function of WSA >0.25 mm and (d) 'ae' as a function of WSA >0.25 mm (from Wang et al., 2016).

irrigation and crop residue management will all affect soil properties, which in turn determine soil erodibility.

2.3 Soil amendments/conditioners

Soil properties affecting erodibility can be affected by artificial and natural soil conditioners, such as biochar (Bayabil et al., 2015; Jeffery et al., 2015; Kumar et al., 2019; Verheijen et al., 2014), anaerobic digestate (Voelkner et al., 2015), manures (Guo et al., 2019) and composts (Annabi et al., 2007).

3 Assessment of soil erodibility

Sampling and analysing all the factors affecting erodibility is time-consuming and very costly. Qian et al. (2019) propose the use of robots to collect soil data to address some of these issues. As a result, erodibility indices have been developed where only the most salient soil properties are measured, using a variety of methods. Ideally, these indices should be:

- Highly correlated with soil loss;
- Derived from easily available data;
- Simple to use;
- Universal in application and accuracy;
- Inexpensive;
- Easily replicated;
- Sensitive enough to reflect differences that discriminate between soils or treatments; and
- Representative of a site, where replication is limited.

Erodibility indices have a number of applications. They allow erodibility to be mapped at local, regional, national and global scales, as an indicator of spatial variation in soil erosion risk (Fig. 5). Panagos et al. (2012; 2014) used a pan-European soil dataset, of around 20 000 points across 25 EU Member States to produce a harmonised high-resolution soil erodibility map.

Mapping erodibility helps target areas in need of soil protection, although other factors affect soil erosion too (e.g. rainfall erosivity; slope length, steepness and curvature; land use; crop type; land management etc.). Soil erodibility indices have been added to other engineering properties of soils to predict the risk of scouring around river bridge foundations and associated costs (Bones et al., 2017).

Indices of soil erodibility are used as input data for soil erosion prediction models, such as the 'K' factor in the Universal Soil Loss Equation (Ostovari et al., 2019; Schmidt et al., 2018; Wischmeier and Mannering, 1969; Wischmeier and Smith, 1978).

In the Water Erosion Prediction Project (WEPP; (Nearing et al., (1989)), soil erodibility is considered at the sub-process level. K_i is the interrill erodibility

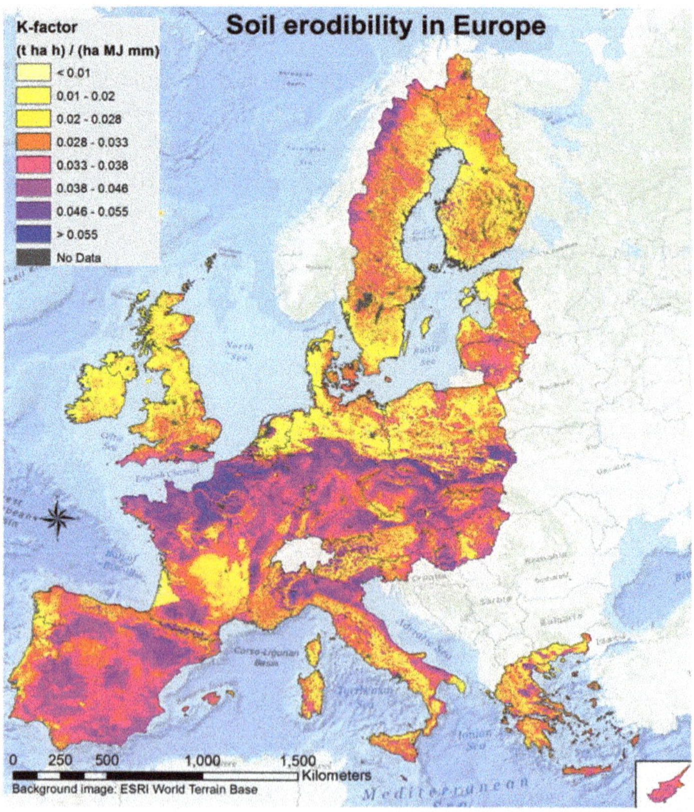

Figure 5 High-resolution (500 m grid cell size) map of soil erodibility estimated as K-factor in the European Union (from Panagos et al., 2014).

factor (for unconcentrated flow) and K_r is the erodibility of the soil within rill features, where flow is concentrated (Laflen et al., 1991). These erodibility factors are then compared with critical hydraulic shear stresses of the flow at which detachment of soil material occurs. Laflen et al. (1991) state that these factors vary widely as they are dependent on soil properties, climate and land management. In the EUROSEM model, soil erodibility is represented by the 'EROD' and 'COH' factors (Morgan et al., 1998).

Different approaches have been used to develop indicators of soil erodibility: there is no one standard test (Bryan et al., 1989) or scale of testing (Mhaske et al., 2019). These approaches can be classified as static or dynamic tests, based either in the laboratory or in the field (Bryan, 1968). Some indices are measured directly under natural or simulated rainfall conditions, and others are based on empirical regression models (Saygın et al., 2011).

3.1 Static laboratory techniques of assessing soil erodibility

Many static assessments of soil erodibility assessments are based on particle size distribution (Bouyoucos, 1935).

For example,

$$\frac{(\text{Sand} + \text{silt }\%)}{\text{Clay }\%} \tag{1}$$

For these assessments (and others based on particle size e.g. the USLE K factor (Wischmeier and Mannering (1969); Section 3.4.2)), the way particle sizes are measured will influence erodibility estimates (Dumbrovský et al., 2019). Centeri et al. (2015) conducted soil particle size measurements in different institutions with a variety of methodologies (laser, aerometer and pipette methods) on various soil materials (sandy, loamy and clay). Statistical analyses of the eight soil samples examined showed differences among the particle size measurements. Static tests have also been criticised for lacking simulation of erosion processes.

3.2 Static field techniques of assessing soil erodibility

Field soil surveys can measure soil characteristics linked to erodibility (Section 2.1). For example, Chorley (1959) studied soil physical properties near Oxford, UK, in an attempt to explain local relief patterns. The properties were moisture content, bulk density, grain size, range of grain size, permeability and shearing resistance. Moisture content, bulk density and range of grain size were used to determine shearing resistance, and were combined in an index of resistance (I_r):

$$I_r = \frac{(\text{soil bulk density} \times \text{range of grain size})}{\text{soil moisture content}} \quad (2)$$

Combining mean shearing resistance with permeability gave an index of erodibility:

$$\text{Erodibility} \approx \frac{1}{(\text{mean shear resistance} \times \text{permeability})} \quad (3)$$

3.3 Dynamic laboratory techniques of assessing soil erodibility

Laboratory tests on small soil samples (<100 g) are easy to carry out and are far less expensive and time-consuming than field experiments (Barthès and Roose, 2002). Simple laboratory determinations can generate data that is significantly correlated with more expensive or time-consuming field investigations (Imeson and Vis, 1984; Mentler et al., 2004; Fristensky and Grismer, 2008; Zhu et al., 2009, 2017; Rawlins et al., 2013).

Soil erodibility is usually determined by measuring aggregate stability (the ability of soil aggregates, or a size range of aggregates to withstand disaggregation when moistened under laboratory conditions) (Almajmaie et al., 2017). A number of workers have found aggregate stability to be the best indicator of erodibility (Jaiyeoba and Ologe (1990); Barthes and Roose (2002); Bryan (1968); Nciizah and Wakindiki (2015)).

Aggregate stability is determined by a number of procedures, usually by exposing aggregates to various methods of dispersion; wet sieving; rainfall simulation; or ultrasonic vibration. Measurement of aggregate stability has attracted numerous reviews and comparative studies between procedures (e.g. Amézketa et al., 1996, 2003; Le Bissonnais, 1996; Rohošková and Valla, 2004).

3.3.1 Clay dispersion

Clay dispersion is measured by comparing the amount of suspended material resulting from immersion in different fluids, dispersion tests or levels of agitation (Rengasamy et al., 1984). For example, the ASTM has a number of tests to identify dispersive clays, including the double hydrometer test (D4221); pinhole dispersion (D4647/D4647M), crumb or aggregate coherence test (D6572) and the analysis of pore water extraction (D4542). According to Almajmaie et al. (2017), dispersion is influenced by many factors including; (i) clay charge characteristics, (ii) cation ratios, especially the exchangeable sodium percentage (ESP), (iii) the electrolyte concentration of infiltration water and (iv) attractive forces between colloidal particles.

Middleton (1930) developed an 'erosion ratio' that was significantly correlated with erodibility. It compared the amount of silt + clay in an undispersed sample with that in a sample previously treated with a dispersing agent. These two results were combined as a ratio, expressed as a percentage, termed as the 'dispersion ratio'. The dispersion ratio was then divided by the ratio of colloid to moisture equivalent (Equation 4). A value of 10 was thought to be a significant boundary between erodible and non-erodible soils (Bryan, 1968).

$$\text{Erosion ratio} = \frac{\text{dispersion ratio}}{(\text{colloid content} / \text{moisture content})} \tag{4}$$

Mutter (2018) used portable turbidity meters to develop a quick dispersion ratio test (DR, %) as an index of soil stability. This method was quick, easy and cost-effective. It gave results that were highly correlated with other erodibility parameters.

3.3.2 Wet sieving

Wet sieving simulates the disintegrating forces found in sheet erosion and flooding (Kemper and Rosenau, 1986; Loch and Pocknee, 1995; Valmis et al., 2005). Valmis et al. (1988) describe the method used in wet sieving. Two brass sieves are oscillated in a cylindrical vessel containing water at the rate of 72 vertical strokes per minute. The sieves are positioned so that the mesh of sieves is 13 mm above the surface of the water at the top of each stroke. To determine the proportion of water stable aggregates, 10 g of size-graded (2.0–4.7 mm), dry-stable aggregates are placed in each of the two sieves and lowered into the water until covered to a depth of about 5 mm for a 3 min pre-soaking. After the pre-soaking treatment, the sieves are oscillated for 4 min.

An instability index, β, is determined according to the following formula:

$$(8)\beta = 12[\log(Wa - Ws) - \log(Wa - Wk - Ws)] \tag{5}$$

where Wa is the dry mass of the initial soil sample (g), Ws is the dry mass of the sand-sized mineral particle (≥0.25 mm) (g), Wk is the dry mass of the material, which has passed through the sieve during the preliminary soaking (3 min) and the oscillation period (4 min) (g).

However, there is no standard technique and the intensity of agitation and duration of immersion (Russell and Feng, 1947), the prewetting treatments prior to immersion, and whether the aggregates are sieved in water or alcohol (Amezketa et al., 1996) all affect aggregate stability results. Determination of aggregate stability by wet sieving has also been criticised, because it does not take into account all mechanisms responsible for aggregate breakdown in the field and tends to overemphasise slaking (Kemper and Rosenau, 1986).

Almajmaie et al. (2017) concur that aggregate breakdown during wet sieving is largely due to slaking and, to a lesser extent, to the physical effects of water movement during lifting of the sieve and abrasion on the sieve surface (Emerson, 1967). Furthermore, the wet sieving procedures developed by Kemper and Rosenau (1986) and by Le Bissonnais (1996) do not consider the effect of raindrop bombardment on disaggregation, and are usually conducted over too short a duration to allow for dispersion.

3.3.3 Laser granulometry

Rawlins et al. (2013) point out the limitations of aggregate stability testing by wet sieving and propose the use of laser granulometry instead. Two measurements of the continuous size distribution (<2000 μm) of a sample of aggregates are taken. The first is made on the WSA after these have been added to circulating water (initial air-dried aggregate size range from 1000 μm to 2000 μm). The second measurement is made on the disaggregated material (DM) after the circulating aggregates have been disrupted with ultrasound (sonication). The difference between the mean weight diameters (MWD) of these two size distributions is the disaggregation reduction (DR; μm). Soils with aggregates resistant to both slaking and mechanical breakdown by the hydrodynamic forces during circulation have larger values of DR (and lower erodibility).

3.3.4 Rainfall simulation

Rainfall simulators have become a significant tool for assessing soil erodibility (Fig. 6; Grismer, 2012; Mhaske et al., 2019), using a wide range of rainfall properties (Almajmaie et al., 2017; McCalla, 1944; Low, 1954; Morin and Benyamini, 1977; Norton, 1987; Bruce-Okine and Lal, 1975). Ideally raindrop sizes, temperature and energies should be realistic. McCalla (1944) and Imeson and Vis (1984) used the number of drops needed to break down a proportion of tested aggregates into fragments of a predetermined size to be the index of erodibility.

According to Bryan (1968), the water-drop testing technique proposed by McCalla (1944), modified by Smith and Cernuda (1951) and Pereira (1956) may be a more efficient index of soil erodibility than those derived from wet-sieving. Martínez-Mena et al. (1999) measured aggregate stability using a laboratory rainfall simulator inside an environmental chamber (Ternan et al., 1996). Twenty-five air-dry aggregates of 4.5–5 mm diameter were subjected to rainfall simulation at 45 mm h^{-1} intensity with a mean drop size of 580 μm. The percentage of water-stable aggregates surviving at 0.5-min intervals was determined and a mean rainfall simulation survival index (RSSI) was calculated based on the number of aggregates surviving at 5, 10, 15 and 20 min during the test.

Figure 6 Rainfall simulator apparatus; needle tray (top left), rainfall catch area (top right), and four sieves with saturated aggregates ready for testing (bottom left) (from Cooper (2006)).

However, Agassi and Bradford (1999) are critical of rainfall simulation studies to assess erodibility, due to variations in:

- The rainfall representativeness of natural conditions (e.g. intensity, kinetic energy etc.).
- Rain simulator types, for example, methods to form drops and energy levels.
- Chemical properties of the simulated rainstorm water.
- Plot size.
- Soil preparation methods.
- Soil surface changes during rainfall.

3.3.5 Flow shear stresses (runoff simulation)

A number of erodibility tests compare inherent soil shear strength with flow shear stress applied. The ASTM pinhole test (ASTM, 2013) identifies the dispersive characteristics of clay soils by simulating the action of water flowing along a crack in an earth embankment. The turbidity of flow, final size of the pinhole and computation of flow rates through the pinhole are used to calculate the dispersive characteristics of the soil. The results of the tests classify soils into six categories of dispersiveness: dispersive (D1, D2), slight to moderately dispersive (ND4, ND3) and nondispersive (ND2, ND1). Comparison of results from the pinhole test and other indirect tests of clay dispersion indicates that the pinhole test has the best correlation with the erosional performance of clay soils in nature (ASTM, 2013).

Briaud et al. (2001) and Briaud and Chen (2005) described an erosion function apparatus (EFA) which investigated the relationship between the hydraulic shear stress of water flowing over the soil and the soil erosion rate. Water at velocities between 0.1 m/s and 6 m/s flows over a soil sample, protruding 1 mm into the flow, which then erodes. The time taken to erode the 1 mm soil is recorded and then the flow velocity is increased to 0.6 m/s to erode the next 1 mm protrusion. Typical results are shown in Fig. 7. The plot of erosion rate versus shear stress indicates the critical shear stress at which erosion starts and the rate of erosion beyond that shear stress.

The Jet Erosion Test (JET) and Jet Erosion Device (JEd) (Hanson and Cook, 2004; Ibrahim et al., 2016) quantify the erodibility of soil and other cohesive sediments based on theory derived for circular impinging jets (Ghaneeizad et al., 2015). Water jets flow over the soil by destroying the soil structure, and then detaching and transporting soil particles out of the cavity (Maksimov et al., 2019). Mutter (2018) describes the use of a 'Mini' Jet Erosion Test to determine scouring depth (SD, in mm), critical shear stress (τc, pa) and erodibility coefficient (kd, cm^3/kN.s). According to Hanson and Cook (2004), the test is simple, quick and relatively inexpensive to perform. It is repeatable and gives consistent results.

Ribeiro et al. (2013) describe the 'Inderbitzen test' as the most commonly used erodibility assessment for geotechnical soils. It consists of a flume in which the soil sample is placed at the same inclination of the channel. Soil losses, caused by the controlled laminar flow of water, are measured to assess soil resistance to erosion.

3.3.6 Rainfall and run-off combined

Sheridan et al. (2000) developed and tested laboratory-scale measurements of erodibility using a laboratory tilting flume and rainfall simulator to determine rill and interrill erodibility for soils and overburdens from open-cast coal mines. The laboratory determinations of erodibility were compared with field erosion rates from 12 m-long plots under simulated rainfall. Regression analysis demonstrated a strong relationship between predicted and measured erosion rates, demonstrating that rill and interill erosion

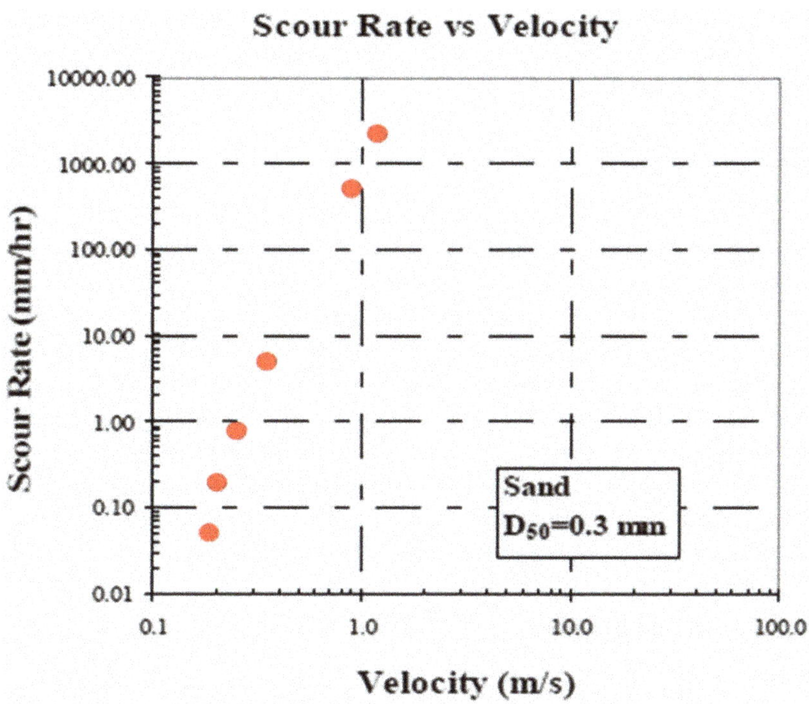

Figure 7 Scour rate vs. velocity to express erodibility of a sandy soil (from Briaud and Chen, 2005).

at the hillslope scale can be predicted successfully from laboratory-scale measurements of erodibility.

3.3.7 Sonication or ultrasonic vibration

Sonication creates cavitation, the formation of an empty space or tiny bubbles within soil aggregates (Almajmaie et al., 2017; Zhu et al., 2017). As these bubbles expand, they break up the aggregate along the lines of weakness, into smaller sub-aggregates or primary particles. The ultrasonic energy required to liberate and disperse soil aggregates is the index of erodibility (Schmidt et al., 1999; Schomakers et al., 2011; Zhu et al., 2009, 2010; Fristensky and Grismer, 2008; Imeson and Vis, 1984; Mentler et al., 2004; Lehtinen et al., 2014).

3.4 Dynamic field tests of assessing soil erodibility

Prediction of hillslope-scale soil erosion rates requires extensive data collection from field erosion plots under natural or simulated rainfall. The high costs of installing and maintaining these plots, and the inconvenience associated with fieldwork favours the development of much simpler and cost-effective (yet still field based) soil erodibility indices.

3.4.1 Field test kits

Herrick et al. (2001) describe a soil stability kit that is inexpensive and easy to fabricate (Fig. 8). Up to 18 samples can be tested in less than 10 min on-site, minimizing damage to soil samples during transfer to the lab. Soil samples are rated on a scale from one to six, based on visual observations of slaking during the first 5 min following immersion in distilled water, and the per cent remaining on a 1.5-mm sieve after five dipping cycles at the end of the 5-min period. The field stability classes correlate well with per cent aggregate stability using laboratory-based techniques. Herrick et al. (2001) found the method to be highly sensitive, and although it is no substitute for careful laboratory-based measurements of soil aggregate stability, it can provide valuable information when these more intensive procedures are not possible.

3.4.2 Soil erosion plots

Borselli et al. (2012) argue that one of the best estimates of the soil erodibility factor still remains the Wischmeier et al.'s (1971) formula (or nomograph), which is largely based on field-data for medium-textured soils in the United States. Wischmeier and Mannering (1969) used field soil erosion plots measuring 1.8 m × 22 m, on 9% slopes, to measure and then correlate soil properties with soil loss measurements. Of all the soil properties measured, highest correlations with observed soil losses involved: silt + very fine sand %; sand %; soil organic

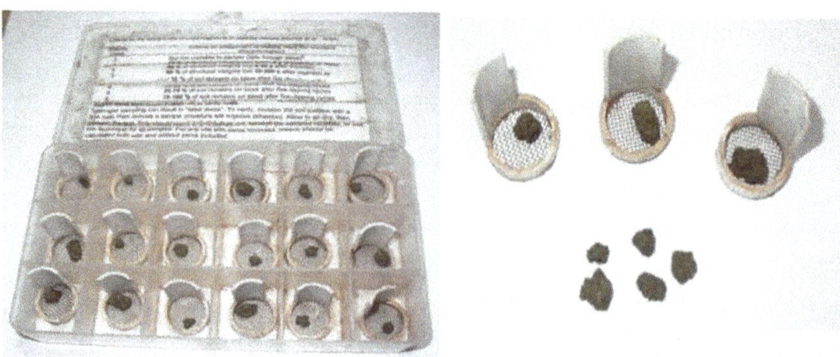

Figure 8 Field test kit (FTK) apparatus. Test box dimensions 21 × 10.5 × 3.5 cm. Sieve size 2 cm diameter and 1.65 cm aperture (from Cooper, 2006).

matter %; soil structure and permeability. An equation and nomograph were derived from these empirical findings:

$$100\,K = 2.1 \times 104 \times (2 - OM) \times M1.14 + 3.25 \times (St - 2) + 2.5 \times (Pt - 3) \quad (6)$$

Where

K = erodibility factor, expressed in soil loss (ton / tonne) per unit of rainfall erosivity (EI_{30})
(Wischmeier and Smith, 1978)
OM = Organic matter content (%)
M = Silt plus fine sand content (%)
St = Soil structure code (very fine granular = 1, fine granular = 2, coarse granular = 3, blocky, platy or massive = 4),
Pt = Permeability class (rapid = 1, moderate to rapid = 2, moderate = 3, slow to moderate = 4, slow = 5, very slow = 6).

The K factor is used in the Universal Soil Loss Equation (USLE; Wischmeier and Smith, (1978)) and is probably the most well-known index of erodibility. It has the advantage of being based on soil properties that are relatively easy to measure. However, it derives from a limited soil database, so may not apply to other geographical locations. Parsons et al. (1994) found field-derived K-factors using a portable field rainfall simulator were 38% of that of the average K-factor from the Wischmeier soil erodibility nomograph. Jaiyeoba and Ologe (1990) tested soil erodibility in the Zaria area, Nigeria, where values were consistently lower than those predicted by the erodibility nomograph. Vanelslande et al. (1987) tested the erodibility of 55 surface soils in different regions of Nigeria using a rainfall simulator, and compared results with K estimates from the Wischmeier nomograph (Wischmeier et al., 1971). The actual measured erodibility was greater than that estimated, often by several orders of magnitude. In soils with high silt content, however, the estimated K was slightly higher than the measured K.

Despite its continuing popularity, the K-factor has been criticised. Vanelslande et al. (1987) noted that erosion processes such as slaking and surface sealing were not accounted for in the K nomograph. Song et al. (2005) found the K-factor did not relate rainsplash or run-off. Centeri (2002) warns of extrapolating empirical relations from one set of climatic and soil conditions to other geographical areas. The K-factor has been shown to be poorly correlated with aggregate stability. As with all USLE factors, it is a mean annual estimate of erodibility, so any seasonal variability is not taken into account.

Anache et al. (2015) argue that direct assessment of erodibility in the field is important to validate models, but is expensive. To overcome this expense, some researchers have developed 'pedo-transfer functions' (PTFs) to predict the K-factor, using regression tree (RT) and multiple linear regression (MLR). The results showed that the mean K value predicted by the USLE and RUSLE was 2.08 and 2.84 times more than the measured K, respectively (Ostovari et al., 2019).

Another example of deriving soil erodibility indices from field plots can be found in Laflen et al. (1991). Here, erodibility indices for interrill (K_i) and rill erosion (K_r) were derived, showing consideration of the different processes operating. Under simulated rainfall and overland flow (the latter to increase the shear stress of the water), sites on cropland and rangeland were selected. Interrill erodibility was determined by measuring soil loss rates, taking account of the rainfall intensity and a slope factor. Rill erodibility and the critical hydraulic shear stress needed to detach material in the rills were determined by subjecting the plot surface to varying levels of rainfall and surface flow, and measuring the resulting erosion. It is interesting that there was virtually no correlation between the USLE K-factor and the interrill erodibility values (K_i). This is explained by the fact that the USLE K-factor encapsulates many processes (e.g. infiltration, detachment and transport), whereas by separating the different erosion sub-processes, WEPP gives very different and unrelated soil erodibility values (Laflen et al., 1991).

3.5 Critical analysis of different methods of assessing soil erodibility

Numerous methods have been used to determine erodibility, but the lack of standardised procedures and instrumentation makes comparison of results challenging (Almajmaie et al., 2017; Loveland and Webb, 2003). Tests are performed at various spatial scales from individual aggregates, microplots, larger run-off plots to hillsides (Barthès and Roose, 2002). Detachment processes dominate in rainfall/raindrop-based tests: slaking dominates in immersion tests such as wet sieving. Immersing soil under water (either intermittently or throughout the test) is not representative of field conditions. Sonication is a very effective means of disaggregating soils, but field soils are rarely exposed to ultrasonic vibration. Different types and levels of disruptive energy are used in erodibility tests, which may or may not represent field conditions. Many techniques deal with disturbed soil (e.g. air dried, destoned and sieved) whose behaviour may not represent field conditions and processes (Song et al., 2005). Loveland and Webb (2003) report that there is little uniformity in sample pre-treatment such as the water content prior to measurement. There is no general agreement as to the size range or size intervals of aggregates used in erodibility tests.

A number of studies have compared different methods of assessing soil erodibility to determine consistency of results. Legout et al. (2005) compared different aggregate stability measurement methods under rainfall and found similar results. Imeson and Vis (1984) compared the water-drop, wet sieving and ultrasonic dispersion methods for a wide range of soils from Colombia, Kenya, Lesotho, Luxembourg and Spain. The water drop and ultrasound methods gave consistent and reproducible results. The former gave best resolution for the most unstable soils, but was unsuitable for the more stable soils.

Cooper (2006) compared three methods of aggregate stability test to compare the erodibility of soils under different intensities of tillage (conventional, a minimum tillage treatment 'SOWAP'; and a farmers' preference treatment (with moderate cultivation intensity)). Cooper (2006) used rainfall simulation, wet sieving and a field test kit (Fig. 9). The relationship between tillage treatment and aggregate stability varied between the methods used. This raises the important question of which method is the most appropriate to assess treatment differences in terms of aggregate stability. The field test kit (FTK) and the wet sieving methods produced similar results. This is not surprising as the FTK method is a simplified version of wet sieving, and both are immersion-based methods. However, the rain-drop impact method generated no treatment differences, unlike the other methods.

Bryan (1968) compared different soil erodibility indices, noting the validity and limitations of different tests, and some inconsistent results. The indices were compared to soil loss from samples subjected to simulated rainfall. The most efficient indices of soil erodibility were: percentage weight of water-stable aggregates (WSA) >3 mm; percentage-weight of WSA > 0.5 mm; erosion ratio;

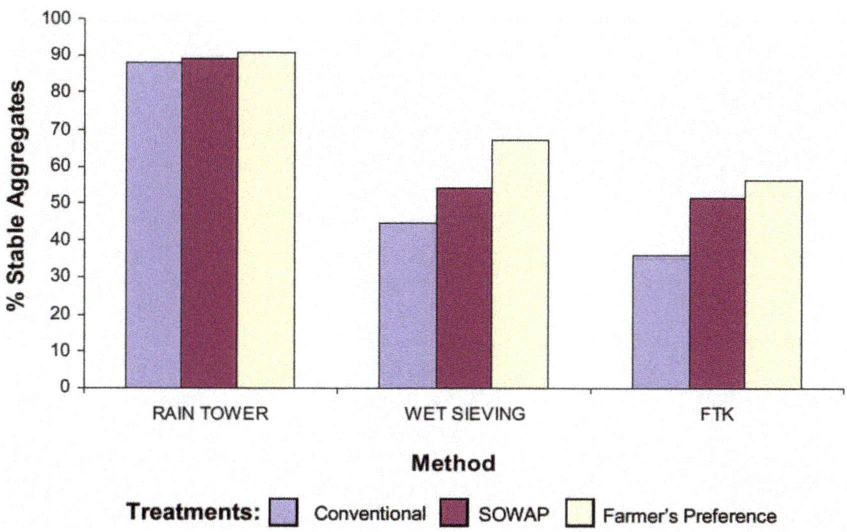

Figure 9 Comparing tillage treatment effect with three different methods of aggregate stability testing (Cooper, 2006).

surface-aggregation ratio; modified surface-aggregation ratio; and clay ratio. Further statistical analysis concluded that the percentage weight of WSA > 3 mm was 'the only completely reliable index of erodibility'.

Deviren Saygin et al. (2012) compared six different aggregate stability approaches (wet sieving, dry sieving, pre-wetting, slow wetting, fast wetting and mechanical breakdown) by calculating a detachment index (DI) for each approach. DI was the ratio of mean weight diameter (MWD) of aggregates after wet sieving (MWDW) to MWD before wet sieving (MWDD). The results varied significantly, depending on the method used. This was explained by the magnitude and extent of the disruptive forces applied during the different tests (Table 1).

Obalum et al. (2019) compared eight different erodibility assessments for their ability to discriminate changes in erodibility due to soil treatments (tillage type and compost additions). The indices included: mean-weight diameter after dry (MWDd) and wet (MWDw) sieving; sand-corrected water-stable aggregates (%WSAcfs); an aggregate stability index (ASI); percolation stability (PS); sealing index (SI); and tensile strength of 24–16, 16–8 and 8–4 mm air-dry (TSad) and re-moistened (TSrm) aggregates. Four of the indices discriminated between different tillage systems. The indices PS, %WSAcfs and MWDd performed best at predicting erosion. The inclusion of MWDd was unexpected as this is

Table 1 Summary of the applied methodologies for evaluating aggregate stability (from Deviren Saygin et al., 2012)

Type of methodology	Initial aggregate sizes before wet sieving	Treatments before wet sieving	Sub-sieve sizes used to obtain MWD value
Wet sieving with single sieve (M_I)	1-2 mm	Pre-wetting with distilled water by spraying to protect slaking effect	0.25 mm
Modified multiple wet sieving (M_{II})	5-8 mm	Pre-wetting with distilled water by spraying to protect slaking effect	5, 4, 2, 1, 0.50, 0.25, 0.125 and 0.063 mm
Multiple wet sieving (M_{III})	4.45 mm	Pre-wetting with distilled water up to field capacity	4.76, 2.83, 2.00, 1.00, 0.5 and 0.3 mm
Fast wetting (M_{IV})	3-5 mm	Direct immersion of soil aggregates into water	2, 1, 0.5, 0.2, 0.1 and 0.05 mm
Slow wetting (M_V)	3-5 mm	Pre-wetting of soil aggregates at matric potential	2, 1, 0.5, 0.2, 0.1 and 0.05 mm
Mechanical breakdown (M_{VI})	3-5 mm	Immersion of soil aggregates into ethanol to test their mechanical cohesion without slaking	2, 1, 0.5, 0.2, 0.1 and 0.05 mm

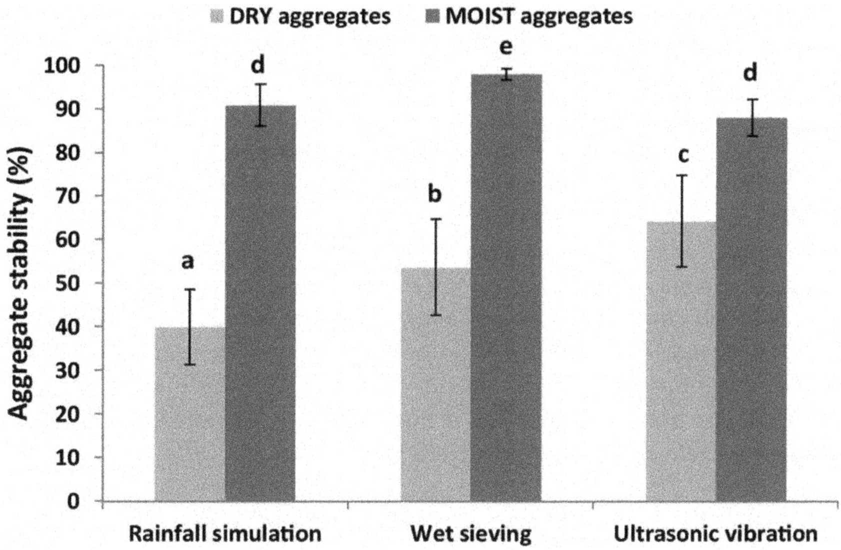

Figure 10 Influence of initial moisture content and method of disaggregation on measured values of aggregate stability. Error bars represent ± standard deviation (From Almajmaie et al., 2017).

usually associated with wind erosion. The inconsistent results were explained by the different disruptive forces and mechanisms of aggregate breakdown (Almajmaie et al., 2017).

Almajmaie et al. (2017) also considered the results of different aggregate stability tests, relating these to the mechanisms responsible for aggregate breakdown. Soil aggregates were exposed to rainfall simulation (RS), wet sieving (WS), ultrasonic vibration (UV) and clay dispersion (CD). Aggregate stability was significantly different for the different methods (Fig. 10), again because of the different mechanisms of aggregate breakdown, and thus the type of disruptive energy applied to the aggregates significantly differed between procedures. Consequently, selection of procedures to measure aggregate stability need to consider the type of disruptive forces and breakdown processes to which field aggregates are subjected.

4 Future trends in research

4.1 Accounting for spatial variability in erodibility

The traditional concept of a spatially universal index of erodibility is flawed. Roose and Sarrailh (1989) showed tropical soil erodibility is different from that in temperate areas, namely ferrallitic and vertisols are less erodible than some alfisols in loessial, temperate regions. Veihe (2002) found difficulty in selecting a suitable erodibility index for soils in the savanna ecosystems, because the majority of existing indices are developed for temperate soils. In West Africa, where kaolinite is the dominant clay mineral, the 'K' erodibility index did not work for vertisols characterised by expanding clay minerals and for soils with a high-coarse sand fraction (Roose, 1977; Vanelslande et al., 1987). Often the 'K' factor is underestimated outside the United States (Vanelslande et al., 1987; Jaiyeoba and Ologe, 1990; Igwe and Mbagwu, 1995; Torri et al., 1997). This implies that erodibility indices have to be regionally specific.

There are very few studies on urban soil erodibility (Kant et al., 2018) or on technosols (artificially created soils) (Deeb et al., 2017; Vidal-Beaudet et al., 2018), and whether these vary from erodibility of agricultural soils. Most assessments have been done on agricultural soils, where sheetwash and rainsplash processes are often more dominant than on non-agricultural soils (Almajmaie et al., 2017; Bryan et al., 1989).

4.2 Accounting for temporal variability in erodibility

Some studies try to capture the variability of erodibility over time, often a result of changing soil moisture conditions (Rejman et al., 2008; Young et al., 1990). Valmis et al. (2005) found aggregate stability increases from April to July, whereas after September it starts to decrease, resulting in an increased susceptibility to erosion especially during the period of the intensive autumn rains.

4.3 Accounting for different erosion processes

There are many processes leading to aggregate breakdown; however, no one method adequately simulates the four main mechanisms (slaking; differential swelling; mechanical breakdown by raindrop impact; and physico-chemical dispersion (Le Bissonnais, 1996, 2016)). For this reason, Almajmaie et al. (2017) do not support efforts to standardise aggregate stability tests. They recommend the use of multiple methods and moisture pre-treatments, with adjustment of test durations to achieve 40–60% aggregate breakdown in order to best reveal differences between sites, treatments and/or erosion processes operating. They recommend that if the processes of aggregate breakdown at a particular site are caused by rainfall or irrigation, then rainfall simulation methods should be used. If soils are likely to be eroded by flood waters, then wet sieving techniques should be used. Bryan et al. (1989) also argue that different properties determine soil erodibility for each erosional sub-process and erodibility can only be defined for precisely identified processes and erosive forces. Obalum et al. (2019) argue that the methodology should be chosen to mimic the processes of aggregate breakdown in the field.

In places, a combination of erosion by water and by wind may be operating. There are no erodibility indices or methodologies for assessing soil erodibility to both processes simultaneously. However, Obalum et al. (2019) were surprised to find one index (mean weight diameter, MWDd) was able to reflect erodibility for both water and wind.

It has been argued that the products of aggregate breakdown are more geomorphologically important than the remaining stable aggregates after breakdown (Farres, 1987). The size and mean weight diameter of the detached material affects its transportability (Fig. 2), resulting in higher erosion rates (and by implication higher erodibility).

4.4 Climate change and soil erodibility

Ongoing global warming causes increases in air and soil temperatures (IPCC, 2019). Shifts in climate could cause changes in soil properties such as soil surface roughness, aggregate stability, sealing and crusting, and thus soil erodibility (Nearing et al., 2004). The analysis of global soil erodibility data by Sanchis et al. (2008) showed how predicted prevailing cool or warm conditions (defined using the Köppen climate classification) would affect soil erodibility values.

It is expected that clay soils will be less erodible in a warm climate, as they tend to have much larger infiltration rates where cracks remain open for longer. At the same time, soil organic matter will be mineralised more quickly resulting in a lower soil aggregate stability. Soil biological composition and activity in warmer winters may be increased, strengthening aggregate bonds and stability. Evidence suggests that shifts in microbial composition affect enzyme production and activity, influencing the degradative capability of microbes and the decomposition rates of organic compounds (i.e. different types of organic matter) (Waldrop et al., 2000; Zogg et al., 1997). Even so, few studies directly show how changing microbial community composition and decomposition rates of specific compounds affect aggregate stability. It is also unclear how the varying decomposition rates of organic compounds mechanistically reduce aggregate stability. Despite much research, a consensus on the temperature sensitivity of soil organic matter decomposition is yet to be reached (Davidson and Janssens, 2006).

Drying and wetting cycles have been reported to both increase and decrease soil aggregate stability (Bullock et al., 1988; Dagesse, 2013; Denef et al., 2001). The number of cycles has been shown to influence the response and development of the soil microbial community (Evans and Wallenstein, 2012; Henry, 2007). Cosentino et al. (2006) investigated the effects of four drying-wetting cycles on microbial properties and aggregate stability. Or et al. (2007) shows microbial activity (and thus contributions to aggregate stability) is limited by organism motility, in turn linked to moisture content.

Drying and wetting cycles mimic the predicted increase in the frequency of extreme rainfall events. The effect of repeated rainfall events on aggregate stability is lacking in the literature (Alegbeleye, 2019). Aggregates deemed 'stable' after one rainfall event may become unstable when subjected to subsequent rainfall events. These gaps in mechanistic understanding of processes may limit accurate prediction of erodibility and erosion risk on a field scale using conventional aggregate stability test results obtained from a single rainfall event. Alegbeleye (2019) tested the aggregate stability of a sandy loam soil subjected to temperature and moisture content treatments (representing different climate change scenarios). The aggregates were subjected to three consecutive, simulated rainfall events. Weight of aggregates before and after rainfall impact was taken to indicate aggregate stability, expressed as a coefficient of vulnerability (Kv). Mean values of Kv varied under the different experimental variables of moisture content, temperature and rainfall scenarios (Fig. 11).

Özgül et al. (2011) report on the increased frequency of soil freezing and thawing cycles during the winter in cool-temperate and other high-latitude regions, and the impact on wet aggregate stability (WAS). WAS increased with increases in freeze thaw cycles (FFTC) from 3 cycles to 6, by 18% to 113%, but decreased after that in all the soil freeze-thaw treatments by 2% to 25%, depending on soil type. An accompanying field study showed that global climate changes of recent decades affected WAS.

Changes in temperature, combined with soil moisture, will influence aggregate stability directly through physical and chemical processes, but may also affect biological processes, by stimulating or limiting the size and activity

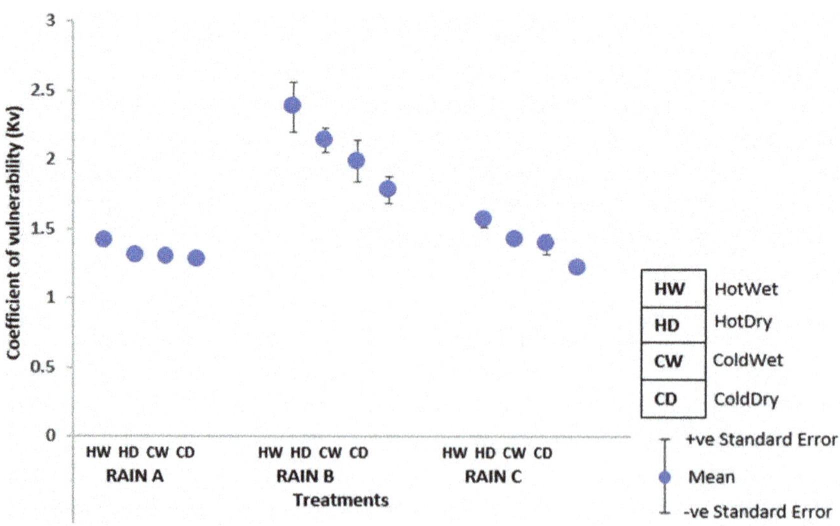

Figure 11 Mean index of vulnerability (Kv) for different climate change scenarios under three rainfall events (Rain A = after 5 days; Rain B = after 5 and 15 days; Rain C = only after 15 days) (from Alegbeleye, 2019).

of the microbial community. Dowdeswell et al. (2018) investigated the effects of different temperature and moisture content regimes on soil aggregate stability at the microcosm scale. The study sought to investigate the effects of climate on aggregate stabilising and destabilising processes (physical, chemical and biological). The research demonstrated how changing climate conditions, and future climate change scenarios, alter soil aggregate stabilisation processes, soil structure and erodibility. Despite these examples, there are many research opportunities in this study area (Borselli et al., 2012).

4.5 Experimental methodologies: 2D versus 3D aggregate layers

Six et al. (2004) notes the different scales and dimensions at which erodibility mechanisms operate and the natural heterogeneity of the soil as a three-dimensional, porous system. Even so, little work has focussed on the variations in aggregate stability when the aggregates are arranged in single or multiple layers. Alegbeleye (2019) hypothesised that aggregates' stability will vary between 2D and 3D layers. This is because the spaces between aggregates may reduce in the 3D arrangement due to smaller aggregates filling them. Also, aggregates in multiple layers have more surrounding aggregates than when arranged in a single layer, so they may experience more protection from the disruptive effect of external forces such as rainfall.

Using an index of vulnerability (Kv), the results from the tests showed the 3D aggregates were on average 1.7 times more stable than the 2D aggregates and results were much less variable between replicates (Fig. 12). The change in aggregate size distribution before and after the rainfall event also revealed differences in stability due to arrangement of the aggregates. The correlation between pre- and post-rainfall aggregate size distribution for the 3D aggregates was on average +0.6, while it was −0.8 for the 2D aggregates. The results suggest the stability of individual soil aggregates in a single layer may not scale up to the field condition. This is a significant finding as many estimates of erodibility have been based on individual aggregates in isolation or multiple aggregates arranged in a 2D layer. This may undermine the accuracy of erosion models and erosion risk assessments using these estimates of erodibility.

4.6 New applications of analytical techniques for soil erodibility assessment

Oades (1993) describes a number of novel techniques to study aggregate structure and stability, including optical and fluorescence microscopy; scanning electron microscopy including EDAX (energy dispersive spectroscopy); transmission electron microscopy using heavy metals or other electron-dense staining techniques for specific chemical compounds; and computer-aided

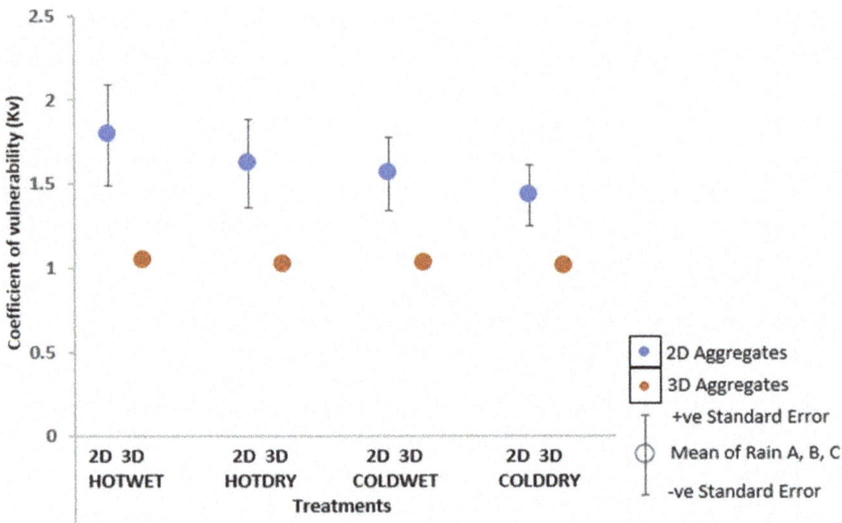

Figure 12 Differences in index of vulnerability (Kv) for 2D and 3D aggregates under different climate change scenarios (from Alegbeleye, 2019).

tomography. These microscopic techniques can be used on individual aggregates, stabilised soils, sections or separates of soils.

4.6.1 Computer-aided tomography (CT) scanning of aggregates

The application of X ray CT-scanning in plant and soil science is flourishing (Lafond et al., 2015) and offers the potential to document soil structure temporally in a non-destructive manner. For example, Grevers et al. (1989) used CT scanning to characterise soil macroporosity, and Garbout et al. (2013) used X-ray CT scanners to monitor the volume, surface area and sphericity of aggregates. Juyal et al. (2019) used CT tomography and fluorescence microscopy to observe the spatial distribution of microbes across 2D and 3D sites.

However, few studies consider the use of CT scanning to assess aggregate stability or erodibility explicitly. Yu et al. (2017) used the technique to investigate the interactions between pore structure and cementing substances within aggregates. Ashenaabad et al. (2019) used CT scanning to observe aggregate micro-cracking, slumping and coalescence following wetting and drying cycles. Papadopoulos et al. (2009) used X-ray μCT to compare stable and unstable soil aggregates from organically and conventionally managed soils that have similar organic matter contents to determine the role of soil management in aggregate stability. They found cracks and elongated pores connected to the surface of stable soil aggregates. These fissures provided an escape route for entrapped air, thus reducing pressure build-up and potential destabilisation of aggregates through slaking.

4.6.2 Visible and near-infrared reflectance spectroscopy

Visible-near-infrared (Vis-NIR) spectroscopy has been used to evaluate a number of physical, biological and chemical soil properties. It is fast, cost-effective, and non-invasive (Ostovari et al., 2018). A number of studies have measured soil properties related to erodibility using spectral reflectance values. Ostovari et al. (2019) predicted the soil erodibility index 'K' from spectral reflectance values for soil organic matter (SOM) and calcium carbonate equivalent ($CaCO_3$) in calcareous soils in southern Iran. A derived 'Spectrotransfer Function' (STF) performed better than the USLE K factor in estimating soil erodibility.

Wang et al. (2016) used hyperspectral visible and near-infrared reflectance spectroscopy to measure factors affecting erodibility (SOM, WSA and the geometric mean radius (D_g) of aggregates). Using a quantitative wavelet analysis model, erodibility (measured in laboratory simulation experiments) was estimated for two contrasting soils from multiple spectral models of these factors affecting erodibility. The study led to a new equation for soil erodibility, derived from the spectral models, providing 'new insights into emerging dynamic variations between erodibility estimations'.

4.6.3 Fractals

A fragmentation fractal dimension, D_f, was used to characterise aggregate-size distribution, as an index of soil erodibility, of five soils with contrasting properties from semi-arid and temperate areas (Martínez-Mena et al., 1999). The D_f was the slope of the regression of aggregate number versus log of aggregate diameter/diameter of the largest aggregates, using scale-variant bulk density and assuming a constant shape for all the aggregates sizes. D_f values correlated with a 'rainfall simulation survival index' (RSSI) and soil loss suggesting that the fragmentation fractal dimension was an useful index of erodibility. Furthermore, the authors argued that fragmentation fractal parameters can be more useful than stability parameters in describing soil erodibility, because they can describe other factors affecting erodibility such as surface roughness. The results related to only a few soil types, so additional experiments are still needed to generate Df values for a wider range of soils. They also need to be compared to aggregate stability results from other methodologies.

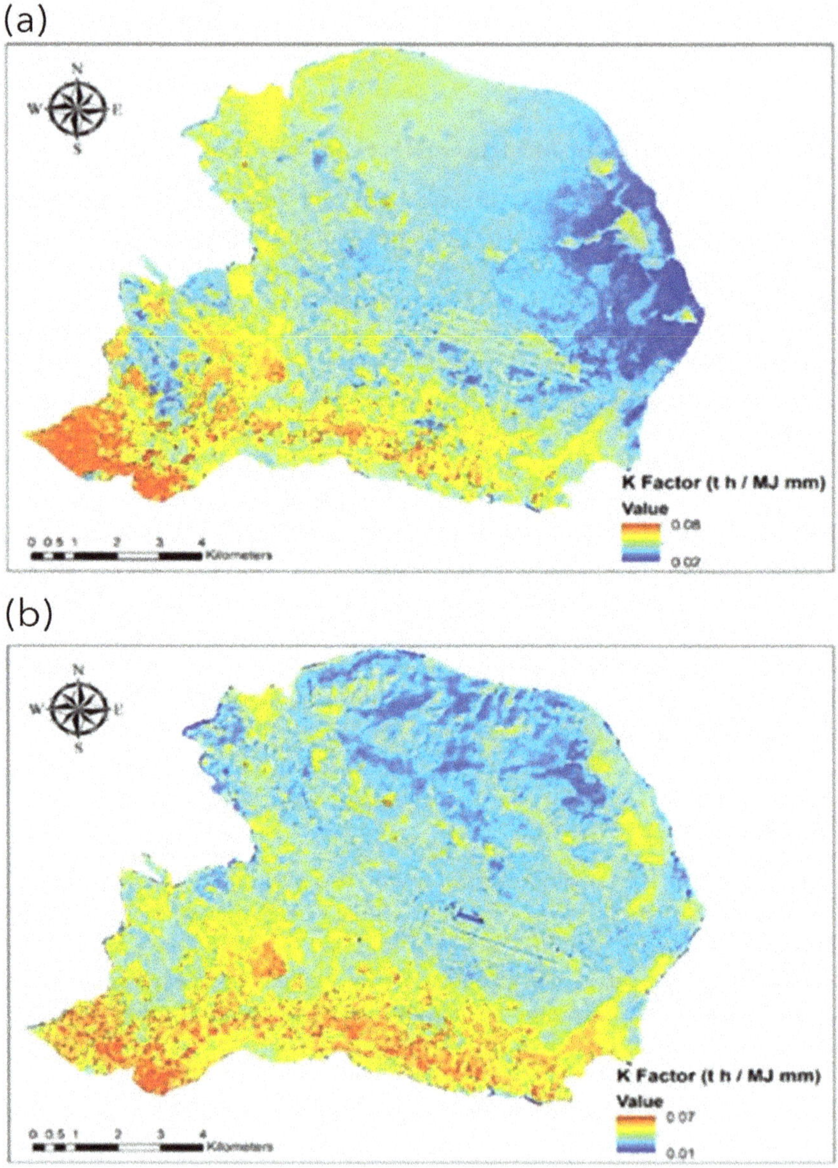

Figure 13 Spatial distribution of K-factor as estimated from the integrated use of ANN and (a) Landsat 8 and (b) Sentinel-2 imageries (from Alexakis et al., 2019).

4.6.4 Use of artificial neural networks (ANNs) in erodibility assessments

Artificial neural networks were used to integrate several innovative methods (satellite remote sensing, field spectroscopy, soil chemical analysis and GIS) to assess soil erodibility and the factors affecting it, namely SOM and $CaCO_3$ (Alexakis et al., 2019). The study presents the potential of an ANN inversion model for the estimation of the soil erosion phenomenon with the use of earth observation data from satellite sensors, Sentinel-2 and Landsat 8 and field spectral VIS-NIR data. The output (the 'K' factor of the USLE) was mapped (Fig. 13). The authors proposed that this will replace conventional methods of soil data collection that are often impractical, expensive and time-consuming. Park and Egbert (2005) also combine digital elevation models, GIS and remote sensing to assess soil erodibility indices for conservation reserve programme lands in Southwestern Kansas.

5 Conclusion

Soil erodibility plays a significant role in soil erosion processes that have been identified as major threats to soil resources, and the ecosystem goods and services they deliver to society (European Commission, 2006). However guidelines for selection or standardisation of procedures to assess erodibility are lacking (Le Bissonnais et al., 1993; Herrick et al., 2001)).

Current methods for assessing soil erodibility are often poorly correlated (Almajmaie et al., 2017). This is because there is no consensus as to the salient factors affecting erodibility, or how to measure them. There are a vast number of potential factors, including soil physical, chemical and biological properties, land management and climate. Also, erodibility has to reflect a soil's resistance/susceptibility to a range of processes operating in the field, including rainsplash detachment, clay dispersion, aggregate breakdown and run-off shear forces.

As a result, Bryan (1968) suggests 'it seems unlikely that a universal index of soil erodibility can be developed'. Instead, it is recommended that the choice of procedure should aim to mimic the types and levels of disruptive energy experienced under field conditions (Almajmaie et al., 2017). It is also recognised that as well as soil properties, erodibility will be influenced by complicated and complex interactions, and interdependencies between soil fauna, microbes, roots, inorganic binding agents, environmental factors, and the factors affecting soil hydrology and erosive forces (Almajmaie et al., 2017). Simple correlations cannot capture these, or inform about the importance of single factors and the interactive effects between factors (Six et al., 2004).

Geeves et al. (2000) and Six et al. (2004) argue that erodibility should be viewed as a dynamic continuum operating over three dimensions, rather than a stable property of the soil surface. Finding a better assessment of soil erodibility is essential because 'the greatest obstacle to soil erosion modelling at larger spatial scales is the lack of data on soil characteristics' (Panagos et al., 2014).

6 Where to look for further information

An introduction to the topic of soil erodibility can be found in an excellent review of erodibility indices in Bryan (1968). Song et al. (2005) extend this further by including the susceptibility of soil to wind erosion. Amézketa (1999) provides a comprehensive discussion of soil aggregate stability. The most widely used soil erodibility indicator is the 'K' factor of the Universal Soil Loss Equation. While originally devised in the United States, east of the Rocky Mountains, it is still frequently used throughout the world. The genesis of the 'K-factor' is given in Wischmeier and Mannering (1969), but readers will notice that there is no consideration of the biological soil properties that are known to influence aggregate stability and soil erodibility.

Soil erodibility studies and recent scientific and methodological developments are reported in a number of international, peer reviewed journals. These include (but are not limited to) Geoderma; Catena; Soil and Tillage Research; Pedosphere; European Journal of Soil Science; Soil Science Society of America Journal; Transactions of the American Society of Agricultural and Biological Engineers; Soil Use and Management; and Earth Surface Processes and Landforms. Advances in the mechanistic understanding of soil erodibility are regularly presented in papers and posters at international conferences such as the European Geosciences Union (EGU) General Assembly and the World Congress of Soil Science, the latter is held every four years.

The Food and Agricultural Organisation of the United Nations (FAO) supports the Global Soil Partnership (GSP; http://www.fao.org/global-soil-partnership/en/), founded in December 2012 as a mechanism to develop a strong interactive partnerships, enhanced collaboration and synergy of efforts between soil scientists. Regional and national soil science societies with interests in soil erodibility research are listed on the GSP pages (http://www.fao.org/global-soil-partnership/pillars-action/2-awareness-raising/soil-science-societies/en/).

7 References

Abiven, S., Menasseri, S. and Chenu, C. (2009). The effects of organic inputs over time on soil aggregate stability–A literature analysis, *Soil Biology and Biochemistry*, 41(1) Elsevier, 1–12.

Agassi, M. and Bradford, J. M. (1999). Methodologies for interrill soil erosion studies, *Soil and Tillage Research*, 49(4) Elsevier, 277–287.

Alegbeleye, G. (2019). *Assessing and Comparing the Effects of Climate Change on Soil Aggregate Stability in Single and Multiple Soil Layers*. Unpublished M Sc thesis, Cranfield University, Bedfordshire, UK.

Alexakis, D. D., Tapoglou, E., Vozinaki, A. K. and Tsanis, I. K. (2019). Integrated use of satellite remote sensing, artificial neural networks, field spectroscopy, and GIS in estimating crucial soil parameters in terms of soil erosion, *Remote Sensing*, 11(9), 1106.

Algayer, B., Wang, B., Bourennane, H., Zheng, F., Duval, O., Li, G., Le Bissonnais, Y. and Darboux, F. (2014). Aggregate stability of a crusted soil: differences between crust and sub-crust material, and consequences for interrill erodibility assessment: an example from the Loess Plateau of China, *European Journal of Soil Science*, 65(3), 325–335.

Almajmaie, A., Hardie, M., Acuna, T. and Birch, C. (2017). Evaluation of methods for determining soil aggregate stability, *Soil and Tillage Research*, 167, 39–45.

Amézketa, E. (1999). Soil aggregate stability: a review, *Journal of Sustainable Agriculture*, 14(2–3), 83–151.

Amezketa, E., Aragüés, R., Carranza, R. and Urgel, B. (2003). Macro-and micro-aggregate stability of soils determined by a combination of wet-sieving and laser-ray diffraction, *Spanish Journal of Agricultural Research*, 1(4), 83–94.

Amezketa, E., Singer, M. J. and Le Bissonnais, Y. (1996). Testing a new procedure for measuring water-stable aggregation, *Soil Science Society of America Journal*, 60(3) Soil Science Society of America, 888–894.

Anache, J. A. A., Bacchi, C. G. V., Panachuki, E. and Alves Sobrinho, T. (2015). Assessment of methods for predicting soil erodibility in soil loss modeling, *Geociencias*, 34(1), 32–40.

Annabi, M., Houot, S., Francou, C., Poitrenaud, M. and Le Bissonnais, Y. L. (2007). Soil aggregate stability improvement with urban composts of different maturities, *Soil Science Society of America Journal*, 71(2), 413–423.

Ashenaabad, R. H., Larsbo, M., Isekog, D., Jarvis, N., Hallett, P., Dathe, A. and Keller, T. (2019). Quantification of post-tillage soil structure dynamics via dynamic micro-CT imaging. *Geophysical Research Abstracts*, 21 EGU General Assembly, EGU2019-16213.

ASTM (2013). *Standard Test Methods for Identification and Classification of Dispersive Clay Soils by the Pinhole Test*. ASTM International, West Conshohocken, PA, 2020, www.astm.org. http://www.astm.org/cgi-bin/resolver.cgi?D6572.

Barbosa, R. S., Marques Júnior, J., Barrón, V., Martins Filho, M. V., Siqueira, D. S., Peluco, R. G., Camargo, L. A. and Silva, L. S. (2019). Prediction and mapping of erodibility factors (USLE and WEPP) by magnetic susceptibility in basalt-derived soils in northeastern, São Paulo state, Brazil, *Environmental Earth Sciences*, 78(1), 12.

Barthes, B. and Roose, E. (2002). Aggregate stability as an indicator of soil susceptibility to runoff and erosion; validation at several levels, *CATENA*, 47(2), 133–149.

Bayabil, H. K., Stoof, C. R., Lehmann, J. C., Yitaferu, B. and Steenhuis, T. S. (2015). Assessing the potential of biochar and charcoal to improve soil hydraulic properties in the humid Ethiopian Highlands: the Anjeni watershed, *Geoderma*, 243–244 Elsevier, 115–123.

Bearden, B. N. and Petersen, L. (2000). Influence of arbuscular mycorrhizal fungi on soil structure and aggregate stability of a vertisol, *Plant and Soil*, 218(1–2), 173–183.

Blankinship, J. C., Fonte, S. J., Six, J. and Schimel, J. P. (2016). Plant versus microbial controls on soil aggregate stability in a seasonally dry ecosystem, *Geoderma*, 272, 39-50.

Bones, E. J., Garrow, L. A. and Sturm, T. W. (2017). Incorporating soil erodibility properties into scour risk-assessment tools using HYRISK, *Journal of Infrastructure Systems*, 23(3). doi:10.1061/(ASCE)IS.1943-555X.0000340.

Borselli, L., Torri, D., Poesen, J. and Iaquinta, P. (2012). A robust algorithm for estimating soil erodibility in different climates, *CATENA*, 97, 85-94.

Bossuyt, H., Denef, K., Six, J., Frey, S. D., Merckx, R. and Paustian, K. (2001). Influence of microbial populations and residue quality on aggregate stability, *Applied Soil Ecology*, 16(3) Elsevier, 195-208.

Bouyoucos, G. J. (1935). The clay ratio as a criterion of susceptibility of soils to erosion, *Agronomy Journal*, 27(9), 738-741.

Briaud, J. L. and Chen, H.-C. (2005). The EFA, erosion function apparatus: an overview. Proceedings of the 16th International Conference on Soil Mechanics and Geotechnical Engineering: Geotechnology in Harmony with the Global Environment 2, 479-481.

Briaud, J. L., Ting, F. C. K., Chen, H. C., Cao, Y., Han, S. W. and Kwak, K. W. (2001). Erosion function apparatus for scour rate predictions, *Journal of Geotechnical and Geoenvironmental Engineering*, 127(2) American Society of Civil Engineers, 105-113.

Bronick, C. J. and Lal, R. (2005). Soil structure and management: a review, *Geoderma*, 124(1-2) Elsevier, 3-22.

Bruce-Okine, E. and Lal, R. (1975). Soil erodibility as determined by raindrop technique, *Soil Science*, 119(2), 149-157.

Bryan, R. B. (1968). The development, use and efficiency of indices of soil erodibility, *Geoderma*, 2(1), 5-26.

Bryan, R. B., Govers, G. and Poesen, J. (1989). The concept of soil erodibility and some problems of assessment and application, *CATENA*, 16(4-5), 393-412.

Bullock, M. S., Nelson, S. D. and Kemper, W. D. (1988). Soil cohesion as affected by freezing, water content, time and tillage, *Soil Science Society of America Journal*, 52(3) Soil Science Society of America, 770-776.

Centeri, C. (2002). Importance of local soil erodibility measurements in soil loss prediction, *Acta Agronomica Hungarica*, 50(1), 43-51.

Centeri, C., Szalai, Z., Jakab, G., Barta, K., Farsang, A., Szabó, S. and Bíró, Z. (2015). Soil erodibility calculations based on different particle size distribution measurements, *Hungarian Geographical Bulletin*, 64(1), 17-23.

Chepil, W. S., Siddoway, F. H. and Armbrust, D. V. (1962). Climatic factor for estimating wind erodibility of farm fields, *Journal of Soil and Water Conservation*, 17 (4), 162-165.

Chepil, W. S. and Woodruff, N. P. (1954). Estimations of wind erodibility of field surfaces, *Journal of Soil and Water Conservation*, 9(257-265), 285.

Chorley, R. J. (1959). The geomorphic significance of some Oxford soils, *American Journal of Science*, 257(7), 503-515.

Cooper, S. (2006). *The Role of Conservation Soil Management on Soil and Water Protection at Different Spatial Scales*. Unpublished PhD thesis, Cranfield University, Amended.

Cosentino, D., Chenu, C. and Le Bissonnais, Y. (2006). Aggregate stability and microbial community dynamics under drying-wetting cycles in a silt loam soil, *Soil Biology and Biochemistry*, 38(8) Elsevier, 2053-2062.

Dagesse, D. F. (2013). Freezing cycle effects on water stability of soil aggregates, *Canadian Journal of Soil Science*, 93(4) NRC Research Press, 473–483.

Davidson, E. A. and Janssens, I. A. (2006). Temperature sensitivity of soil carbon decomposition and feedbacks to climate change, *Nature*, 440(7081) Nature Publishing Group, 165–173.

de Oro, L. A., Colazo, J. C., Avecilla, F., Buschiazzo, D. E. and Asensio, C. (2019). Relative soil water content as a factor for wind erodibility in soils with different texture and aggregation, *Aeolian Research*, 37 Elsevier, 25–31.

Deeb, M., Desjardins, T., Podwojewski, P., Pando, A., Blouin, M. and Lerch, T. Z. (2017). Interactive effects of compost, plants and earthworms on the aggregations of constructed Technosols, *Geoderma*, 305, 305–313.

Degens, B. P. (1997). Macro-aggregation of soils by biological bonding and binding mechanisms and the factors affecting these: a review, *Soil Research*, 35(3), 431–460.

Degens, B. P., Sparling, G. P. and Abbott, L. K. (1994). The contribution from hyphae, roots and organic carbon constituents to the aggregation of a sandy loam under long-term clover-based and grass pastures, *European Journal of Soil Science*, 45(4) Wiley Online Library, 459–468.

Demenois, J., Carriconde, F., Rey, F. and Stokes, A. (2017). Tropical plant communities modify soil aggregate stability along a successional vegetation gradient on a ferralsol, *Ecological Engineering*, 109, 161–168.

Denef, K., Six, J., Bossuyt, H., Frey, S. D., Elliott, E. T., Merckx, R. and Paustian, K. (2001). Influence of dry–wet cycles on the interrelationship between aggregate, particulate organic matter, and microbial community dynamics, *Soil Biology and Biochemistry*, 33(12–13) Elsevier, 1599–1611.

Deviren Saygin, S., Cornelis, W. M., Erpul, G. and Gabriels, D. (2012). Comparison of different aggregate stability approaches for loamy sand soils, *Applied Soil Ecology*, 54, 1–6.

Dowdeswell, E., Grabowski, R. and Rickson, R. (2018). Aggregate stability, climate conditions and the underlying physical, chemical and biological (de)stabilising mechanisms, European Geosciences Union, Geophysical Research Abstracts Vol. 21, EGU2019-9565, 2019 EGU General Assembly 2019.

Dumbrovský, M., Larišová, L., Sobotková, V. and Kulihová, M. (2019). Comparison of different texture analysis for soil erodibility calculations of loamy and sandyloam soils in Moravian regions, *Acta Univ. Agric. Silvic. Mendelianae Brun* 67(2), 383–393.

Emerson, W. W. (1967). A classification of soil aggregates based on their coherence in water, *Soil Research*, 5(1) CSIRO, 47–57.

European Commission (2006). Communication from the Commission to the Council, the European Parliament, the European Economic and Social Committee and the Committee of the Regions - Thematic Strategy for Soil Protection [SEC(2006)620] [SEC(2006)1165] /* COM/2006/0231 final */. Brussels.

Evans, S. E. and Wallenstein, M. D. (2012). Soil microbial community response to drying and rewetting stress: does historical precipitation regime matter?, *Biogeochemistry*, 109(1–3) Springer, 101–116.

Farres, P. J. (1987). The dynamics of rainsplash erosion and the role of soil aggregate stability, *CATENA*, 14(1–3), 119–130.

Fristensky, A. and Grismer, M. E. (2008). A simultaneous model for ultrasonic aggregate stability assessment, *CATENA*, 74(2) Elsevier, 153–164.

Garbout, A., Munkholm, L. J. and Hansen, S. B. (2013). Temporal dynamics for soil aggregates determined using X-ray CT scanning, *Geoderma*, 204 Elsevier, 15-22.

Geeves, G. W., Leys, J. F. and McTainsh, G. H. (2000). Soil erodibility. In: *Soils, Their Properties and Management*. Eds. P. E. V. Charman and B. W. Murphy, Oxford University Press, New York, pp. 205-220.

Ghaneeizad, S. M., Karamigolbaghi, M., Atkinson, J. F. and Bennett, S. J. (2015) Hydrodynamics of confined impinging jets and the assessment of soil erodibility, World Environmental and Water Resources Congress 2015: Floods, Droughts, and Ecosystems - Proceedings of the 2015 World Environmental and Water Resources Congress, pp. 1742-1751.

Gheytasi, F., Mohammadian Behbahani, A. L. I., Hosseinalizadeh, M. and Asgari, H. R. (2019). Comparative analysis of soil physico-chemical properties in erodibility of various desert crusts (case study: hills around Aji-Gol lake Golestan Province, Iran), *Desert Ecosystem Engineering Journal*, 7(21), 33-44.

Greenland, D. J., Rimmer, D. and Payne, D. (1975). Determination of the structural stability class of English and Welsh soils, using a water coherence test, *Journal of Soil Science*, 26(3) Wiley Online Library, 294-303.

Grevers, M. C. J., Jong, E. D. E. and St. Arnaud, R. J. (1989). The characterization of soil macroporosity with CT scanning, *Canadian Journal of Soil Science*, 69(3) NRC Research Press, 629-637.

Grismer, M. (2012). Standards vary in studies using rainfall simulators to evaluate erosion, *California Agriculture*, 66(3) University of California, Agriculture and Natural Resources, 102-107.

Guo, Z., Zhang, J., Fan, J., Yang, X., Yi, Y., Han, X., Wang, D., Zhu, P. and Peng, X. (2019). Does animal manure application improve soil aggregation? Insights from nine long-term fertilization experiments, *Science of the Total Environment*, 660, 1029-1037.

Hanson, G. J. and Cook, K. R. (2004). Apparatus, test procedures, and analytical methods to measure soil erodibility in situ, *Applied Engineering in Agriculture*, 20(4) American Society of Agricultural and Biological Engineers, 455.

Henry, H. A. L. (2007). Soil freeze-thaw cycle experiments: trends, methodological weaknesses and suggested improvements, *Soil Biology and Biochemistry*, 39(5) Elsevier, 977-986.

Herrick, J. E., Whitford, W. G., De Soyza, A. G., Van Zee, J. W., Havstad, K. M., Seybold, C. A. and Walton, M. (2001). Field soil aggregate stability kit for soil quality and rangeland health evaluations, *CATENA*, 44(1) Elsevier, 27-35.

Ibrahim, S. L., Ariffin, J., Abdullah, J. and Muhamad, N. S. (2016). Jet erosion device (JED) - measurement of soil erodibility coefficients, *Jurnal Teknologi*, 78(5), 63-67.

Igwe, C. A. and Mbagwu, J. S. C. (1995). The use of some soil aggregate indices to assess potential soil loss in soils of south-eastern Nigeria, *International Agrophysics*, 9(2), 95-100.

Imeson, A. C. and Vis, M. (1984). Assessing soil aggregate stability by water-drop impact and ultrasonic dispersion, *Geoderma*, 34(3), 185-200.

IPCC (2019). *Climate Change and Land: An IPCC Special Report on Climate Change, Desertification, Land Degradation, Sustainable Land Management, Food Security, and Greenhouse Gas Fluxes in Terrestrial Ecosystems*. P. R. Shukla, J. Skea, E. Calvo Buendia, V. Masson-Delmotte, H.-O. Pörtner, D. C. Roberts, P. Zhai, R. Slade, S. Connors, R. van Diemen, M. Ferrat, E. Haughey, S. Luz, S. Neogi, M. Pathak, J. Petzold,

J. Portugal Pereira, P. Vyas, E. Huntley, K. Kissick, M. Belkacemi and J. Malley (Eds). © 2019 Intergovernmental Panel on Climate Change. www.ipcc.ch.

Jaiyeoba, I. A. and Ologe, K. O. (1990). Soil erodibility measurements in Nigeria, *Zeitschrift fur Geomorphologie*, 34(3), 307-312.

Jeffery, S., Meinders, M. B. J., Stoof, C. R., Bezemer, T. M., van de Voorde, T. F. J., Mommer, L. and van Groenigen, J. W. (2015). Biochar application does not improve the soil hydrological function of a sandy soil, *Geoderma*, 251-252, 47-54.

Johannsen, U., Schäfer, W. and Düwel, O. (1998). Estimation of soil erodibility by wind from parameters of the German soil productivity assessment, *Zeitschrift fur Kulturtechnik und Landentwicklung*, 39(1), 23-26.

Juyal, A., Otten, W., Falconer, R., Hapca, S., Schmidt, H., Baveye, P. C. and Eickhorst, T. (2019). Combination of techniques to quantify the distribution of bacteria in their soil microhabitats at different spatial scales, *Geoderma*, 334 Elsevier, 165-174.

Kant, C., Majumder, M., Kumar Tyagi, D. and Prabhat Singh, A. (2018). Estimation of the soil erodibility factor of a peri-urban catchment by group method of data handling and enhanced particle swarm optimization method, *International Journal of Engineering and Technology*, 7(2), 41-45.

Kemper, W. D. and Rosenau, R. C. (1986). Aggregate stability and size distribution, *SSSA Book Series*, 9(9), 425-442.

Kumar, H., Ganesan, S. P., Bordoloi, S., Sreedeep, S., Lin, P., Mei, G., Garg, A. and Sarmah, A. K. (2019). Erodibility assessment of compacted biochar amended soil for geo-environmental applications, *Science of the Total Environment*, 672, 698-707.

Laflen, J. M., Elliot, W. J., Simanton, J. R., Holzhey, C. S. and Kohl, K. D. (1991). WEPP: soil erodibility experiments for rangeland and cropland soils, *Journal of Soil and Water Conservation*, 46(1) Soil and Water Conservation Society, 39-44.

Lafond, J. A., Han, L. and Dutilleul, P. (2015). Concepts and analyses in the CT scanning of root systems and leaf canopies: a timely summary, *Frontiers in Plant Science* 6 Frontiers, 1111.

Le Bissonnais, Y. (1996). Aggregate stability and assessment of soil crustability and erodibility: I. Theory and methodology, *European Journal of Soil Science*, 47(4), 425-437.

Le Bissonnais, Y. (2016). Aggregate stability and assessment of soil crustability and erodibility: I. Theory and methodology, *European Journal of Soil Science*, 67(1), 11-21.

Le Bissonnais, Y., Singer, M. J. and Bradford, J. M. (1993). Assessment of soil erodibility: the relationship between soil properties, erosion processes and susceptibility to erosion, Farm land erosion in temperate plains environment and hills. Proceedings Symposium, Saint-Cloud, Paris, 1992, 87-96.

Legout, C., Leguédois, S. and Le Bissonnais, Y. (2005). Aggregate breakdown dynamics under rainfall compared with aggregate stability measurements, *European Journal of Soil Science*, 56(2), 225-238.

Lehtinen, T., Lair, G. J., Mentler, A., Gísladóttir, G., Ragnarsdóttir, K. V. and Blum, W. E. H. (2014). Soil aggregate stability in different soil orders quantified by low dispersive ultrasonic energy levels, *Soil Science Society of America Journal*, 78(3) The Soil Science Society of America, Inc., 713-723.

Leys, J. F. and McTainsh, G. H. (2002). Soil erodibility to wind. In: *Soils: Their Properties and Management*. Eds P. E. V Charman and B. W. Murphy, Oxford University Press, Melbourne, 213-221.

Loch, R. J. and Pocknee, C. (1995). Effects of aggregation on soil erodibility: Australian experience, *Journal of Soil and Water Conservation*, 50(5) Soil and Water Conservation Society, 504–506.

Loveland, P. and Webb, J. (2003). Is there a critical level of organic matter in the agricultural soils of temperate regions: a review, *Soil and Tillage Research* 70(1), 1–18.

Low, A. J. (1954). The study of soil structure in the field and the laboratory, *Journal of Soil Science*, 5(1) Wiley Online Library, 57–74.

Maksimov, I., Alekseev, V. and Chuchkalov, S. (2019). Erosion resistance potential as a soil erodibility characteristic based on energy approach, *IOP Conference Series: Earth and Environmental Science*, 226 IOP Publishing 2067.

Martínez-Mena, M., Deeks, L. K. and Williams, A. G. (1999). An evaluation of a fragmentation fractal dimension technique to determine soil erodibility, *Geoderma*, 90(1), 87–98.

McCalla, T. M. (1944). Water-drop method of determining stability of soil structure, *Soil Science*, 58(2) LWW, 117–122.

Mentler, A., Mayer, H., Strauß, P. and Blum, W. E. H. (2004). Characterisation of soil aggregate stability by ultrasonic dispersion, *International Agrophysics*, 18(1) Catholic University of Lublin, 39–46.

Mhaske, S. N., Pathak, K. and Basak, A. (2019). A comprehensive design of rainfall simulator for the assessment of soil erosion in the laboratory, *CATENA* 172, 408–420.

Middleton, H. E. (1930). *Properties of Soils Which Influence Soil Erosion. Soil Science Society of America Journal*, B11(2001), 119–21.

Morel, J. L., Habib, L., Plantureux, S. and Guckert, A. (1991). Influence of maize root mucilage on soil aggregate stability, *Plant and Soil* 136(1), 111–119.

Morgan, R. P. C., Quinton, J. N., Smith, R. E., Govers, G., Poesen, J. W. A., Auerswald, K., Chisci, G., Torri, D., Styczen, M. E. and Folly, A. J. V. (1998). *The European Soil Erosion Model (EUROSEM): Documentation and User Guide*, Silsoe College, Cranfield University, p. 124.

Morin, J. and Benyamini, Y. (1977). Rainfall infiltration into bare soils, *Water Resources Research*, 13(5) Wiley Online Library, 813–817.

Mutter, G. M. (2018). Utilization of water turbidity meter devices in estimating the aggregate stability of artificially stabilized soils, *International Journal of Integrated Engineering*, 10(1), 9–16.

Nciizah, A. D. and Wakindiki, I. I. C. (2015). Physical indicators of soil erosion, aggregate stability and erodibility, *Archives of Agronomy and Soil Science*, 61(6) Taylor & Francis, 827–842.

Nearing, M. A., Foster, G. R., Lane, L. J. and Finkner, S. C. (1989). A process-based soil erosion model for USDA-Water Erosion Prediction Project technology, *Transactions of the American Society of Agricultural Engineers*, 32(5), 1587–1593.

Nearing, M. A., Pruski, F. F. and O'Neal, M. R. (2004). Expected climate change impacts on soil erosion rates: a review, *Journal of Soil and Water Conservation*, 59(1), 43–50.

Norton, L. D. (1987). Micromorphological study of surface seals developed under simulated rainfall, *Geoderma*, 40(1–2) Elsevier, 127–140.

Oades, J. M. (1993). The role of biology in the formation, stabilization and degradation of soil structure, *Geoderma*, 56(1–4), 377–400.

Obalum, S. E., Uteau-Puschmann, D. and Peth, S. (2019). Reduced tillage and compost effects on soil aggregate stability of a silt-loam luvisol using different aggregate stability tests, *Soil and Tillage Research*, 189, 217–228.

Or, D., Smets, B. F., Wraith, J. M., Dechesne, A. and Friedman, S. P. (2007). Physical constraints affecting bacterial habitats and activity in unsaturated porous media-a review, *Advances in Water Resources*, 30(6-7) Elsevier, 1505-1527.

Ostovari, Y., Ghorbani-Dashtaki, S., Bahrami, H.-A., Abbasi, M., Dematte, J. A. M., Arthur, E. and Panagos, P. (2018). Towards prediction of soil erodibility, SOM and CaCO3 using laboratory Vis-NIR spectra: a case study in a semi-arid region of Iran, *Geoderma*, 314 Elsevier, 102-112.

Ostovari, Y., Ghorbani-Dashtaki, S., Kumar, L. and Shabani, F. (2019). Soil erodibility and its prediction in semi-arid regions, *Archives of Agronomy and Soil Science*, 65(12), 1688-1703.

Özgül, M., Aksakal, E. L., Günes, A., Angn, İ, Turan, M. and Öztaş, T. (2011). Influence of global warming on aggregate stability and hydraulic conductivity under highland soil order in Turkey, *Soil Science*, 176(10), 559-566.

Panagos, P., Meusburger, K., Alewell, C. and Montanarella, L. (2012). Soil erodibility estimation using LUCAS point survey data of Europe, *Environmental Modelling and Software*, 30 Elsevier Ltd, 143-145.

Panagos, P., Meusburger, K., Ballabio, C., Borrelli, P. and Alewell, C. (2014). Soil erodibility in Europe: a high-resolution dataset based on LUCAS, *Science of the Total Environment*, 479-480(1) Elsevier B.V., 189-200.

Papadopoulos, A., Bird, N. R. A., Whitmore, A. P. and Mooney, S. J. (2009). Investigating the effects of organic and conventional management on soil aggregate stability using X-ray computed tomography, *European Journal of Soil Science*, 60(3), 360-368.

Park, S. and Egbert, S. L. (2005). Assessment of soil erodibility indices for conservation reserve program lands in Southwestern Kansas using satellite imagery and GIS techniques, *Environmental Management*, 36(6), 886-898.

Parsons, T. S., Burney, J. R. and Edwards, L. (1994). Field measurement of soil-erodibility and cover-management factors in Prince Edward Island using simulated rainfall, *Canadian Agricultural Engineering*, 36(3), 127-133.

Pereira, H. C. (1956). Soils, *Journal of Soil Science*, 7(1) Wiley Online Library, 68-75.

Poesen, J. W., Torri, D. and Bunte, K. (1994). Effects of rock fragments on soil erosion by water at different spatial scales: a review, *CATENA* 23(1-2) Elsevier, 141-166.

Puscas, C., Stremtan, C. and Kristály, F. (2010). Past surface conditions and speleogenesis as inferred from cave sediments in the Great Cave of Salitrari Mountain (SW Romania), *Studia Universitatis Babes-Bolyai, Geologia*, 55(2), 51-57.

Qian, F., Lee, D., Nikolich, G., Koditschek, D. and Jerolmack, D. (2019). Rapid in situ characterization of soil erodibility with a field deployable robot, *Journal of Geophysical Research: Earth Surface*, 124(5) John Wiley & Sons, Ltd, 1261 -1280.

Rawlins, B. G., Wragg, J. and Lark, R. M. (2013). Application of a novel method for soil aggregate stability measurement by laser granulometry with sonication, *European Journal of Soil Science*, 64(1) Wiley Online Library, 92-103.

Rejman, J., Brodowski, R. and Iglik, I. (2008). Annual variations of soil erodibility of silt loam developed from loess based on 10-years runoff plot studies, *Annals of Warsaw University of Life Sciences - SGGW. Land Reclamation*, 39 Versita, 77-83.

Rengasamy, P., Greene, R. S. B., Ford, G. W. and Mehanni, A. H. (1984). Identification of dispersive behaviour and the management of red-brown earths, *Soil Research*, 22(4) CSIRO, 413-431.

Ribeiro, A. I., Gonzalez, L. F., Longo, R. M., de Medeiros, G. A., Rosa, A. H., Lourenço, R. W. and Fengler, F. H. (2013). Soil erodibility assessment in a pasture and forest remnant using the inderbitzen device, *WIT Transactions on Ecology and the Environment*, 170, 49–55.

Rohošková, M. and Valla, M. (2004). Comparison of two methods for aggregate stability measurement - a review, *Plant, Soil and Environment*, 50(8), 379–382.

Roose, E. J. (1977). Use of the universal soil loss equation to predict erosion in West Africa. In: *Soil Erosion: Prediction and Control*. Soil Conservation Society of America, Ankeny, IA, 60–74.

Roose, E. and Sarrailh, J.-M. (1989). Erodibility of tropical soils. Twenty years of measurements on runoff plots under natural rainfall, *Cahiers - ORSTOM, Serie Pedologie*, 25(1-2), 7–30.

Russell, M. B. and Feng, C. L. (1947). Characterization of the stability of soil aggregates, *Soil Science*, 63(4), 299–304.

Sanchis, M. P. S., Torri, D., Borselli, L. and Poesen, J. (2008). Climate effects on soil erodibility, *Earth Surface Processes and Landforms: The Journal of the British Geomorphological Research Group*, 33(7) Wiley Online Library, 1082–1097.

Saygın, S. D., Basaran, M., Ozcan, A. U., Dolarslan, M., Timur, O. B., Yilman, F. E. and Erpul, G. (2011). Land degradation assessment by geo-spatially modeling different soil erodibility equations in a semi-arid catchment, *Environmental Monitoring and Assessment*, 180(1-4), 201–215.

Schmidt, M. W. I., Rumpel, C. and Kögel-Knabner, I. (1999). Evaluation of an ultrasonic dispersion procedure to isolate primary organomineral complexes from soils, *European Journal of Soil Science*, 50(1) Wiley Online Library, 87–94.

Schmidt, S., Ballabio, C., Alewell, C., Panagos, P. and Meusburger, K. (2018). Filling the European blank spot–Swiss soil erodibility assessment with topsoil samples, *Journal of Plant Nutrition and Soil Science*, 181(5), 737–748.

Schomakers, J., Mentler, A., Steurer, T., Klik, A. and Mayer, H. (2011). Characterization of soil aggregate stability using low intensity ultrasonic vibrations, *International Agrophysics* 25(2), 165–172.

Sheridan, G. J., So, H. B., Loch, R. J. and Walker, C. M. (2000). Estimation of erosion model erodibility parameters from media properties, *Australian Journal of Soil Research*, 38(2), 265–284.

Six, J., Bossuyt, H., Degryze, S. and Denef, K. (2004). A history of research on the link between (micro)aggregates, soil biota, and soil organic matter dynamics, *Soil and Tillage Research*, 79(1), 7–31.

Smith, R. M. and Cernuda, C. F. (1951). Some applications of water-drop stability testing to tropical soils of Puerto Rico, *Soil Science*, 71(5) LWW, 337–346.

Song, Y., Liu, L., Yan, P. and Cao, T. (2005). A review of soil erodibility in water and wind erosion research, *Journal of Geographical Sciences*, 15(2) Springer, 167–176.

Ternan, J. L., Elmes, A., Williams, A. G. and Hartley, R. (1996). Aggregate stability of soils in central Spain and the role of land management, *Earth Surface Processes and Landforms*, 21(2) Wiley Online Library, 181–193.

Tisdall, J. M., Smith, S. E. and Rengasamy, P. (1997). Aggregation of soil by fungal hyphae, *Australian Journal of Soil Research*, 35(1), 55–60.

Torri, D. (1987). Theoretical study of soil detachability. In: Ahnert, F. (Ed.), *Geomorphological Models: Theoretical and Empirical Aspects*. Catena, Cremlingen, W. Germany, c1987.

Torri, D., Poesen, J. and Borselli, L. (1997). Predictability and uncertainty of the soil erodibility factor using a global dataset, *CATENA*, 31(1-2) Elsevier, 1-22.

Valmis, S., Dimoyiannis, D. and Danalatos, N. G. (2005). Assessing interrill erosion rate from soil aggregate instability index, rainfall intensity and slope angle on cultivated soils in central Greece, *Soil and Tillage Research*, 80(1), 139-147.

Valmis, S., Kerkides, P. and Aggelides, S. (1988). Soil aggregate instability index and statistical determination of oscillation time in water, *Soil Science Society of America Journal*, 52(4) Soil Science Society of America, 1188-1191.

Vanelslande, A., Lal, R. and Gabriels, D. (1987). The erodibility of some Nigerian soils: a comparison of rainfall simulator results with estimates obtained from the Wischmeier nomogram, *Hydrological Processes*, 1(3) Wiley Online Library, 255-265.

Veihe, A. (2002). The spatial variability of erodibility and its relation to soil types: a study from northern Ghana, *Geoderma*, 106(1-2), 101-120.

Verheijen, F. G. A., Graber, E. R., Ameloot, N., Bastos, A. C., Sohi, S. and Knicker, H. (2014). Biochars in soils: new insights and emerging research needs, *European Journal of Soil Science*, 65(1), 22-27.

Vidal-Beaudet, L., Rokia, S., Nehls, T. and Schwartz, C. (2018). Aggregation and availability of phosphorus in a Technosol constructed from urban wastes, *Journal of Soils and Sediments*, 18(2), 456-466.

Voelkner, A., Holthusen, D. and Horn, R. (2015). Determination of soil dispersion caused by anaerobic digestates: interferences of pH and soil charge with regard to soil texture and water content, *Journal of Soils and Sediments*, 15(7), 1491-1499.

Waldrop, M. P., Balser, T. C. and Firestone, M. K. (2000). Linking microbial community composition to function in a tropical soil, *Soil Biology and Biochemistry*, 32(13) Elsevier, 1837-1846.

Wang, G., Fang, Q., Teng, Y. and Yu, J. (2016). Determination of the factors governing soil erodibility using hyperspectral visible and near-infrared reflectance spectroscopy, *International Journal of Applied Earth Observation and Geoinformation*, 53, 48-63.

Webb, N. P. and Strong, C. L. (2011). Soil erodibility dynamics and its representation for wind erosion and dust emission models, *Aeolian Research*, 3(2), 165-179.

Wischmeier, W. H., Johnson, C. B. and Cross, B. V. (1971). Soil erodibility nomograph for farmland and construction sites, *Journal of Soil and Water Conservation*, 26(5), 189-93.

Wischmeier, W. H. and Mannering, J. V. (1969). Relation of soil properties to its erodibility, *Soil Science Society of America Journal*, 33(1), 131-137.

Wischmeier, W. H. and Smith, D. D. (1978). *Predicting Rainfall Losse: A Guide to Conservation Planning*. U.S Department of Agriculture, Agriculture Handbook, Maryland.

Young, R. A., Römkens, M. J. M. and McCool, D. K. (1990). Temporal variations in soil erodibility, *Catena, Supplement*, 17, 41-53.

Yu, X., Fu, Y. and Lu, S. (2017). Characterization of the pore structure and cementing substances of soil aggregates by a combination of synchrotron radiation X-ray micro-computed tomography and scanning electron microscopy, *European Journal of Soil Science*, 68(1), 66-79.

Zhu, B., Li, Z., Li, P., Shen, Z. and Lu, J. (2009). Dynamic changes of soil erodibility during process of land degradation and restoration, *Transactions of the Chinese Society of Agricultural Engineering*, 25(2) Editorial Office of Transactions of the Chinese Society of Agricultural, 56-61.

Zhu, Z., Angers, D. A., Field, D. J. and Minasny, B. (2017). Using ultrasonic energy to elucidate the effects of decomposing plant residues on soil aggregation, *Soil and Tillage Research* 167, 1–8.

Zhu, Z., Field, D. J. and Minasny, B. (2010). Measuring and modelling the actual energy involved in aggregate breakdown, *CATENA*, 82(1) Elsevier, 53–60.

Zogg, G. P., Zak, D. R., Ringelberg, D. B., White, D. C., MacDonald, N. W. and Pregitzer, K. S. (1997). Compositional and functional shifts in microbial communities due to soil warming, *Soil Science Society of America Journal*, 61(2) Soil Science Society of America, 475–481.

Chapter 8

Advances in measuring mechanical properties of soil in relation to soil health

Muhammad Naveed, University of West London, UK

1 Introduction

Soil health determines its ability to carry out diverse and essential functions that support human life and well-being. These functions include producing food, storing water, carbon and nutrients, protecting our buried cultural heritage and providing a habitat for flora and fauna. Therefore, it is important to quantify soil health and how it changes over space and time using sensitive soil indicators in response to natural factors (such as changing weather patterns) and human factors (land use and management practices). Soil health indicators are based on physical, chemical and biological properties of soils with certain cut-off values beyond which soil functions significantly impedes (Horn et al., 2018; Rickson et al., 2012). The physical, chemical and biological properties of soils are often interlinked and dependent to each other. The mechanical properties of soils always coincide with a link to capacity attributes in soil physical properties, that is, porosity, pore-size distribution, pore connectivity and the derived parameters such as hydraulic conductivity, soil water retention (Horn et al., 2018), and thus linked with plant growth and yield, climate change aspects as well as groundwater recharge and water quality (Fig. 1).

http://dx.doi.org/10.19103/AS.2020.0079.15

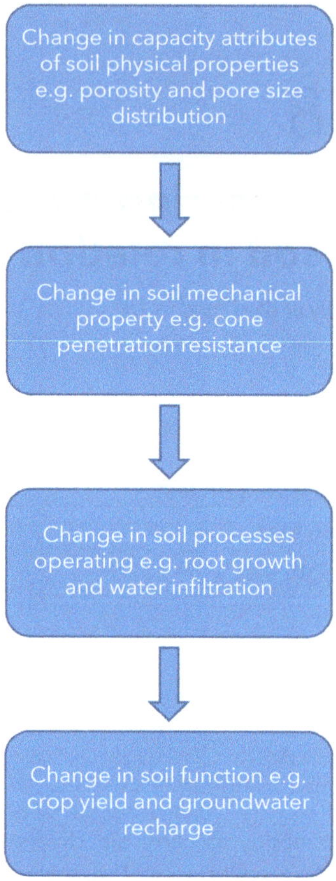

Figure 1 Relationship of soil mechanical properties with processes and functions.

Mechanical properties of soils are typically referred to as the strength of the soil. Soil strength means the maximum amount of stress a soil can withstand before it fails, thus exceeding soil strength results in soil failure or yield. Stress is defined as the force per unit area of the soil. Stress can be induced by external or internal forces which lead to a change in the soil's volume and/or shape expressed as deformations or strains if the body is nonrigid. The mechanical behavior of soil can, therefore, be characterized by its stress-strain relationship (Horn and Fleige, 2011). The yield point is the point on a stress-strain curve that indicates the limit of elastic behavior and the beginning plastic behavior. Yield strength or yield stress is the material property defined as the stress at which a material begins to deform plastically. Prior to the yield point, the material will deform elastically and will return to its original shape when the applied stress is removed. Once the yield point is passed, some fraction of the deformation will be permanent and nonreversible (Fig. 2).

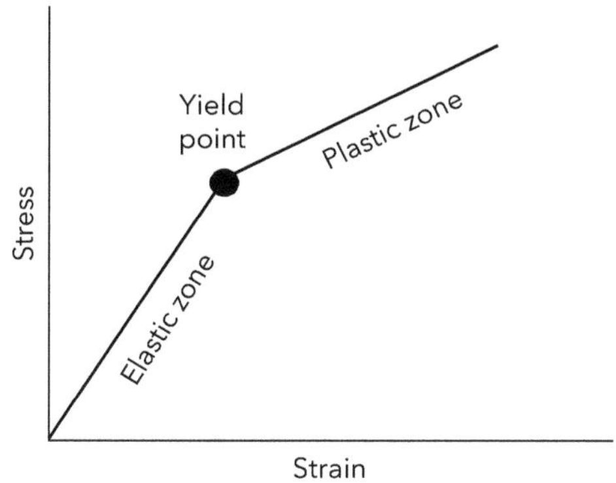

Figure 2 Idealized stress-strain behavior of soils.

Mechanical properties of homogenized soils are higher for gravelly soils compared to sandy or finer-textured soils (Horn, 1988). Granular soils with angular particles possess higher shear strength compared to that with spherical particles providing the same density. Soil strength also depends on the type and content of clay minerals. At the same bulk density and pore water pressure, compressibility of homogenized soil increases with the clay content.

Agriculture soils are structured soils. Soil structure refers to how particles of soil are grouped together into aggregates (also called peds). Soil aggregation results from the interaction of many factors including the environment, soil management, plant influences and soil properties. For a given soil, soil aggregation can be directly altered by means of management strategies (Kay, 1998). The stress-strain behavior of structured soils depends on the degree of soil aggregation and history of hydraulic stresses (degree of maximum pre-drying), as well as shape and internal arrangement of soil particles and aggregates and pores in-between. Along with the history of hydraulic stresses, mechanical properties of soils also depend on the dynamic changes in hydraulic stresses, that is, the frequency of swell/shrink and wet/dry cycles and on the actual pore water pressure at that time. Soils with a well-developed vertical pore system are stronger than those with randomized or horizontal pore arrangements since pores in the direction of the major principle stress are more stable. Therefore, vertical (bio-) pores are less prone to compression than oblique or horizontally oriented pores.

Mechanical properties of soils are useful and quick to measure soil health indicators. Mechanical properties of soils depend on the matric potential of the soils and, therefore, different parameters are suitable to measure at different water potentials (i.e. water contents) and are indicators of specific soil processes and functions (Fig. 3).

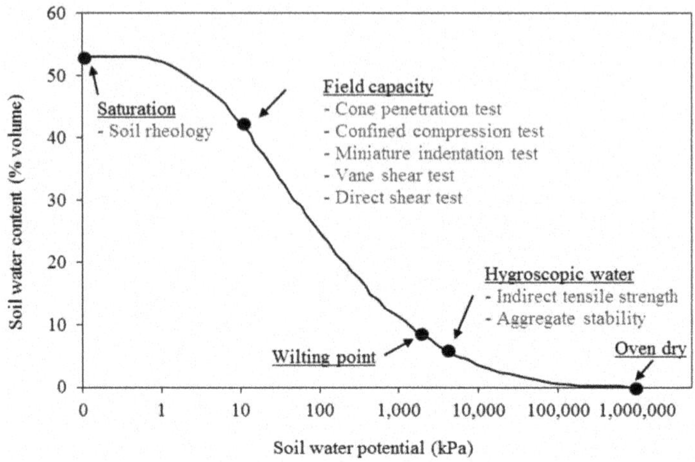

Figure 3 Mechanical measurements of soils are listed at recommended matric potentials.

At or near saturation rheological techniques are an appropriate tool to investigate and quantify microstructural stability of soil on the particle-particle scale. Furthermore, it can be combined with and extended by other methods that are intertwined in micromechanics and physicochemical investigations on the microscale, that is, ζ potential measurements, scanning electron microscopy and X-ray diffractometry, or compared to well-known soil mechanical parameters, that is, shear strength, angle of internal friction, cohesion and soil hydraulics, that is, shrinkage and swelling processes on the mesoscale (Markgraf, 2011). Most mechanical properties of soils are generally measured at field capacity, for example, cone penetration test, confined compression test, miniature indentation test, vane shear test and direct shear test. At air-dry condition (at hygroscopic water), aggregate stability and indirect tensile strength of soil aggregates or remolded soil disks are measured (Fig. 3). Although a huge number of studies have been carried out on mechanical properties of soils measured at air condition, these measurements may not be a true representation of field soils.

2 Soil rheology

2.1 Introduction

Rheology is the part of science dealing with the flow of matter: mainly liquids but also soft solids or solids in states where they flow rather than deforming elastically. The experimental characterization of the rheological properties of a material is termed rheometry. In short, rheology is an extension of the 'classical' disciplines of elasticity (Hook's law) and (Newtonian) fluid mechanics to materials whose mechanical behavior cannot be described by the classical theories. The task of rheometry is to experimentally establish the relationship between deformations and stresses (stress-strain relations), and their derivatives. Strain consists of elastic and plastic parts. According to the fundamental definition, the yield point differentiates the state of soil between elastic and plastic (viscous).

At macroscale soil structure develops as a result of shrinkage and swelling, plant roots and macrofauna, while at microscale it depends on electrostatic forces and cementing substances like cations and organic substances. Quality and quantity of organic substances affect the microscale soil stability and its resilience, that is, ability to recover from any destructive impact. Rheological tests such as the amplitude sweep test quantify microstructural stability of soil on the particle-particle scale. According to Ghezzehei and Or (2001), the specific bonds between particles at the microscale can be considered the origin of processes and phenomena observed at much larger scales. But there is no much effort done to correlate microscale soil stability to the soil processes observed at the macroscale and soil functions.

2.2 Amplitude sweep test

A rotational rheometer with a profiled parallel-plate measuring device can be used to perform an amplitude sweep test, which defines soil as a viscoelastic material according to Maxwell's model. In Maxwell's model both Hooke's law of ideal elastic bodies and Newton's law of ideal viscous fluids are considered (Mezger, 2002). Most amplitude sweep tests under oscillatory condition are conducted in controlled strain mode with a shear strain of $\gamma = 0.0001$ to 100% on log scale at an angular frequency of 1 Hz with 30 measuring points, which lead to an average test duration of 15 min. An oscillatory sweep test stresses and then relaxes a specimen under shear; at each step, an increased shear strain is applied. A plate gap of at least twice the maximum soil particle size should be used to avoid interlocking of soil particles. Author conducted tests at a constant temperature of 20°C with a Discovery Hybrid Rheometer HR-3 (TA Instruments, New Castle, DE, USA). Normal force should be maintained <10 N during the tests (Naveed et al., 2017).

From the stress-strain relationship in oscillation, with preset strain γ (%), specific parameters such as the storage modulus G' (Pa), representing elasticity, stiffness, or rigidity of a material, the loss modulus G'' (Pa), representing viscosity, the loss factor tan δ (= G''/G') and the linear viscoelastic (LVE) range (phase I) are calculated (Fig. 4). The intersection of G'' and G' is called flow point beyond which soil behaves as a fluid. The elastic stress ($G' \gamma$) was plotted as a function of oscillation strain (Naveed et al., 2017). The peak elastic stress was denoted as yield stress and the corresponding strain is denoted as yield strain (Fig. 4). The blue shaded area is calculated as an integral z (Fig. 4), which is referred to

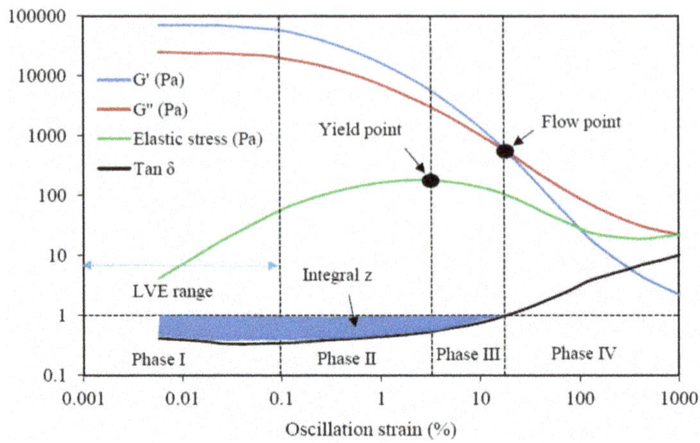

Figure 4 Analysis of rheological data obtained from an amplitude sweep test performed on sandy loam soil at a gravimetric moisture content of 37.5% (unpublished data).

as the amount of energy required to reach at the flow point. The G′ and G″ can be calculated using equations 1 and 2. Plots of storage and loss modulus (G′ and G″) are generated automatically during a test.

$$G' = \frac{\sigma_0}{\varepsilon_0} \cos\delta \tag{1}$$

$$G'' = \frac{\sigma_0}{\varepsilon_0} \sin\delta \tag{2}$$

where σ_0 and ε_0 are the amplitudes of stress and strain, respectively, and δ is the phase shift between them.

Four phases of elasticity loss can be identified in Fig. 4. In phase I, the initial or plateau phase, G′ > G″, a quasielastic behavior can be observed. This can be described by an analog of spring for ideal elastic substances according to Hooke's law. The quasielastic stage of soil as viscoelastic solid is defined by the LVE range (Fig. 4). At this stage of low strain, a full recovery of the microstructure can be assumed on unloading. In phase II, a stage of pre-yielding occurs due to higher strain - soil particles reorientate and microstructural stability of soil decreases. At the end of phase II, a peak elastic stress is reached and is denoted as the yield point. Yield point marked the point on stress-strain curve beyond which no recovery of the microstructure of the soil can be achieved on unloading. Phase III lies between the yield point and the flow point. At the flow point, the soil starts behaving as fluid instead of solid. The intersection of tanδ with the tanδ = 1-line also indicates a complete loss of stiffness and rigidity of the soil microstructure (Markgraf and Horn, 2009) that can be used to calculate integral z (Fig, 4). At the end of phase III, an intersection of G′ and G″ indicates the flow point and the start of phase IV. Phase IV defines the final stage of structural collapse, G′ < G″ - a viscous character predominates and substances creep or run. This behavior can be represented by a dashpot, an analog for ideal fluids according to Newton's law.

2.3 Results

Some of the results of an amplitude sweep test are shown in Figs 5 and 6. Water content is the primary parameter that impacts the microstructural stability of the soil. Yield stress of the soil rapidly decreased with increasing water content as shown in Fig. 5. Yield stress of soil at a water content of 35% was 177 Pa that decreased to only 4.48 Pa at a water content of 47.5%. Figure 6 shows how organic substances (in this case chia seed mucilage) impact microstructural stability of the soil. Yield stress of soil has increased from 8.91 Pa to 199.63 Pa at a water content of 45% by the addition of chia seed mucilage at a concentration of 4.5% by the dry weight of soil (Fig. 6). Studies carried out to date, for example, Naveed et al., 2017, Deng et al., 2013, and Barré and Hallett, 2009 show that rheometry is a sensitive technique in measuring microstructural stability of the soil impacted by different land use and management practices or addition of organic substances. However, there is no framework available to monitor soil health in terms of rheological measurements. Although partial knowledge is available on the interaction between rheology and cations (kind and amount) salt, organic effects, hydraulic effects, bulk density effects, and so on, there is a great scope of predicting macroscale soil physicochemical properties,

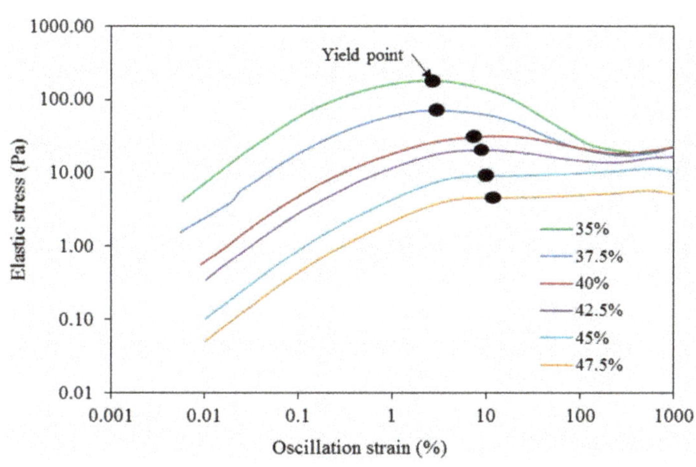

Figure 5 Effect of water content on the yield stress of sandy loam (A-Horizon) using an amplitude sweep test (unpublished data).

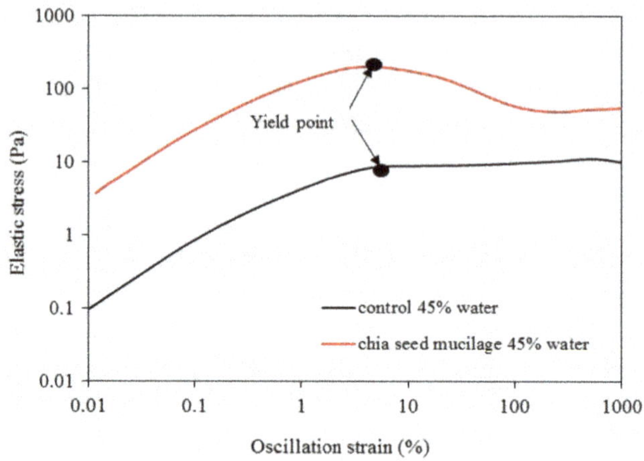

Figure 6 Impact of chia seed mucilage at a concentration of 4.5% by the dry weight of soil on microstructural stability of the sandy loam soil (A-horizon) using an amplitude sweep test (unpublished data).

processes and soil functions as a function of rheological measurements carried out at the microscale.

3 Cone penetration resistance

3.1 Introduction

Cone penetration resistance, a function of several soil properties, provides a rapid method to characterize the variability of soil strength within the soil profile. Studies in the past have used cone penetration resistance to screen the soil for spatial changes in specific soil properties (Carter, 1988), to quantify the depth and intensity of soil compaction and loosening (O'Sullivan et al., 1987) and to estimate the extent of the potential rooting depth and degree of soil structure regeneration (Bengough et al., 2011; Carter and White, 1986).

Originally, cone penetration resistance tests were conducted to determine soil load-bearing capacity for engineering purposes. They have been used extensively in geotechnical engineering to describe the stratigraphy of geological layers. In agriculture, penetrometers are traditionally used to measure soil compaction, although other uses, including soil mapping, have been implemented. Cone penetration resistance is a measure of the capacity of soil to resist deformation and refers to the amount of energy that is required to break apart aggregates or move implements through the soil. It is measured in megapascals (MPa).

Cone penetration resistance is influenced by several factors such as soil water content, bulk density, texture and structure. Cone penetration resistance generally increases with increase in depth from natural surface level. Cone penetration resistance increases as soil dries as much force is required to break up aggregates (Lomeling and Lasu, 2015). Cone penetration resistance values are directly correlated with soil bulk density at comparable soil matric potentials (Lowery and Schuler, 1994; Grunwald et al., 2001). Inconsistent correlations were observed between cone penetration resistance and soil textures (Grunwald et al., 2001). Cone penetration resistance is smaller in fine aggregated soils compared to large aggregates with a stable macrostructure and microstructure. Cone penetration resistance can be modified by inputs of organic matter such as mulches, composts or cover crops which cause aggregate macrostructure to become more stable (Celik et al., 2010). The application of gypsum to soil stabilizes aggregate microstructure and prevents clay dispersion (Shainberg et al., 1989). However, gypsum will restore or improve soil structure only on areas that have been flooded with sea water. In SEPA's opinion, therefore, gypsum can be applied to land to improve the soil structure of land that has been flooded by seawater at 5 tonne/ha. Excessive tillage can break down both the macrostructure and microstructure of aggregates leading to hardsetting and crusting of surface soils.

There is a consensus that a penetrometer resistance greater than 2 MPa will significantly impede root growth when measured at or near field capacity (Table 1). Generally, the proportion of roots penetrating the soil was inversely related to penetration resistance (Martino and Shaykewich, 1994). A survey by MAFF (Ministry for Agriculture, Food and Fisheries) recommended the following categories for cone penetration resistance: loose <0.5 MPa; medium 0.5–1.25 MPa; dense 1.25–2.0 MPa; and very dense >2.0 MPa (MAFF, 1982). Penetrometer resistance in excessively wheeled areas of arable soils was between 2.0 MPa and 2.5 MPa (Ball and O'Sullivan, 1982). Cone penetration resistance of 2.5 MPa was an upper limit for cotton taproots to grow through soil, with rooting density reduced at 1.0–2.0 MPa (Taylor, 1971), while more recent studies of UK soils found that at >2.0 MPa root elongation was slowed to <50% of its unimpeded rate (Valentine et al., 2012).

3.2 Equipment

There are several types of tools and methods used to measure cone penetration resistance. They all rely on determining the resistance of the soil to penetration and are best used when the soil is sufficiently moist. Cone penetrometer has a stainless steel cone on the end of a shaft. It is inserted into the soil and pushed through the soil profile at a steady rate. A pressure sensor records the pressure (units of MPa) needed to push it through the soil. Pocket penetrometers are convenient since, as the name implies, they can be easily carried in a pocket or bag. This would be useful to determine the presence of hardpans and depth

Table 1 Cut-off values of cone penetration resistance (MPa) related to soil health and root growth.

Cone penetration resistance (MPa)	Relevance
<1	• Roots grow through the soil without difficulty. • Soil physical health is good.
1-3	• Moderate adverse effects on plant root growth. • Soil physical health is moderate.
>3	• Root growth is retarded except through cracks and old root channels. For conservation/zero tillage this may not be true. • Soil physical health is poor.

of subsoiling required to alleviate this condition. Soil penetrometers used for most research purposes are designed to allow for the recording of the data digitally as the rod/tip is pushed into the soil/material, as well as the soil depth associated with the reading. The force is typically measured with a load cell and the depth with a sonic depth indicator, an encoder or a potentiometer. The dynamic cone penetrometer (Herrick and Jones, 2002) is yet another very simple device for measuring cone penetration resistance. According to Herrick and Jones (2002), the dynamic penetrometer method does not require one to push the penetrometer through the soil, nor does it need one to apply continuous force to the penetrometer. Dynamic penetrometers supply a known amount of kinetic energy (KE) to the penetrometer, which causes the penetrometer to move some distance through the soil. Penetration distance depends on the KE applied, the geometry of the penetrometer tip and the soil penetration resistance. Soil penetrometers have also been designed to be driven hydraulically, thus allowing for constant insertion rates. Soil penetrometer with multiple probes was developed by Raper et al. (1999) consisting of an array of five rods, each one with its own independent load-cell. This setup provides cone penetration resistance at five locations simultaneously. Other penetrometer designs have incorporated simultaneous determination of soil water content with some success as the soil water content is the key parameter on which cone penetration resistance depends (Morrison et al., 2000; Young et al., 2000).

The dimensions and geometry of the cone tip and the rod itself can affect penetration measurements. For agricultural purposes, most penetrometers consist of a stainless steel rod with a threaded, replaceable tip attached to one end and a handle on the other. The tip is constructed of stainless steel with a 30° angle and can have two different base diameters, 12.8 mm or 20.3 mm (ASAE, 1999). It is important to note that penetrometers with a cone that has an angle greater than 30° will exhibit lower penetrability (i.e. greater cone index values), while decreasing the diameter of its base will have the opposite effect.

If one desires to make comparisons between treatments or soil conditions, it is important that the same type of penetrometer is used.

3.3 Location and timing

It is important to assess the soil strength of each soil layer that will impact on root growth and water infiltration. Ideally, you should measure cone penetration resistance in each soil layer and at three positions in a soil pit. The best time to carry out cone penetration resistance measurement is when the soil is at field capacity. This is when the soil matric potential is approximately 10 kPa. This can be measured using a tensiometer. As a general rule, field capacity usually occurs approximately 24–48 h after soaking rain or penetrating irrigation.

3.4 Results

In various studies, cone penetration resistance is proved to be a strong predictor of root elongation (Valentine et al., 2012) and crop yield (Bölenius et al., 2018; Schjønning et al., 2016; Naderi-Boldaji et al., 2013). There are two very obvious responses of plant roots to the mechanical impedance of soil (measured by cone penetration resistance): slowing of the rate of extension and an increase in root diameter immediately behind the root tip. The cone penetration resistance of most soils increases as they dry, so shortage of soil water and compaction of soils are commonly interlinked and contribute to mechanical impedance. It was reported that the root elongation rate decreased to half with the increase of cone penetration resistance from nearly 0–1 MPa at comparable water contents (Valentine et al., 2012). Cone penetration resistance also proved to be a strong predictor of grain yield as shown by Schjønning et al. (2016) (Fig. 7), Naderi-Boldaji et al. (2013) (Fig. 8) and Bölenius et al. (2018) (Fig. 9). Schjønning et al. (2016) showed a linear relationship between grain yield and cone penetration resistance on compacting the soil with different wheel loads. Contour maps shown in Figs 8 and 9 for grain yield and cone penetration resistance look like mirror images, reflecting the fact that cone penetration resistance is the key predictor of grain yield as well of soil health.

4 Uniaxial confined compression test

4.1 Introduction

Soil compaction due to agricultural or forest traffic adversely impacts soil structure and as a result several soil functions, including crop production, water and nutrient cycling, and habitat for soil organisms. Prevention of compaction in the subsoil is of critical importance because of the low recovery potential

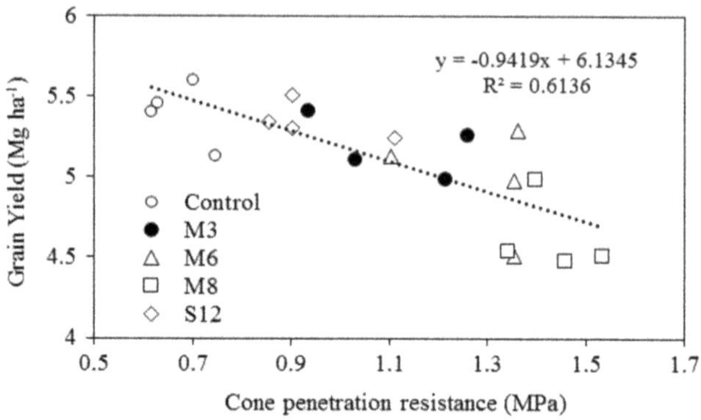

Figure 7 Cone penetration resistance is a strong predictor of grain yield. Control: no compaction; M3, M6, M8: multiple wheel passage of 3, 6 and 8 Mg of wheel load; S12: single wheel passage of 12 Mg of wheel load (Schjønning et al., 2016).

of subsoil. Prediction of soil compaction risks prior to field traffic, for example, using decision-support tools (Horn and Fleige, 2009; Stettler et al., 2014), can help avoid soil compaction. Prediction of compaction risks requires knowledge of compressive properties of soil (recompression or swelling index, compression index and precompression stress). Compressive properties of arable and forest soils are typically determined from data obtained in uniaxial-confined compression tests. In uniaxial confined compression test a uniaxial force is applied and corresponding soil displacement was measured. This test is generally fast to simulate field traffic ignoring changes in pore water pressure. The soil core used for the test should have a diameter (D) at least twice the height (h) of the soil core to minimize the wall friction effect. The relative impact of soil-wall friction on sample average behavior decreased with increasing the ratio of D/h. De Lima and Keller (2019) suggested that the effect of soil-wall friction on sample-average bulk density cannot be neglected unless $D/h > 8$.

4.2 Semi-logarithmic diagram for determination of compressive properties

Soil compression properties are traditionally derived from the semi-logarithmic diagram between log-applied stress (σ) and void ratio (e), the compression curve, and measured on cylindrical soil samples, typically in uniaxial compression equipment. The compressive behavior of the soil apparently changes when analyzing the data in a log σ−e diagram as compared to analysis in a σ−e diagram. For example, for sandy loam soil shown in Fig 10a and b, the calculation of precompression stress changes from 150 kPa to 85 kPa when data was analyzed in log σ−e diagram. Similarly, for silty clay loam soil, the calculation of precompression stress becomes impossible as compression

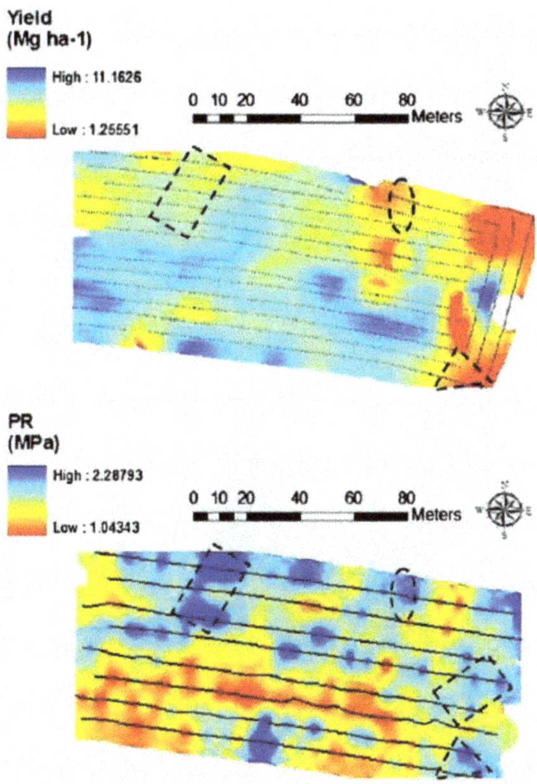

Figure 8 Maps of crop yield and penetration resistance (PR) at Hardrain field with a plot area of 1.5 ha, Bern, Switzerland (Naderi-Boldaji et al., 2013).

data becomes very linear in log σ–e diagram (Fig. 10d) compared to the actual trend of data in σ–e diagram (Fig. 10c). For this reason, the use of the logarithmic stress scale in compression studies has been criticized by Janbu (1998) and Wesley (2010). Therefore I recommend the use of σ–e diagram for the derivation of compressive properties of the soils.

4.3 Derivation of compression properties of soils

In a σ–e diagram, the soil compression characteristic has two distinct regions that can be approximated by two lines (Fig. 11): the swelling (or recompression) line (SL) and the virgin compression line (VCL). This approximation represents an elastoplastic model and the intersection between SL and VCL is referred to as the precompression stress, σ_{pc} (Fig. 11). It is assumed that the soil behaves elastically (reversible deformation) along SL, and plastically (irreversible deformation) along the VCL. The compression curve as shown in Fig. 10 yields

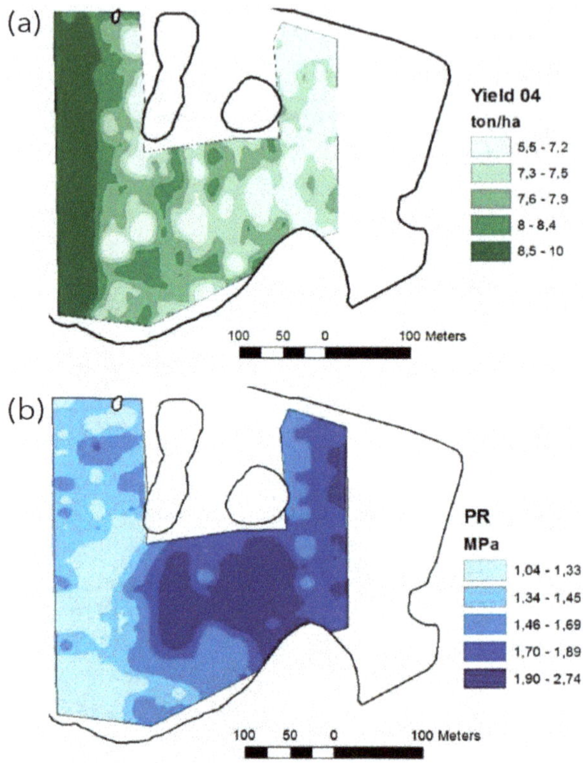

Figure 9 Maps of (a) spring barley yield 2004, (b) penetration resistance at 50 cm in spring 2004 at Kvarnbo farm, Sweden (Bölenius et al., 2018).

three important soil compression properties: the swelling or recompression index, C_S; the compression index, C_C; and the precompression stress, σ_{pc}. The change in void ratio during the confined compression test is calculated using equation 3. The C_S and C_C are calculated using equations 4 and 5, respectively (Fig. 11).

$$e = \frac{\rho_s}{\rho_d}\frac{H-d}{H} - 1 \tag{3}$$

where ρ_s is the particle density, ρ_d the initial bulk density, d is the recorded displacement during the experiment, and H is the initial height of the soil core.

$$C_S = -\frac{\Delta e_s}{\log \Delta \sigma_s} \tag{4}$$

$$C_C = -\frac{\Delta e_v}{\log \Delta \sigma_v} \tag{5}$$

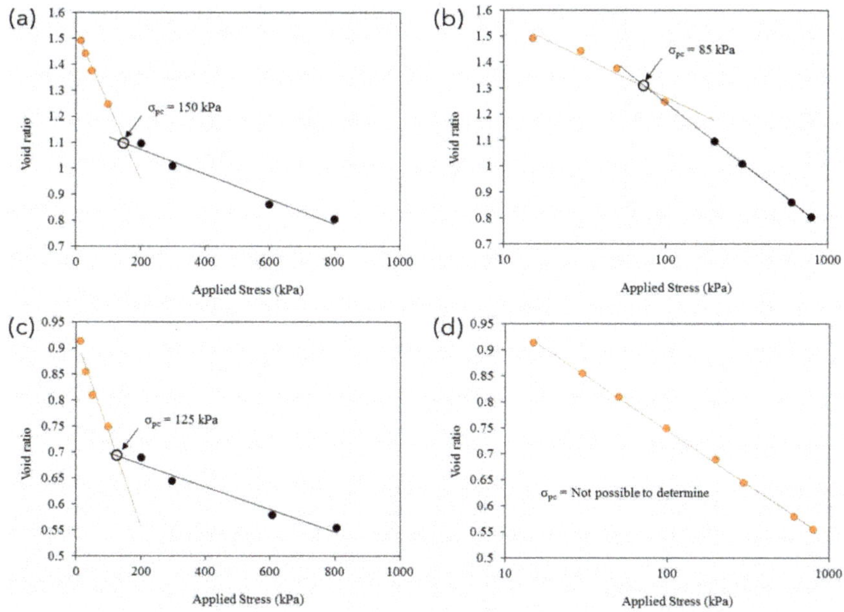

Figure 10 (a) and (b) compression curves for sandy loam soil as σ–e and log.σ–e diagrams, respectively, (c) and (d) compression curves for silty clay loam soil as σ–e and log.σ–e diagrams, respectively. Both soils are from 0–30 cm layer (data from Tang et al., 2009).

where Δe_s is the change in void ratio on swelling line, $\Delta\sigma_s$ is the change in stress on swelling line, Δe_v is the change in void ratio on virgin compression line, $\Delta\sigma_v$ is the change in stress on the virgin compression line, C_S is the swelling index and C_C is the compression index.

The compression index (C_C) is considered an indicator of soil susceptibility to compaction or soil resistance to compression. The swelling index is used as a measure of rebound and soil mechanical resilience. Kuan et al. (2007) defined recovery from compression as the ratio of C_S to C_C. As it is assumed that soil behaves elastically as long as the applied stress does not exceed σ_{pc} σ_{pc} is considered the soil load support capacity or soil compression strength. Consequently, soil compaction due to agricultural field traffic could in principle be avoided by limiting the stresses exerted by agricultural machinery to below σ_{pc} (Alakukku et al., 2003; van den Akker, 2004).

4.4 Factors affecting compression properties of soils

Soil mechanical properties are strongly influenced by soil water potential or soil water content: generally, the soil becomes more resistant to compaction with decreasing (more negative) soil water potential or soil water content.

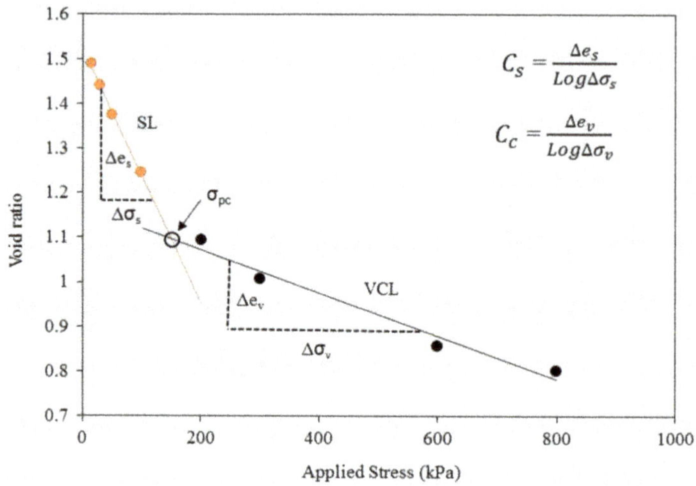

Figure 11 Soil compression characteristics, SL is the swelling line, VCL is the virgin compression line, σ_{pc} is the precompression stress, Δe_s is the change in void ratio on swelling line, $\Delta\sigma_s$ is the change in stress on swelling line, Δe_v is the change in void ratio on virgin compression line, $\Delta\sigma_v$ is the change in stress on virgin compression line, C_S is the swelling index and C_C is the compression index.

The influence of soil texture on soil compressive properties has been widely investigated, but the results are contradictory. The compression index and σ_{pc} have been reported to increase with increasing clay content by Gregory et al. (2006) and Imhoff et al. (2004), while Keller and Arvidsson (2007) could not find any correlations with soil texture. Soil organic matter affects soil strength, but the mechanisms are complex: soil organic matter increases soil aggregate stability and soil mechanical strength, but reduces bulk density and improves water retention, both of which negatively affect soil strength. Kuan et al. (2007) reported that soils with high organic matter are less resistant to compaction, but are able to recover better from compaction. Soil mechanical properties are affected by soil structure because soil strength is a function of the contact properties of soil aggregates. Bulk density or initial void ratio is a universally good predictor of compression index as well as of precompression stress (Saffih-Hdadi et al., 2009; Keller and Arvidsson, 2007; Imhoff et al., 2004). Keller et al. (2011) reported a positive correlation between compression index and initial void ratio while a negative correlation between precompression stress and initial void ratio. However, bulk density or initial void ratio is a poor measure of soil structure (Dexter, 1988, 2004; Horn, 1990). Soil structure parameters other than bulk density such as porosity, macroporosity, pore-size distribution, pore connectivity are little investigated in relation to soil compression properties. Ajayi et al. (2009a,b) found that σ_{pc} and C_C were influenced by soil structure, which was associated with clay mineralogy. Horn (2004) and da Veiga et al. (2007) measured differences in σ_{pc} and C_C between different tillage systems.

5 Miniature indentation test

Miniature indentation test is very useful to carry out at the field capacity when mechanical properties of soil are required to measure at the mm scale.

5.1 Measurements of soil hardiness and elasticity

The packed soil cores amended with plant root exudates were first saturated and then drained to −10 kPa matric potential on a suction plate (ecoTech, Umwelt-Meßsysteme) at 4°C to suppress microbial decomposition of the exudates. One indentation measurement was performed on each soil core using a 3-mm-diameter spherical indenter. Figure 12 shows a typical load-displacement curve obtained from a soil indentation test. A corresponding schematic cross-section of such an indentation is depicted in Fig. 13. The soil hardness parameter (Oliver and Pharr, 1992) was obtained using equation 6;

$$H = \frac{F_{max}}{A_c} \quad (6)$$

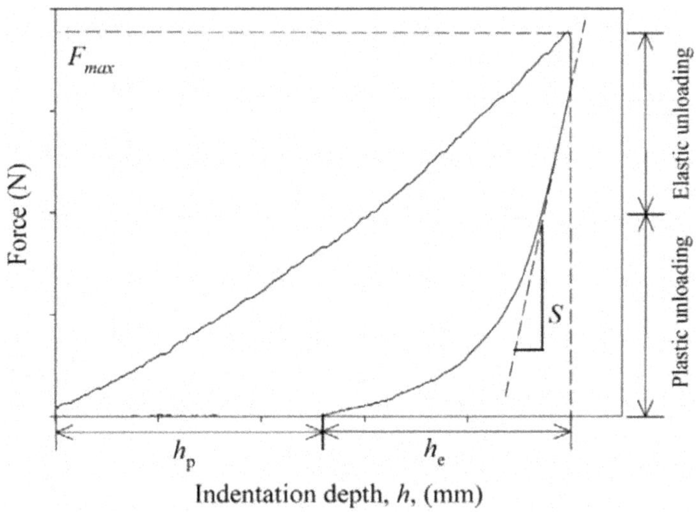

Figure 12 Typical load-displacement curve obtained from an indentation test; F_{max} = maximum measured force for the required indentation, h_e = elastic indentation, that is, deformation recovered on unloading, h_p = plastic indentation, that is, permanent deformation and not recovered on unloading and *S* = stiffness, that is, slope of the initial linear part of the unloading curve (Naveed et al., 2018).

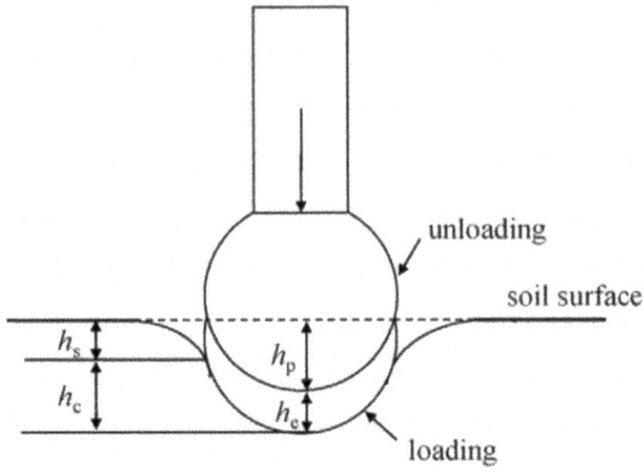

Figure 13 A schematic of an indentation test; h_e = elastic indentation, that is, deformation recovered on unloading, h_p = plastic indentation, that is, permanent deformation and not recovered on unloading, h_c = contact depth at maximum load, h_s = displacement of the surface at the perimeter of the contact (Naveed et al., 2018).

where F_{max} is the maximum force applied during an indentation test as shown in Fig. 12 and A_c is the projected contact area at F_{max}, obtained using equation 7

$$A_c = \pi \cdot \left(2h_c R - h_c^2\right) \tag{7}$$

where R is the radius of the spherical indenter (i.e. 1.5 mm) and h_c is the contact depth, as shown in Fig. 13 and obtained using equation 8

$$h_c = h_{max} - h_s \tag{8}$$

where h_{max} is the indentation depth and h_s is the displacement of the surface at the perimeter of the contact at F_{max} before the unloading. The value of h_s was determined by equation 9

$$h_s = \varepsilon \frac{F_{max}}{S} \tag{9}$$

where ε is a geometric constant, which for a spherical indenter is theoretically equal to 0.75 (Fischer-Cripps, 2011), and S is the stiffness, that is, the initial slope of the unloading curve as shown in Fig. 12. It was determined by taking the derivative of F with respect to h for the initial linear part of the unloading curve. The reduced modulus of elasticity (E_r) was obtained following Oliver and Pharr (1992) using equation 10

$$E_r = \frac{S\sqrt{\pi}}{2\beta\sqrt{A_c}} \tag{10}$$

where β is the tip geometry correction factor, which is equal to 1 for a spherical indenter. Because the indenter's (steel) modulus of elasticity is much higher than that of the soil, the indenter can be treated as a rigid body and the modulus of elasticity can be determined using equation 11

$$E = (1 - \nu_s^2)E_r \quad (11)$$

where v_s is the Poisson ratio of the soil, assumed to be 0.30. The value of E was determined in addition to hardness because the soil at −10 kPa matric potential behaves as an elastoplastic material, as evident from the initial elastic followed by plastic unloading of the indenter shown in Fig. 12.

5.2 Results

The sandy loam soil (from A-horizon) hardness was not significantly impacted by barley root exudate at 0.46 mg g^{-1} concentration, whereas at 4.6 mg g^{-1} concentration soil hardness significantly increased ($p < 0.05$) by 28% compared with the control (Fig. 14). Soil hardness was increased by 20% for maize root exudate at 0.46 mg g^{-1} concentration ($p < 0.05$) and 62% at 4.6 mg g^{-1} concentration ($p < 0.05$) for the sandy loam soil (Fig. 14). The clay loam soil showed significantly greater hardness than the sandy loam soil. However, this difference was overcome by the addition of barley root exudate at 4.6 mg g^{-1} concentration and maize root exudate at both 0.46 and 4.6 mg g^{-1} concentrations. Neither barley nor maize root exudates at any of the studied concentrations improved soil hardness for the clay loam soil (Fig. 14).

6 Indirect tensile strength test

The tensile strength (TS) of air-dried soil aggregates or remolded soil disks (Fig. 15) is determined by the indirect crushing method, which involved crushing soil aggregates/disks between two flat plates at a displacement rate of 2 mm min^{-1} and recording the maximum force (F) required to break them (brittle failure). The TS (kPa) of soil aggregates was then computed using equation 12

$$TS = 0.576\,F/D^2 \quad (12)$$

where F (N) is the polar force required to rupture the aggregate and D (m) is the mean aggregate diameter. The 0.576 value is a constant based on the assumption of spherical form and incompressible behavior of aggregates (Poisson ratio = 0.5). The failure of air-dried soil aggregates starts with plastic deformation followed by brittle failure. The TS (kPa) of air-dried remolded soil disks was computed using equation 13

$$TS = 2F/\pi Dh \quad (13)$$

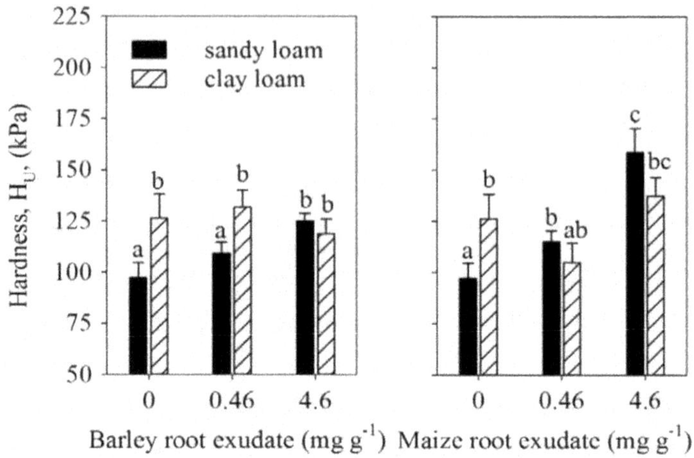

Figure 14 Hardness and modulus of elasticity of sandy loam and clay loam soils treated with barley root, maize root and chia seed exudates at different concentrations (in mg exudate g^{-1} dry soil) measured at −10 kPa matric potential. Different letters indicate a significant difference between treatments at $p < 0.05$ (Naveed et al., 2018).

where F (N) is the polar force required to rupture the soil disks, D (m) is the diameter of soil disk, and h (m) is the height of the soil disk.

The tensile strength of the soil is generally measured at air-dried condition so it is independent of the water content of the soil. Studies have also been

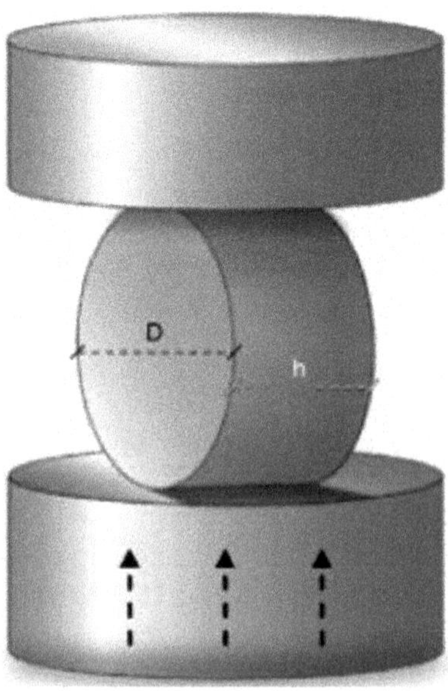

Figure 15 Diagram of the tensile strength of air-dried soil disks.

carried out in the past on the tensile strength of soil aggregates as a function of soil matric potential or water content (Causarano, 1994). As shown in Fig. 16, the tensile strength of soil disks increased with increasing chia seed exudate concentration from 240 kPa to 400 kPa with chia seed exudate concentration of 2.3 mg g^{-1} of dry soil. Tensile strength of soil aggregates is generally increased with decreasing aggregate size. Large organic matter content appeared to strengthen wet soil aggregates and weaken dry ones (Causarano, 1994).

7 Conclusion

Although mechanical properties of soil are fast and easy to measure, they have not been used as indicators of soil health apart from cone penetration resistance. The confined compression test is traditionally used for the prediction of soil compaction risks. The confined compression test data is even used to prepare soil maps, including computation of stress-induced changes in hydraulic soil functions down to depth in the soil profile. Other mechanical tests such as soil rheometry, miniature indentation test and tensile strength test are used for assessing the impact of certain treatments/amendments on the stability of the soil. Rheological techniques are an appropriate tool to investigate and quantify microstructural stability and stiffness degradation of soil on a particle-particle scale. Next-generation and hybrid discovery rheometers allow temperature-controlled measurements. Miniature indentation test is very useful to perform when mechanical properties of soil are required to measure at the mm scale

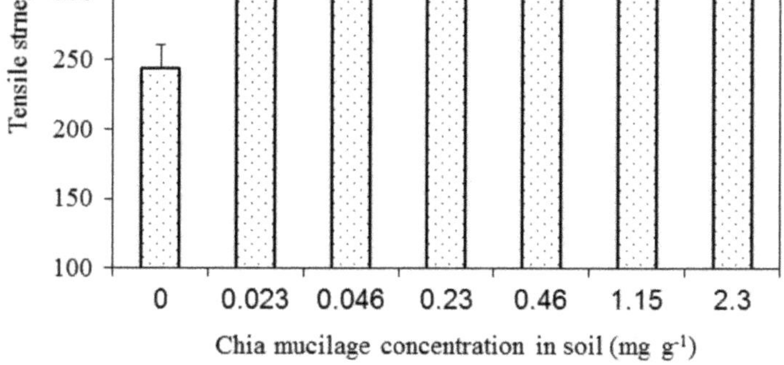

Figure 16 Tensile strength of air-dried sandy loam soil (A-horizon) remolded disks impacted by chia seed exudate.

particularly in rhizpshpere. Measurement of the tensile strength of soil has an advantage as it removes the effect of water content because it is generally measured at the air-dried condition. There is clearly a need for more practical and rigorous testing on comparing different mechanical properties of the soil to test how they perform relative to each other and they might be interlinked. As cone penetration resistance of 1 MPa is considered as a cut-off value for maintaining soil health, cut-off values for other mechanical parameters of soil are also required to establish. Further studies are needed to establish correlations between crop yield and different mechanical properties of the soils in order to develop a soil health framework based on mechanical properties of the soil.

Soil texture, structure, organic carbon and matric potential are the primary parameters that control mechanical properties of soils. In the laboratory, soil texture, structure and organic carbon are measured and matric potential is controlled during the measurements. In-situ measurements of mechanical properties of soils are generally fast, but their main flaw is that they were not developed to account for the variations in matric potential. Therefore, when spatial variation in mechanical properties of soils has to be assessed, they must be accompanied by the measurements of soil matric potential.

8 Future research

Mechanical properties of soils have generally been measured using only one of the above-mentioned tests in any particular study. To assess the soil health holistically using mechanical measurements, a set of mechanical tests should be performed. For example, soil rheology should be measured to assess the microstructural stability of the soil. Soil rheological parameters are highly influenced by the clay content, quantitiy and quality of the organic matter. Cone penetration test assess the soil resistance at the mm scale and highly suitable to assess the rootability of the soil. Confined compression test measures the macro-stability of the soil, as is primarily used to assess the soil compaction risks due to farm vehicles. Confined compression test can also be used to determine the resilience of the soil under soil compaction. Aggregate size distribution coupled with aggregate tensile strength at air dry condition are useful in assessing the soil structure and soil structural stability. Future research should be carried out on inter-linking various mechanical tests to establish a soil health framework based on mechanical properties of soil.

9 References

Ajayi, A. E., Dias Junior, M. S., Curi, N., Gontijo, I., Araujo Junior, C. F. and India Junior, A. V. 2009a. Relation of strength and mineralogical attributes in Brazilian latosols. *Soil and Tillage Research* 102(1), 14–8. doi:10.1016/j.still.2008.05.013.

Ajayi, A. E., Dias Junior, MdS., Curi, N., Araujo Junior, C. F., Souza, T. T. T. and Inda Junior, A. V. 2009b. Strength attributes and compaction susceptibility in Brazilian latosols. *Soil and Tillage Research* 105(1), 122–7. doi:10.1016/j.still.2009.06.004.

Alakukku, L., Weisskopf, P., Chamen, W. C. T., Tijink, F. G. J., van der Linden, J. P., Pires, S., Sommer, C. and Spoor, G. 2003. Prevention strategies for field traffic-induced subsoil compaction: a review. Part 1. Machine/soil interactions. *Soil and Tillage Research* 73(1–2), 145–60. doi:10.1016/S0167-1987(03)00107-7.

American Society of Agricultural Engineering (ASAE) 1999. Soil cone penetrometer. *ASABE Standard* S313(3, February) (rev 2009).

Ball, B. C. and O'Sullivan, M. F. 1982. Soil strength and crop emergence in direct drilled and ploughed cereal seedbeds in seven field experiments. *Journal of Soil Science* 33(4), 609–22. doi:10.1111/j.1365-2389.1982.tb01793.x.

Barré, P. and Hallett, P. D. 2009. Rheological stabilization of wet soils by model root and fungal exudates depends on clay mineralogy. *European Journal of Soil Science* 60(4), 525–38. doi:10.1111/j.1365-2389.2009.01151.x.

Bengough, A. G., McKenzie, B. M., Hallett, P. D. and Valentine, T. A. 2011 Root elongation, water stress, and mechanical impedance: a review of limiting stresses and beneficial root tip traits. *Journal of Experimental Botany* 62(1), 59–68. doi:10.1093/jxb/erq350.

Bölenius, E., Wetterlind, J. and Keller, T. 2018. Can within field yield variation be explained using horizontal penetrometer resistance and electrical conductivity measurements? Results from three Swedish fields. *Acta Agriculture Scandinavica, Section B - Soil & Plant Science* 68(8), 690–700. doi:10.1080/09064710.2018.1464201.

Carter, M. R. and White, R. P. 1986. Determination of variability in soil physical properties and microbial biomass under continuous direct-planted corn. *Canadian Journal of Soil Science* 66(4), 747–50. doi:10.4141/cjss86-075.

Carter, M. R. 1988. Penetration resistance to characterize the depth and persistence of soil loosening in tillage studies. *Canadian Journal of Soil Science* 68(4), 657–68. doi:10.4141/cjss88-063.

Causarano, H. 1994. Factors affecting tensile strength of soil aggregates. *Soil and Tillage Research* 28(1), 15–25.

Celik, I., Gunal, H., Budak, M. and Akpinar, C. 2010. Effect of long-term organic and mineral fertilizers on bulk density and penetration resistance in semi-arid Mediterranean soil conditions. *Geoderma* 160(2), 236–43. doi:10.1016/j.geoderma.2010.09.028.

da Veiga, M., Horn, R., Reinert, D. J. and Reichert, J. M. 2007. Soil compressibility and penetrability of an oxisol from southern Brazil, as affected by long-term tillage systems☆. *Soil and Tillage Research* 92(1–2), 104–13. doi:10.1016/j.still.2006.01.008.

de Lima, R. P. and Keller, T. 2019. Impact of sample dimensions, soil-cylinder wall friction and elastic properties of soil on stress field and bulk density in uniaxial compression tests. *Soil and Tillage Research* 189, 15–24. doi:10.1016/j.still.2018.12.021.

Deng, W., Iannetta, P. P. M., Hallett, P. D., Toorop, P. E., Squire, G. R. and Jeng, D. S. 2013. The rheological properties of the seed coat mucilage of Capsella bursa-pastoris L. Medik. (shepherd's purse). *Biorheology* 50(1–2), 57–67. doi:10.3233/BIR-130627.

Dexter, A. R. 1988. Advances in characterization of soil structure. *Soil and Tillage Research* 11(3–4), 199–238. doi:10.1016/0167-1987(88)90002-5.

Fischer-Cripps, A. C. 2011. *Nano Indentation* (3rd edn). Springer, New York.

Ghezzehei, T. A. and Or, D. 2001. Rheological properties of wet soils and clays under steady and oscillatory stresses. *Soil Science Society of America Journal* 65(3), 624-37. doi:10.2136/sssaj2001.653624x.

Gregory, A. S., Whalley, W. R., Watts, C. W., Bird, N. R. A., Hallett, P. D. and Whitmore, A. P. 2006. Calculation of the compression index and precompression stress from soil compression test data. *Soil and Tillage Research* 89(1), 45–57. doi:10.1016/j.still.2005.06.012.

Grunwald, S., Rooney, D. J., McSweeney, K. and Lowery, B. 2001. Development of pedotransfer functions for a profile cone penetrometer. *Geoderma* 100(1–2), 25–47. doi:10.1016/S0016-7061(00)00079-3.

Herrick, J. E. and Jones, T. L. 2002. A dynamic cone penetrometer for measuring soil penetration resistance. *Soil Science Society of America Journal* 66(4), 1320–4. doi:10.2136/sssaj2002.1320.

Horn, R. 1988. Compressibility of arable land. *Catena. Supplement* 11, 53–71.

Horn, R. 1990. Aggregate characterization as compared to soil bulk properties. *Soil and Tillage Research* 17(3–4), 265–89. doi:10.1016/0167-1987(90)90041-B.

Horn, R. 2004. Time dependence of soil mechanical properties and pore functions for arable soils. *Soil Science Society of America Journal* 68(4), 1131–7. doi:10.2136/sssaj2004.1131.

Horn, R. and Fleige, H. 2011. Soil structure and mechanical strength. In: Gliński, J., Horabik, J. and Lipiec, J. (Eds), *Encyclopedia of Agrophysics: Encyclopedia Of Earth Sciences Series*. Springer, Dordrecht.

Horn, R., Fleige, H., Lal, R. and Zimmermann, I. 2018. Soil health and functions as a basic requirement for advancing the SDG´s. In: Lal, R., Horn, R. and Kosaki, T. (Eds), *Soils and Sustainable Development Goals*. Geo Ecology Essay. Catena soil sciences. Stuttgart, pp. 52–60. ISBN: 978-3-510-65425-3.

Imhoff, S., Da Silva, A. P. and Fallow, D. 2004. Susceptibility to compaction, load support capacity, and soil compressibility of Hapludox. *Soil Science Society of America Journal* 68(1), 17–24. doi:10.2136/sssaj2004.1700.

Janbu, N. 1998. Sediment deformations - a classical approach to stress–strain–time behaviour of granular media as developed at NTH over 50 year period. NTNU, Department of Geotechnical Engineering, Trondheim: Bulletin, 35, 86pp.

Kay, B. D. 1998. Soil structure and organic carbon: a review in soil processes and the carbon cycle. In: Lal, R., Kimble, J. M., Follett, R. F. and Stewart, B. A. (Eds), *Soil Processes and the Carbon Cycle*. CRC Press, Boca Raton, pp. 169–97.

Keller, T. and Arvidsson, J. 2007. Compressive properties of some Swedish and Danish structured agricultural soils measured in uniaxial compression tests. *European Journal of Soil Science* 58(6), 1373Y1381. doi:10.1111/j.1365-2389.2007.00944.x.

Keller, T., Lamandé, M., Schjønning, P. and Dexter, A. R. 2011. Analysis of soil compression curves from uniaxial confined compression tests. *Geoderma* 163(1-2), 13-23. doi:10.1016/j.geoderma.2011.02.006.

Kuan, H. L., Hallett, P. D., Griffiths, B. S., Gregory, A. S., Watts, C. W. and Whitmore, A. P. 2007. The biological and physical stability and resilience of a selection of Scottish soils to stresses. *European Journal of Soil Science* 58(3), 811-21. doi:10.1111/j.1365-2389.2006.00871.x.

Lomeling, D. and Lasu, D. M. 2015. Spatial patterns of penetration resistance and soil moisture distribution in a Sandy Loam Soil (eutric leptosol). *International Journal of Soil Science* 10(3), 130–41. doi:10.3923/ijss.2015.130.141.

Lowery, B. and Schuler, R. T. 1994. Duration and effects of compaction on soil and plant growth in Wisconsin. *Soil and Tillage Research* 29(2-3), 205-10. doi:10.1016/0167-1987(94)90058-2.

MAFF 1982. *Techniques for Measuring Soil Physical Properties. MAFF Reference* (Bk. 441). HMSO, London.

Markgraf, W. and Horn, R. 2009. Rheological investigations in soil micro mechanics: measuring stiffness degradation and structural stability on a particle scale. In: Gragg, L. P. and Cassell, J. M. (Eds), *Progress in Management Engineering*. Nova Science Publishers, Hauppauge, pp. 237-79.

Markgraf, W. 2011. Rheology in soils. In: Gliński, J., Horabik, J. and Lipiec, J. (Eds), *Encyclopedia of Agrophysics: Encyclopedia of Earth Sciences Series*. Springer, Dordrecht.

Martino, D. L. and Shaykewich, C. F. 1994. Root penetration profiles of wheat and barley as affected by soil penetration resistance in field conditions. *Canadian Journal of Soil Science* 74(2), 193-200. doi:10.4141/cjss94-027.

Mezger, T. 2002. *The Rheology-Handbook - for Users of Rotational and Oscillatory Rheometers*. Vincentz Verlag, Hannover.

Morrison, Jr., J. E., Lowery, B. and Hart, G. L. 2000. Soil penetrometer for sensing soil water content as in soil characterization for site-specific farming. In: *Proceedings of the Agrophysical and Ecological Problems of Agriculture in the 21st Century*. 6-9 September 1999, St. Petersburg (vol. 2). International Soil and Tillage Research Organization, pp. 121-9.

Naderi-Boldaji, M., Sharifi, A., Alimardani, R., Hemmat, A., Keyhani, A., Loonstra, E. H., Weisskopf, P., Stettler, M. and Keller, T. 2013. Use of a triple-sensor fusion system for on-the-go measurement of soil compaction. *Soil and Tillage Research* 128, 44-53. doi:10.1016/j.still.2012.10.002.

O'Sullivan, M. F., Dickson, J. W. and Campbell, D. J. 1987. Interpretation and presentation of cone resistance data in tillage and traffic studies. *Journal of Soil Science* 38(1), 137-48. doi:10.1111/j.1365-2389.1987.tb02131.x.

Oliver, W. C. and Pharr, G. M. 1992. An improved technique for determining hardness and elastic modulus using load and displacement sensing indentation experiments. *Journal of Materials Research* 7(6), 1564-83. doi:10.1557/JMR.1992.1564.

Naveed, M., Brown, L. K., Raffan, A. C., George, T. S., Bengough, A. G., Roose, T., Sinclair, I., Koebernick, N., Cooper, L. and Hallett, P. D. 2018. Rhizosphere-scale quantification of hydraulic and mechanical properties of soil impacted by root and seed exudates. *Vadose Zone Journal* 17(1), 170083. doi:10.2136/vzj2017.04.0083.

Naveed, M., Brown, L., Raffan, A., George, T., Bengough, A., Roose, T., Sinclair, I., Koebernick, N., Cooper, L. and Hallett, P. 2017. Plant exudates can either stabilise or weaken soils depending on species, origin and time. *European Journal of Soil Science* 68, 806-16, doi:10.1111/ejss.12487.

Raper, R. L., Washington, B. H. and Jarrell, J. D. 1999. A tractor mounted multiple-probe soil cone penetrometer. *Applied Engineering in Agriculture* 15(4), 287-90.

Rickson, R. J., Deeks, L. K., Corstanje, R., Newell-Price, P., Kibblewhite, M. G., Chambers, B., Bellamy, P., Holman, I., James, I. T., Jones, R., Kechavarsi, C., Mouazen, A. M., Ritz, K. and Waine, T. 2012. Indicators of the quality of the physical property of soil - SP1611. Available at: http://randd.defra.gov.uk/Default.aspx?Module=More&Location=None&ProjectID=17595.

Saffih-Hdadi, K., Défossez, P., Richard, G., Cui, Y. -J., Tang, A.-M. and Chaplain, V. 2009. A method for predicting soil susceptibility to the compaction of surface layers as a function of water content and bulk density. *Soil and Tillage Research* 105(1), 96–103. doi:10.1016/j.still.2009.05.012.

Schjønning, P., Lamandé, M., Munkholm, L. J., Lyngvig, H. S. and Nielsen, J. A. 2016. Soil precompression stress, penetration resistance and crop yields in relation to differently-trafficked, temperate-region sandy loam soils. *Soil and Tillage Research* 163, 298–308. doi:10.1016/j.still.2016.07.003.

Shainberg, I., Sumner, M. E., Miller, W. P., Farina, M. P. W., Pavan, M. A. and Fey, M. V. 1989. Use of Gypsum on soils: a review. *Advances in Soil Science* 9, 1–111. doi:10.1007/978-1-4612-3532-3_1.

Stettler, M., Keller, T., Weisskopf, P., Lamandé, M., Lassen, P. and Schjønning, P. 2014. Terranimo® – a web-based tool for evaluating soil compaction. *Landtechnik* 69, 132–7.

Tang, A.-M., Cui, Y.-J., Eslami, J. and Défossez, P. 2009. Analysing the form of confined uniaxial compression curve of various soils. *Geoderma* 148(3–4), 282–90. doi:10.1016/j.geoderma.2008.10.012.

Taylor, H. M. 1971. Effect of soil strength on seedling emergence, root growth and crop yield. In: Barnes, K. K., Taylor, H. M., Throckmorton, R. I. and Carleton, W. M. (Eds), *Compaction of Agricultural Soils*. American Society of Agricultural Engineers Monograph, St. Joseph, Michigan, pp. 292–305.

Valentine, T. A., Hallett, P. D., Binnie, K., Young, M. W., Squire, G. R., Hawes, C. and Bengough, A. G. 2012. Soil strength and macropore volume limit root elengation rates in many UK agricultural soils. *Annals of Botany* 110(2), 259–70. doi:10.1093/aob/mcs118.

van den Akker, J. J. H. 2004. SOCOMO: a soil compaction model to calculate soil stresses and the subsoil carrying capacity. *Soil and Tillage Research* 79(1), 113–27. doi:10.1016/j.still.2004.03.021.

Wesley, L. D. 2010. *Fundamentals of Soil Mechanics for Sedimentary and Residual Soils*. John Wiley & Sons Ltd, New York.

Young, G. D., Adams, B. A. and Topp, G. C. 2000. A portable data collection system for simultaneous cone penetrometer force and volumetric water content measurements. *Canadian Journal of Soil Science* 80(1), 23–31. doi:10.4141/S99-025.

Chapter 9

Advances in near-infrared (NIR) spectroscopy to assess soil health

Francisco J. Calderón, Oregon State University, USA; Andrew J. Margenot, University of Illinois at Urbana-Champaign, USA; and Scarlett Bailey, National Resources Conservation Service - National Soil Survey Center, USA

1 Introduction

Soil health is an emerging conceptual framework that integrates soil chemical and physical properties with biological processes. The concept of soil health is proposed to advance understanding of how soils sustain long-term functions such as crop production. The notion of healthy (and unhealthy) soils holds an intuitive appeal to stakeholders such as producers, researchers and the general public (Arbuckle and Roesch-McNally, 2016). As suggested by Arbuckle and Roesch-McNally (2016), soil health is more explicit than its predecessor concept of soil quality in placing an emphasis on biology as central to the sustainable use of soils, because 'health' connotes a living entity. The United States Department of Agriculture (USDA) Natural Resources and Conservation Service (NRCS) defines soil health as 'the continued capacity of soil to function as a vital living ecosystem that sustains plants, animals, and humans'. The central concept of a healthy soil is a soil that supports high crop productivity while maintaining ecosystem services related to aesthetics, biological diversity and environmental quality.

Soil health is additionally distinguished from its predecessor concepts by its integration of physical, chemical and biological components of soils. Soil

http://dx.doi.org/10.19103/AS.2020.0079.16
Published by Burleigh Dodds Science Publishing Limited, 2021.

health is assessed and monitored through soil health indicators (SHI) that represent these chemical, physical or biological components. Many SHI are also dynamic soil properties, in contrast to static soil properties (NRCS, 2019). Static properties such as texture or mineralogy are relatively unchanged by human activities and often result from pedogenic processes. In contrast, dynamic properties change depending on how soils are managed. For example, management practices can alter soil organic matter content, soil structure and nutrient and water availability. Thus, soil health can be conceptualized as an integration of inherent properties and properties sensitive to how the soil is managed.

Importantly, this implies that the response of soil health to management, and which SHI best capture this, can strongly depend on edaphic context. Corroborating this approach, recent assessments of major US geographic regions finds that certain SHI are more or less suited to broad combinations of soil, climate and cropping system characteristics (Fine et al., 2017). Thus, one of the major focuses of soil health research in the United States is developing context-specific evaluation and interpretation guidelines for SHI. This is reflected by the partnership of NRCS and the Soil Health Institute to collaborate on long-term agricultural experimental research sites to identify sensitive SHI for particular environmental and edaphic conditions (SHI, 2017). A second major focus of soil health research is establishing baselines and monitoring programs for soil health across the United States (NRCS, 2019).

2 Infrared spectroscopy for the analysis of soils and soil health

Infrared spectroscopy measures vibration frequencies of chemical bonds in response to exposure to infrared light (12,000 - 50 cm^{-1}). Molecules with dipole moments, either permanent or induced, are infrared active. Absorption of infrared light occurs when photon transfer to the molecular bond increases the energy state of the bond. In this excited state, the bonded atoms vibrate in various modes, including symmetric and asymmetric stretching, bending, rocking, twisting and scissoring. The utility of infrared spectroscopy arises from the fact that specific molecular bonds absorb infrared light at a characteristic frequency, and often with a distinct vibration behavior, enabling identification - and theoretically quantification - of molecular bonds that correspond to various functional groups. Thus, an infrared spectrum exhibits peaks that represent absorption of infrared light at frequencies (i.e. wavenumbers, expressed as nm or cm^{-1}) specific to the type and excitation behavior of functional groups. Additional reading on principles of infrared spectroscopy and its utility for the study of soil and other environmental samples can be found in Parikh et al. (2014).

Published by Burleigh Dodds Science Publishing Limited, 2021.

The rising adoption of infrared spectroscopy for soil property prediction in the past decades has emerged in parallel with the more recent crescendo of the soil health concept. An explicit evaluation of how these two related efforts may be coupled is therefore merited. Infrared spectroscopy has long been recognized for its ability to enable soil property prediction using chemometrics. The advantage of infrared spectroscopy is that once collected, an infrared spectrum of a soil sample can be used to measure more than one property (Ben-Dor and Banin, 1995), and theoretically all properties. Building spectral libraries offers a way to rapidly store large amounts of information on soils (Shepherd and Walsh, 2002), enabling archiving of soil information in spectral form. These advantages were recognized in the 1990s, and since then have driven increasing interest and application of infrared spectroscopy for assessments of soil quality (Viscarra Rossel et al., 2006), the predecessor of soil health (Arbuckle and Roesch-McNally, 2016). The potential of infrared to displace laborious wet chemistry measurements and routine chemical characterizations of soil fertility was immediately recognized (Janik et al., 1998), though this has not been articulated for soil health until recently (Calderón et al., 2017; Margenot et al., 2020).

Increasing adoption of infrared spectroscopy for mapping and monitoring soils has made rapid, high-throughput prediction of static soil properties (e.g. texture) routine and reliable. Though infrared has been applied for over three decades, the advent of technologies and synchronization of efforts on national (Clairotte et al., 2016; Wijewardane et al., 2018), continental (Shepherd, 2010; Vågen et al., 2016) and even global (Viscarra Rossel et al., 2016) scale efforts have precipitated an exponential use of infrared spectroscopy in soil science. This not only includes benchtop application for ex situ analysis, but also field-deployable or even portable (e.g. hand-held) infrared spectrometers (Kodaira and Shibusawa, 2013). For the purpose of soil survey, routinely measured chemical and physical properties (e.g. total organic carbon (TOC), clay, total carbonates, CEC) except pH have an acceptable level of accuracy (Seybold et al., 2019). In the United States, the accuracy of soil property predictions by infrared spectroscopy using a global model is sufficient for regional use by NRCS field offices (Seybold et al., 2019).

However, less evaluated for infrared spectroscopy prediction are biological SHI as well as less routinely evaluated physical SHI (e.g. aggregate stability). While many of these properties are what separates soil health from soil quality and agronomic assessments, SHI prediction by infrared spectroscopy does not commonly extend to biological or physical SHI. However, many biological properties that are thought to be key components of soil health were first evaluated decades ago for predictability using infrared spectroscopy (Pankhurst et al., 1995). Though not understood as SHI at the time, previous work has also evaluated the less commonly predicted soil variables such as enzymes (Zornoza et al., 2008) or labile carbon fractions (Peltre et al., 2014; Calderón et al., 2017; Margenot et al., 2020).

3 Near-infrared (NIR) spectroscopy for the analysis of soil properties

This chapter reviews and demonstrates the potential and challenges to extending infrared spectroscopy prediction using the near-infrared (NIR) for these components of soil health assessment and monitoring. The NIR as used in this chapter refers to the absorbances between 350 nm and 2500 nm. This encompasses part of the visible spectrum (350–700 nm), the near-NIR and then the region between 1100 nm and 2500 nm which is considered the NIR proper. Though this chapter focuses on laboratory-based NIR measurements, it is important to note that NIR is amenable for direct field measurements as well as remote sensing applications (Gomez et al., 2008). Though the use of NIR for the prediction of forages, grain and soil composition has a relatively long history spanning decades (Reeves III et al., 2002; Nocita et al., 2015), there is still no gold standard about how to perform the calibration development. However, the basic principles applicable for the calibration of chemometric models in general also apply to predictions based on NIR spectra. The reference analyte data, such as TOC, must be obtained with precision and accuracy, since the prediction can at best only match the quality of the calibration data. The analyte data must also bracket the range of measurements expected in the predictions. Though partial least squares (PLS) regression is well established as the common mathematical procedure to develop chemometric calibrations of soil properties using whole spectra or spectral regions, other options exist such as machine learning procedures and principal component regression.

Soil attributes that are predicted accurately with the mid-infrared (MIR) often achieve acceptable to very good calibrations in the NIR. Both infrared regions contain information on the major constituents of soils. The MIR contains the fundamental absorbances due to a variety of bonds between specific atoms, in particular the fundamental vibrations of polar bonds that are common in the inorganic and organic constituents of soils: C-O, C-C, C-H, O-H, Si-O(H), Fe-O(H) and Al-O(H) (Parikh et al., 2014). In contrast, absorbances in the NIR are largely overtones: absorbances due to harmonics and combinations of O-H, N-H and C-H bonds (Workman Jr and Weyer, 2012). Therefore, similar soil constituents have influence on the NIR and MIR spectra. These include clays, organic C and N, carbonates, and metal oxides. Compared to the MIR, however, NIR has lower absorption intensity *and* less diversity of pronounced absorbance bands corresponding to polar bonds in soil. Weak and overlapping absorbances in the NIR absorbances mean that Beer-Lambert's Law does not apply to interpret spectral features, nor can bands be resolved or attributed to particular polar bonds. NIR has distinct advantages over traditional soil analyses of soil organic matter such as Walkley-Black titration or dry combustion for TOC, because NIR is rapid and non-destructive (i.e. does not consume the soil sample). Further, the same set of NIR spectra can potentially be used to develop calibrations for other soil health biological and physical parameters, which can save time and expense.

Several challenges to the prediction of SHI using infrared spectroscopy can be expected. These are discussed below.

3.1 Minor mass component

Indicators of soil health, in particular biological, generally change at small spatiotemporal scales, making them sensitive to management but challenging to monitor consistently using labor-intensive and often prolonged wet chemistry procedures. Additionally, many biological SHI such as soil enzyme activities are the subject of ongoing research (SHI, 2017) and require additional data before being fully integrated into soil health frameworks (NRCS, 2019). To this end, NIR offers advantages of rapidity, high-throughput and low cost.

A likely challenge to infrared spectroscopy prediction of biological SHI is that these are often a minor or even negligible component by mass of a soil sample, making their direct detection by NIR unlikely. While the inability to directly measure absorbances attributed to a sample component is not inherently a challenge to NIR predictions, the low concentration of many biological SHI (e.g. part per million for labile carbon, to below part per trillion for a particular enzyme in a soil sample) means that chemometric models are likely to rely on indirect correlations of these SHI with IR-detectable soil components (e.g. organic matter and mineral absorbances).

Limited evidence exemplifies this likely challenge for perhaps the most minor by-mass component of soils: enzymes. Activities of soil enzymes were proposed as SHI in 2019 by the USDA NRCS (2019). While it is the activity of the enzyme (μmol or nmol product g^{-1} h^{-1}) rather than the concentration of enzyme itself (likely < ng kg^{-1}) that is the SHI, the mass of any particular enzyme in a soil sample is likely to be extremely low. This may challenge chemometric prediction of enzyme activity since direct detection of the enzyme is unlikely. Attempts at predicting soil enzyme activities have had mixed results, but overall indicate low accuracy. For example, enzyme activities and PLFA were poorly predicted across calcareous soils in southeastern Spain using NIR ($n = 343$) (Zornoza et al., 2008).

3.2 Component versus function

Given the focus on processes, SHI include soil variables that are not constituents of soils but rather functions or rates. Here, the biological SHI of soil enzyme

Published by Burleigh Dodds Science Publishing Limited, 2021.

activities are also illustrative. As described previously, enzyme activities are rates, expressed as an amount of substrate hydrolyzed per unit soil and per unit time (Burns et al., 2013). Soil structure is similarly an SHI (physical) that is not directly detectable. Traditional sequential sieve-based fractionation of water stable aggregates (Six et al., 2000) can be expressed as mean weight diameter (MWD), which is the geometric mean of aggregate diameter (mm) (Van Bavel, 1950). While used in some early soil quality or health assessments, MWD is labor-intensive and there is a diversity of more rapid, on-site assessments of aggregate stability (Almajmaie et al., 2017). In the interest of rapidity and low cost of user-friendly SHI (NRCS, 2019), a single measure of aggregate stability as the percent of water-stable aggregates is approximated as the percentage (by mass) of soil aggregates > 4 mm diameter that does not slake. However, these metrics of soil structure are emergent properties of soil constituents that express absorbances in the NIR, and aggregation does not have a unique spectral signature for direct detection.

3.3 Operation vs in situ SHI

In other cases, SHI may not reflect a true in situ component because the indicator is an operational variable: it is defined by the method of measurement. This is commonly the case for extraction-based SHI. For example, past attempts to predict available phosphorus pools (e.g. Mehlich-3 extractable phosphorus) have been largely unsuccessful with low R^2 and poor root mean square error (RMSE; i.e. accuracy) of predictions (He et al., 2007; Morón and Cozzolino, 2007; Zornoza et al., 2008; Abdi et al., 2012). This has been attributed to the measurement of an operationally defined variable (e.g. extractable) as opposed to direct measurement of a pool such as TOC. Though phosphate exhibits absorbance frequencies that are highly sensitive to binding environment (Arai and Sparks, 2001; Kizewski et al., 2011), in soils its detection is more challenging and requires spectral subtractions to isolate P-O frequencies (Margenot et al., 2015). On the other hand, the promise of chemometrics is that information on a particular polar bond contained in the infrared spectrum should be usable. Thus, it may be a combination of low total concentration, and absorbance, and the variable in question that collectively challenges chemometrics.

3.4 Accuracy of the wet chemistry measurement

Biological indicators of soil health can have substantial analytical variability that challenges their use in soil health assessments, requiring interpretation of RMSE. For example, soil respiration, often referred to as mineralizable carbon, measured at 6–48-hour time scales (Franzluebbers et al., 2000; NRCS, 2019), varied substantially by location of analyses and had 2–20-fold greater variability than other common soil measures such as pH or TOC (Wade et al., 2017). This can challenge predictions in two ways. First, the true value requires high replication numbers, which are not necessarily followed (Wade et al., 2017). For example, high variability in permanganate-oxidizable carbon (POXC) would require >18 analytical replicates to have sufficient statistical power, though this undermines the rapidity that is at the heart of many soil health assessments (Haney et al., 2018). Second, the advantage of NIR (or other regions of the infrared) for 'big data' and large spectral libraries with substantial regional coverage (Wijewardane et al., 2016; Wijewardane et al., 2018) can be challenged by this variability. For example, leveraging NIR for a network of field station-deployed spectrometers for region-specific monitoring of soil properties may work well for properties such as TOC and texture with low analytical variability (Seybold et al., 2019), but could be less useful to predict other SHI.

4 Using near-infrared (NIR) spectroscopy in practice: methodology

The National Cooperative Soil Survey Soil Characterization Database is housed at the KSSL in Lincoln, NE, and contains soil samples from over 38000 pedons across the United States that are measured for chemical and physical properties. As described in more detail by Seybold et al. (2019), since 2011 infrared analysis has been conducted on all soils at KSSL (air-dried, <2 mm sieved) prior to archiving. This database exemplifies how NRCS is capitalizing on the potential of large-scale infrared spectral libraries for soil monitoring (Nocita et al., 2015), an effort of increasing importance given growing interest in soil health assessments by NRCS, researchers and land managers in the United States.

NIR spectra (12 500–4000 cm^{-1}) in the NRCS library were obtained using a LabSpec 2500 Visible Near Infrared Spectrometer (ASD Inc., Colorado, USA), equipped with a fiber optic cable, reflectance accessory and sample holder. The NIR spectral collection method is detailed in Wijewardane et al. (2018). Briefly, the soils were ground and dried, sieved (<2 mm) and positioned on a sample holder with a silica window, and scanned with a reflectance accessory (MugLite, ASD Inc.). A Spectralon reference was scanned every 15 min. The resolution was set at 1 nm, and each scan was an average of 100 internal scans.

Four SHI were used to illustrate the utility of NIR for soil health measurement:

- clay content;
- TOC;
- β-glucosidase; and
- aggregate stability.

All methods are described in Soil Survey Staff (2014) by the alphanumeric codes indicated in parentheses, using air-dried (30–35°C, 1B1b2), crushed and sieved (<2 mm, 1B1b2b) soil samples, and are reported on an oven-dry basis (air-dry/oven-dry ratio, 1B1b2b4). Clay content (<0.002 mm) was determined by the pipette method based on Kilmer and Alexander (1949) (3A1a1a). Total soil carbon content was determined by dry combustion. Inorganic carbon was determined using the pressure-calcimeter method. The estimated total organic C (TOC) was obtained by subtracting total carbon minus inorganic carbon (4E1a1a1a1-2). Soil pH was determined 1:1 soil/solution suspensions (4C1a2a2a-b1). Aggregate stability was assessed based on Kemper and Rosenau (1986), which quantifies the retention of air-dry soil aggregates (1-2 mm) on a sieve after the soil has been submerged and agitated. β-glucosidase enzymes are common in soil and catalyze the degradation of cellulosic plant material into glucose. The enzyme activity was determined by incubating a 3 g soil sample with *para*-nitrophenyl-β-D-glucopyranoside substrate in a buffer solution (5 mL of 10 mM modified universal buffer) at pH 6.0 for 1 h at 37 °C, then measuring absorbance at 410 nm. Activities are expressed as mg *para*-nitrophenol kg^{-1} soil h^{-1}.

5 Using near-infrared (NIR) spectroscopy in practice: results and discussion

5.1 Relationship among SHI

Not unexpectedly, there was a strong correlation (R = 0.85) between TOC and β-glucosidase activity (Table 1). TOC was also positively correlated to aggregate stability, albeit weakly. All the other relationships between the variables have weaker Pearson correlation coefficients of -0.22–0.29. The variance of the data in the calibration learning set can govern the extent of the range of samples that can be predicted, although it may come at the expense of a loss in accuracy. In our sample set, the coefficient of variation for the clay content was much smaller than that of TOC, aggregate stability and β-glucosidase activity (Table 2). The skewness or asymmetry of a sample distribution and the kurtosis (i.e. data

Table 1 Pearson correlation coefficients between the soil health indicators. n = 709. All correlations are significant at p = 0.05

	Aggregate Stability, 0.5-2mm	β-Glucosidase	Clay
Total organic carbon	0.46	0.85	-0.22
Aggregate stability, 0.5-2mm		0.29	0.20
β-Glucosidase			-0.20

Published by Burleigh Dodds Science Publishing Limited, 2021.

Table 2 Descriptive statistics for the soil health indicators (n = 709)

Soil health parameter	Range	Max	Min	Mean	Standard Deviation	Coefficient of variation	Skewness	Kurtosis
Total organic carbon (%)	15.0	15.2	0.2	2.1	1.7	0.81	2.76	10.99
Aggregate stability, 0.5–2mm (%)	102.0	102.0	0.0	34.6	27.6	0.80	0.58	-0.95
β-glucosidase (μg *p*NP g^{-1} h^{-1})	882.0	883.0	1.0	140.7	136.7	0.97	1.99	5.20
Clay (%)	43.7	49.9	6.2	26.8	8.6	0.32	0.45	-0.31

frequency toward the extreme ends of the distribution range) can affect the distribution of the prediction errors. The frequency distributions of our data set show that both TOC and enzyme activities can have a more skewed distribution with a higher number of samples with lower values (Fig. 1). The clay content, in contrast, has a distribution that is more even across the data range than the other three variables. Thus, the re-use of an NIR spectral library should consider SHI-specific data distributions, as these may alter the appropriateness of the same spectral-analyte dataset for a given SHI. In our example, the biological SHI (β-glucosidase activity) appears to be more susceptible to prediction issues resulting from data distribution whereas the physical SHI (aggregate stability) does not exhibit this issue.

5.2 Interpretation of infrared absorbance for soil spectra

Due to the fact that most soil absorbances within the NIR range are due to harmonic combinations (Rossel and Behrens, 2010), the assignment of bands to specific soil components is not straightforward. Assigning bands in an NIR spectrum is difficult due to overlap between organic and inorganic absorbances, which is especially true for absorbances above 2000 nm (Stenberg et al., 2010). Thus, there is no basis for using single absorbance bands for the quantification

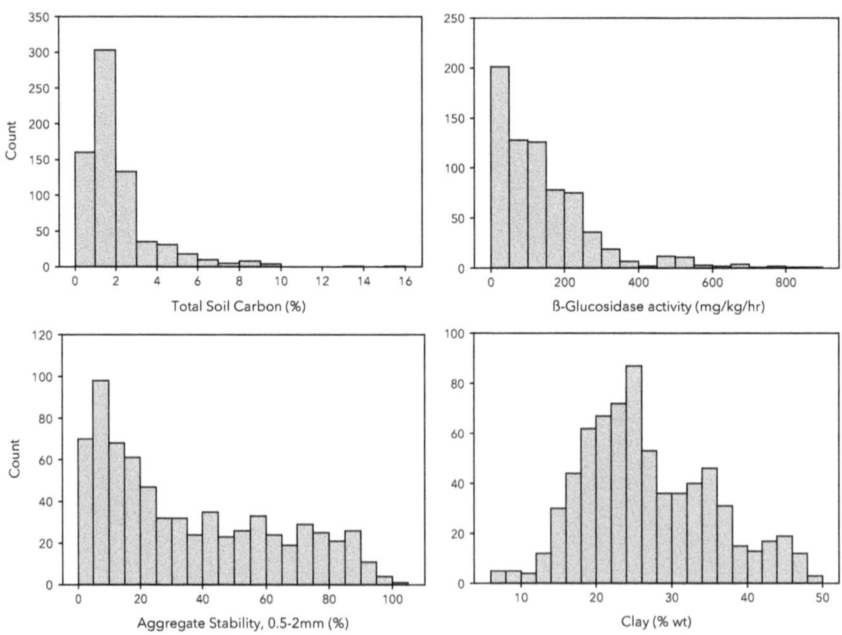

Figure 1 Frequency distribution histograms for the four soil health indicators measured across diverse US soils ($n = 709$).

Published by Burleigh Dodds Science Publishing Limited, 2021.

of TOC or carbonates. This is one reason why NIR spectra in their entirety are often used to develop calibrations.

Figure 2 presents the diffuse reflectance NIR spectrum of a dried and ground sample of a Mollisol. The spectrum contains three prominent absorbance peaks or bands, each resulting from the additive effect of several soil constituents. For example, peak number 1, which ranges from 2150 nm to 2270 nm, includes absorbances attributable to overtones or combinations of fundamental C-O bands from organic materials, as well as Al-OH, Fe-OH, and O-H from clay minerals (Rossel and Behrens, 2010). Peak 2 from 1870 nm to 2000 nm can integrate overtone and combination bands from water O-H, carboxylic C=O, methyl C-H, and phenolic C-OH. Peak number 3 in the figure ranges from 1375 nm to 1460 nm. In this region are the overtones and combination bands for hydroxyls from clays and water, and carboxyl C=O. Besides the prominent absorbance peaks, previous work has shown that other bands near 1100 nm, 1600 nm, 1700–1800 nm, 2000 nm and 2200–2400 nm can have a relevant role during calibration development (Stenberg et al., 2010).

Principal components analysis of NIR spectra shows that for this set of diverse US soils, the clay content has a discernible effect on specific NIR absorbance bands. There is a clear tendency of the soils with higher clay content to have low component 4 scores (Fig. 3). The component 4 loadings indicate that soils with high clay content had high absorbances at 620 nm, 1420 nm, 1920 nm and 2500 nm (Fig. 4). The same principal components analysis shows that TOC content has an effect on the NIR absorbance patterns, but the responses are not as clear as those found for the clay content (Fig. 5). Some of the soils with high TOC content also have relatively high scores for component 1, and loadings

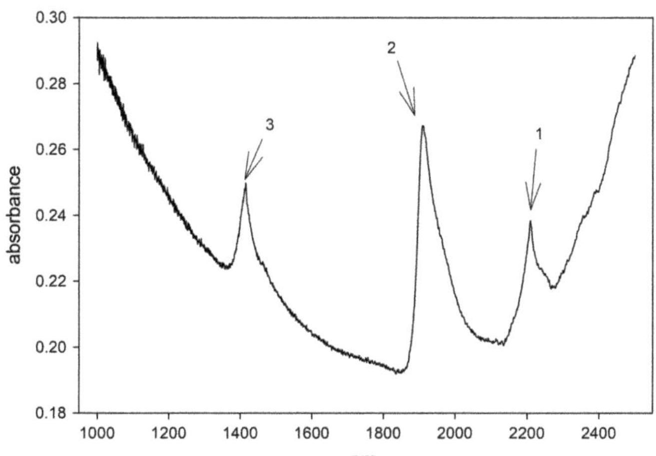

Figure 2 Diffuse reflectance NIR spectrum of a Mollisol from Akron, Colorado, USA.

Published by Burleigh Dodds Science Publishing Limited, 2021.

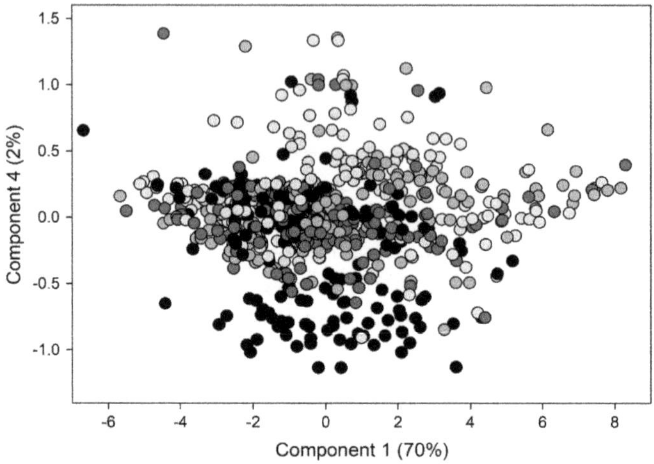

Figure 3 Principal component analysis of the NIR spectral data, color coded by clay percent by weight. Light gray range is from 6.2% to 20.4%, gray range is from 20.5% to 25.1%, dark gray range is from 25.2% to 33.2% and black range is > 33.3%. The portion of the total variation in absorbance explained by each component is in parentheses.

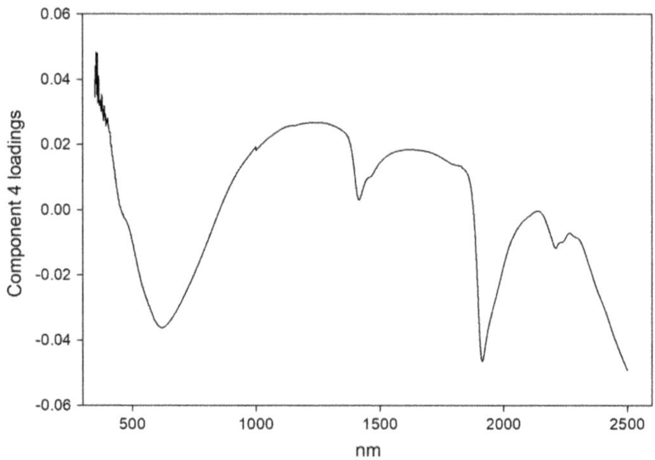

Figure 4 Component 4 loadings of the Principal Components Analysis as shown in Figures 3 and 5.

show that this is due to high absorbance between 590–760 nm (Fig. 6). In contrast, soils with low TOC tend to have high absorbance at 1908 nm.

5.3 Prediction of SHI

The development of calibrations for predicting soil organic matter pools and texture using NIR has been an active area of research since the early

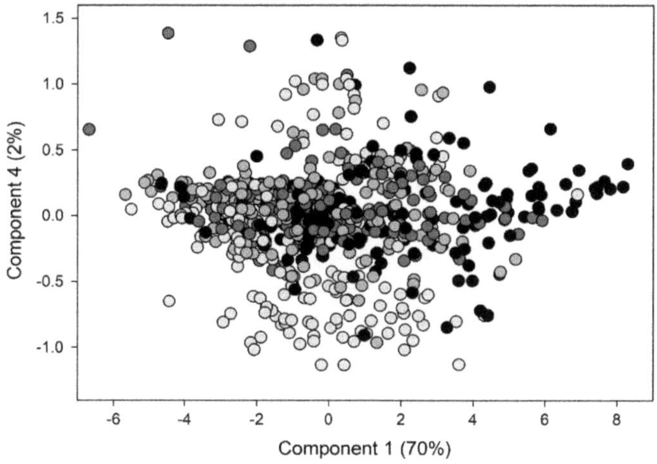

Figure 5 Principal components analysis of the NIR spectra, gray-scaled by total organic C. Light gray range is from 0.20% to 1.06%, gray range is from 1.07% to 1.61%, dark gray range is from 1.62% to 2.39% and black range is > 2.40%. The portion of the total variation in absorbance explained by each component is in parenthesis.

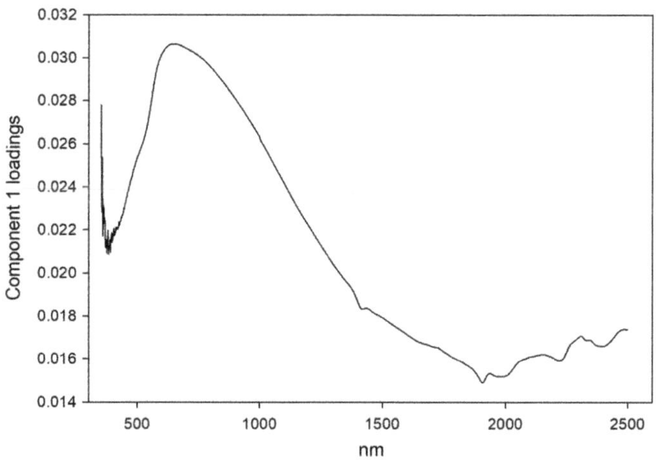

Figure 6 Component 1 loadings of the Principal Components Analysis as shown in Figure 5.

2000s (Viscarra Rossel et al., 2006). With many of these soil variables considered candidate SHI, there is clear potential for NIR for rapid, inexpensive measurement of SHI such as texture and TOC (Vågen et al., 2010). As described above, minerals and organic matter have direct effects on absorbance frequencies in the NIR and provide sound basis for anticipating relatively accurate calibrations that can be developed for SHI related to these soil constituents (e.g. labile carbon fractions). Calibrations for minor mass components or emergent properties, often biological and physical SHI such as enzyme activities or aggregate stability, could also be successful. For example, POXC is a biological SHI with concentrations ranging from 500–1400 mg kg^{-1} across diverse temperate and tropical agroecosystems (Culman et al., 2012; Margenot et al., 2017; Bongiorno et al., 2019). Despite this relatively low mass proportion (<0.14% of total soil mass), POXC was better predicted than TOC by NIR and with moderate accuracy ($RMSE_{val}$ = 170 mg kg^{-1}) across diverse US soils (n = 496). Inspection of component loadings showed similar absorbances were used by the PLS regression model to predict POXC and TOC, consistent with direct detection of POXC in infrared spectra (Calderón et al., 2017; Margenot et al., 2020). However, acceptable prediction accuracies for SHI that are a smaller mass component (e.g. μg kg^{-1} or less) is likely to be challenged by their low to non-existent contribution of absorbance features to an NIR spectrum. Additionally, the calibration might be compromised due to the correlation between soil properties with polar bonds that express frequencies in the NIR and soil variables that have no strong NIR signature. Alternatively, prediction of minor mass components or emergent property SHI may simply be due to their correlations with a major soil constituent such as TOC or clay content – in other words, a calibration by proxy, which will generally be weaker than the correlation between the two soil variables. These types of calibrations by proxy should not be expected to be robust at large geographic scales or across soil types (Stenberg et al., 2010).

Soil organic matter absorbances in the NIR are less marked and intense compared to the MIR. The NIR, however, can include part of the visible range (> 740 nm) depending on the type of detector used. This is important because the visible region does have a more pronounced absorbance response to soil organic matter than the NIR proper, and soils higher in TOC tend to have darker color compared to soils with lower TOC, all else being equal. Because of this, the inclusion of visible frequencies sometimes improves predictive calibrations for TOC (Viscarra Rossel et al., 2006). Quantitative changes in TOC are often accompanied with qualitative changes in the chemistry of TOC (Matamala et al., 2017; Ramírez et al., 2020). This is thought to enable the development of chemometric calibrations. NIR spectra are generally sensitive to changes in organic moieties associated with variations in soil organic matter chemistry (Ben-Dor and Banin, 1995), because the spectral characteristics of high TOC soils are affected by the accumulation of more labile organic moieties. For example, soils that are preserved by freezing, dry conditions or waterlogging can accumulate substantial amounts of organic matter and also have a higher proportion of undecomposed plant residues. Nevertheless, in soils containing relatively low amounts of TOC, any changes in TOC chemistry have only a slight influence on their NIR absorptions because organic matter makes up a small percentage of the soil, and because organic bands are relatively weak absorbing in the NIR (Viscarra Rossel and McBratney, 1998). As a result, soil sample sets that include significant geographical variability and have low SOM can be challenging when developing chemometric calibrations for TOC with infrared spectra (Udelhoven et al., 2003).

For the 709 soils across the United States with SOC ranging from 0.2% to 15.2%, prediction of SOC from NIR using PLS regression was good (Fig. 7) and was the variable for which greatest variance was explained by the PLS model (R^2 = 0.79) (Table 3). RMSE of the calibration and validation set were similar at 0.76 and 0.78, respectively. Though spectral processing treatment of standard normal variate (SNV) detrending did not improve calibration accuracy, there was slightly greater accuracy with multiple scatter correction (MSC) pretreatment (lower RMSE). The activity of β-glucosidase was well predicted and only of slightly less calibration quality than TOC prediction (up to R^2 0.75 and 0.80, respectively) (Table 3). Activity of this hydrolytic enzyme, which tends to be well correlated with TOC in general (Sinsabaugh et al., 2008), was slightly less predicted than TOC by the PLS model with SNV detrending pretreatment (R^2 = 0.77). Clay content was well predicted from NIR spectra by the PLS model (Fig. 8), with no improvement in spectral processing (MSC) or even lower prediction accuracy (SNV) (Table 3). The RMSE of validation of 4.40% indicates an accuracy within < 5% clay content, which is generally comparable or poorer than the accuracy of texture-by-feel. For example, across a set of German soils, clay content estimated using texture-by-feel was found to be within 1.7%

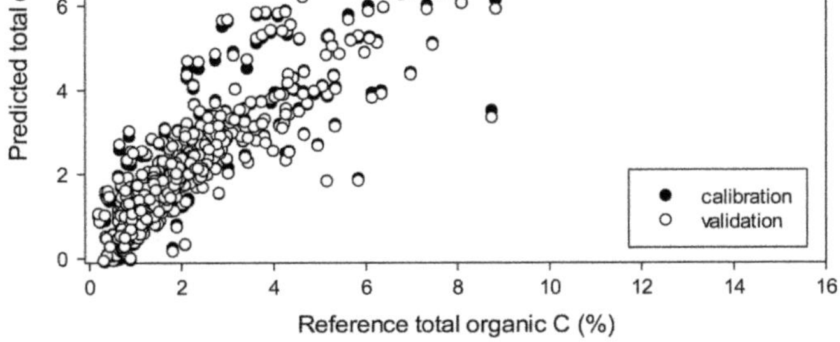

Figure 7 Results of a partial least squares regression used to calibrate a prediction model for total organic carbon (n = 709). NIR spectra were mean centered prior to PLS using the NIPALS algorithm with a 7-factor model. The validation was random cross validation with 20 segments.

Table 3 Calibration and validation performance (as R^2 (RMSE)) of the PLS regression calibrations for the soil health parameters using NIR spectra (n = 709). MSC is multiplicative scatter correction. SNV is standard normal variate

SHI	Mean centering only		Mean centering plus SNV Detrending		Mean centering Plus MSC	
	Cal.	Val.	Cal.	Val.	Cal.	Val.
Total organic carbon (%)	0.80(0.76)	0.79(0.78)	0.80(0.76)	0.79(0.78)	0.85(0.67)	0.83(0.69)
Aggregate stability, 0.5–2mm (%)	0.30(23.1)	0.29(23.30)	0.26(23.6)	0.26(23.9)	0.31(22.8)	0.30(23.1)
β-glucosidase (μg *p*NP g^{-1} h^{-1})	0.66(79.1)	0.64(81.9)	0.77(65.1)	0.75(68.2)	0.77(65.2)	0.75(68.4)
Clay (%)	0.75(4.33)	0.74(4.40)	0.65(5.10)	0.64(5.20)	0.75(4.33)	0.74(4.40)

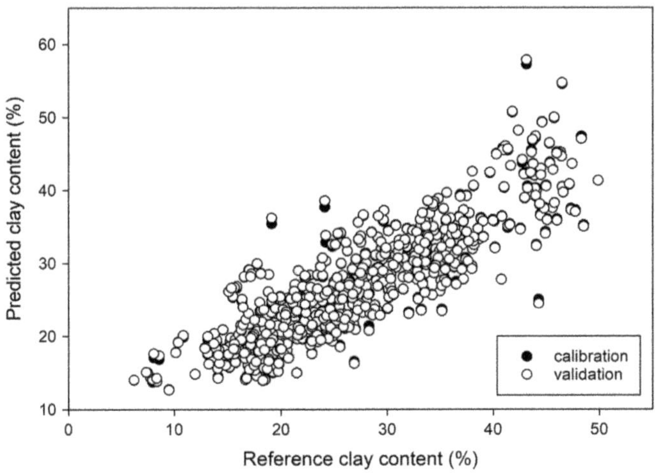

Figure 8 Results of a partial least squares regression used to calibrate a prediction model for soil clay content (n = 709). NIR spectra were mean centered prior to PLS using the NIPALS algorithm with a 7-factor model. The validation was random cross validation with 20 segments.

accuracy, slightly better than NIR PLS at 1.8% accuracy (Vos et al., 2016). In contrast to TOC and clay content, aggregate stability was not well predicted but had some improvement in accuracy with spectral preprocessing of SNV but not MSC (Table 3). A low amount of variability in aggregate stability was explained by the PLS model (R^2 validation 0.26–0.30) with negligible differences follow spectral treatments, and an average RMSE of validation of 23.4% of water-stable aggregates.

6 Conclusion and future trends

Infrared spectroscopy offers cost-effective analysis of soils by enabling rapid, high-throughput prediction of soil properties. To-date, soil properties based on major soil constituents with consequently strong direct or significant influences on infrared absorbances can be accurately predicted by chemometric models, commonly PLS regression, using NIR spectra. Many soil properties for which accurate prediction from spectra are well established, such as TOC and clay content, are considered SHI, meaning benefits of infrared spectroscopy for soil health analyses can already be realized. The evaluation of minor soil constituents or even soil *processes* that are at the heart of many biological and physical SHI such as enzyme activities and soil structure are likely more challenging to predict because their prediction is likely based on indirect relationships with infrared absorbance features. As illustrated using n = 709 soil spectra from the USDA NRCS National Soil Survey Center library, NIR offers accurate prediction of TOC and clay content, though the accuracy of prediction (i.e. RMSE) of β-glucosidase activity and aggregate stability may not be acceptable for soil health assessment and monitoring purposes.

As infrared spectroscopy continues to transform soil science and survey (Nocita et al., 2015; Seybold et al., 2019), prediction of sometimes costly SHI can facilitate greater soil health assessment and monitoring. Future directions of research include: (1) Predicting soil health scores, produced by scoring functions that weigh multiple SHI (Fine et al., 2017); (2) In situ sensing or real-time monitoring, which has been performed using tractor-mounted or mobile infrared spectrometers (Kodaira and Shibusawa, 2013; Schirrmann et al., 2013); (3) transferability of PLS models among spectrometers, spectral libraries (i.e. validation sets) and geographic areas or soil types (Seybold et al., 2019).

7 Where to look for further information

A number of excellent resources exist on fundamental and applied aspects of infrared spectroscopy in soil science and soil health. For a detailed overview of the theory of infrared spectroscopy, the reader is referred to Griffiths and de Haseth (1986) and Johnston and Aochi (1996). Detailed descriptions of additional applications of infrared spectroscopy for soils beyond soil organic matter characterization and quantification are offered by Johnston and Aochi (1996), Wetterlind et al. (2013) and Parikh et al. (2014).

8 References

Abdi, D., Tremblay, G. F., Ziadi, N., Bélanger, G. and Parent, L.-É. 2012. Predicting soil phosphorus-related properties using near-infrared reflectance spectroscopy. *Soil Science Society of America Journal* 76(6), 2318–2326.

Almajmaie, A., Hardie, M., Acuna, T. and Birch, C. 2017. Evaluation of methods for determining soil aggregate stability. *Soil and Tillage Research* 167, 39–45.

Arai, Y. and Sparks, D. L. 2001. ATR-FTIR spectroscopic investigation on phosphate adsorption mechanisms at the ferrihydrite-water interface. *Journal of Colloid and Interface Science* 241(2), 317–326.

Arbuckle, J. and Roesch-McNally, G. 2016. Iowa farmers' perspective on soil health, soil health conference. Iowa State University, Ames, IA.

Ben-Dor, E. and Banin, A. 1995. Near-infrared analysis as a rapid method to simultaneously evaluate several soil properties. *Soil Science Society of America Journal* 59(2), 364–372.

Bongiorno, G., Bünemann, E. K., Oguejiofor, C. U., Meier, J., Gort, G., Comans, R., Mäder, P., Brussaard, L. and de Goede, R. 2019. Sensitivity of labile carbon fractions to tillage and organic matter management and their potential as comprehensive soil quality indicators across pedoclimatic conditions in Europe. *Ecological Indicators* 99, 38–50.

Burns, R. G., DeForest, J. L., Marxsen, J., Sinsabaugh, R. L., Stromberger, M. E., Wallenstein, M. D., Weintraub, M. N. and Zoppini, A. 2013. Soil enzymes in a changing environment: current knowledge and future directions. *Soil Biology and Biochemistry* 58, 216–234.

Calderón, F. J., Culman, S., Six, J., Franzluebbers, A. J., Schipanski, M., Beniston, J., Grandy, S. and Kong, A. Y. Y. 2017. Quantification of soil permanganate oxidizable C (POXC) using infrared spectroscopy. *Soil Science Society of America Journal* 81(2), 277–288.

Clairotte, M., Grinand, C., Kouakoua, E., Thébault, A., Saby, N. P. A., Bernoux, M. and Barthès, B. G. 2016. National calibration of soil organic carbon concentration using diffuse infrared reflectance spectroscopy. *Geoderma* 276, 41–52.

Culman, S. W., Snapp, S. S., Freeman, M. A., Schipanski, M. E., Beniston, J., Lal, R., Drinkwater, L. E., Franzluebbers, A. J., Glover, J. D., Grandy, A. S., Lee, J., Six, J., Maul, J. E., Mirksy, S. B., Spargo, J. T. and Wander, M. M. 2012. Permanganate oxidizable carbon reflects a processed soil fraction that is sensitive to management. *Soil Science Society of America Journal* 76(2), 494–504.

Fine, A. K., van Es, H. M. and Schindelbeck, R. R. 2017. Statistics, scoring functions, and regional analysis of a comprehensive soil health database. *Soil Science Society of America Journal* 81(3), 589–601.

Franzluebbers, A. J., Haney, R. L., Honeycutt, C. W., Schomberg, H. H. and Hons, F. M. 2000. Flush of carbon dioxide following rewetting of dried soil relates to active organic pools. *Soil Science Society of America Journal* 64(2), 613–623.

Gomez, C., Rossel, R. A. V. and McBratney, A. B. 2008. Soil organic carbon prediction by hyperspectral remote sensing and field vis-NIR spectroscopy: an Australian case study. *Geoderma* 146(3–4), 403–411.

Griffiths, P. R. and de Haseth, J. A. 1986. *Fourier Transform Infrared Spectrometry*. John Wiley and Sons, New York.

Haney, R. L., Haney, E. B., Smith, D. R., Harmel, R. D. and White, M. J. 2018. The soil health tool–theory and initial broad-scale application. *Applied Soil Ecology* 125, 162–168.

He, Y., Huang, M., García, A., Hernández, A. and Song, H. 2007. Prediction of soil macronutrients content using near-infrared spectroscopy. *Computers and Electronics in Agriculture* 58(2), 144–153.

Janik, L. J., Merry, R. H. and Skjemstad, J. O. 1998. Can mid infrared diffuse reflectance analysis replace soil extractions? *Australian Journal of Experimental Agriculture* 38(7), 681–696.

Johnston, C. T. and Aochi, Y. O. 1996. Fourier transform infrared and Raman spectroscopy. In: *Methods of Soil Analysis Part 3–Chemical Methods* Sparks, D. L., Page, A. L., Helmke, P. A. and Loeppert, R. H. (Eds). Soil Science Society of America, American Society of Agronomy, Madison, WI, pp. 269–321.

Kemper, W. and Rosenau, R. 1986. Aggregate stability and size distribution: methods of soil analysis: part 1. *Physical and Mineralogical Methods* 5, 425–442.

Kilmer, V. J. and Alexander, L. T. 1949. Methods of making mechanical analyses of soils. *Soil Science* 68(1), 15–24.

Kizewski, F., Liu, Y. T., Morris, A. and Hesterberg, D. 2011. Spectroscopic approaches for phosphorus speciation in soils and other environmental systems. *Journal of Environmental Quality* 40(3), 751–766.

Kodaira, M. and Shibusawa, S. 2013. Using a mobile real-time soil visible-near infrared sensor for high resolution soil property mapping. *Geoderma* 199, 64–79.

Margenot, A., O'Neill, T., Sommer, R. and Akella, V. 2020. Predicting soil permanganate oxidizable carbon (POXC) by coupling DRIFT spectroscopy and artificial neural networks (ANN). *Computers and Electronics in Agriculture* 168, 105098.

Margenot, A. J., Calderón, F. J. and Parikh, S. J. 2015. Limitations and potential of spectral subtractions in fourier-transform infrared spectroscopy of soil samples. *Soil Science Society of America Journal* 80(1), 10–26.

Margenot, A. J., Pulleman, M. M., Sommer, R., Paul, B. K., Parikh, S. J., Jackson, L. E. and Fonte, S. J. 2017. Biochemical proxies indicate differences in soil C cycling induced by long-term tillage and residue management in a tropical agroecosystem. *Plant and Soil* 420(1-2), 315–329.

Matamala, R., Calderón, F. J., Jastrow, J. D., Fan, Z., Hofmann, S. M., Michaelson, G. J., Mishra, U. and Ping, C.-L. 2017. Influence of site and soil properties on the DRIFT spectra of northern cold-region soils. *Geoderma* 305, 80–91.

Morón, A. and Cozzolino, D. 2007. Measurement of phosphorus in soils by near infrared reflectance spectroscopy: effect of reference method on calibration. *Communications in Soil Science and Plant Analysis* 38(15–16), 1965–1974.

Nocita, M., Stevens, A., van Wesemael, B., Brown, D. J., Shepherd, K. D., Towett, E., Vargas, R. and Montanarella, L. 2015. Soil spectroscopy: an opportunity to be seized. *Global Change Biology* 21(1), 10–11.

NRCS 2019. Recommended soil health indicators and associated laboratory procedures. In: *Soil Health* Stott, D. E. (Ed.). USDA NRCS, Washington, DC.

Pankhurst, C., Hawke, B., McDonald, H., Kirkby, C., Buckerfield, J., Michelsen, P., O'Brien, K., Gupta, V. and Doube, B. 1995. Evaluation of soil biological properties as potential bioindicators of soil health. *Australian Journal of Experimental Agriculture* 35(7), 1015–1028.

Parikh, S. J., Goyne, K. W., Margenot, A. J., Mukome, F. N. D. and Calderón, F. J. 2014. Chapter One. Soil chemical insights provided through vibrational spectroscopy. In: *Advances in Agronomy* Donald, L. S. (Ed.). Academic Press, London, pp. 1–148.

Peltre, C., Bruun, S., Du, C., Thomsen, I. K. and Jensen, L. S. 2014. Assessing soil constituents and labile soil organic carbon by mid-infrared photoacoustic spectroscopy. *Soil Biology and Biochemistry* 77, 41–50.

Ramírez, P. B., Calderón, F. J., Fonte, S. J., Santibáñez, F. and Bonilla, C. A. 2020. Spectral responses to labile organic carbon fractions as useful soil quality indicators across a climatic gradient. *Ecological Indicators* 111, 106042.

Reeves III, J., McCarty, G. and Mimmo, T. 2002. The potential of diffuse reflectance spectroscopy for the determination of carbon inventories in soils. *Environmental Pollution* 116 (Suppl. 1), S277–S284.

Rossel, R. A. V. and Behrens, T. 2010. Using data mining to model and interpret soil diffuse reflectance spectra. *Geoderma* 158(1-2), 46–54.

Schirrmann, M., Gebbers, R. and Kramer, E. 2013. Performance of automated near-infrared reflectance spectrometry for continuous in situ mapping of soil fertility at field scale. *Vadose Zone Journal* 12(4), vzj2012.0199.

Seybold, C. A., Ferguson, R., Wysocki, D., Bailey, S., Anderson, J., Nester, B., Schoeneberger, P., Wills, S., Libohova, Z., Hoover, D. and Thomas, P. 2019. Application of mid-infrared spectroscopy in soil survey. *Soil Science Society of America Journal* 83(6), 1746–1759.

Shepherd, K. 2010. Soil spectral diagnostics-infrared, x-ray and laser diffraction spectroscopy for rapid soil characterization in the Africa Soil Information Service.

19th World congress of soil science, soil solutions for a changing world. Brisbane Australia, pp. 1-6.

Shepherd, K. D. and Walsh, M. G. 2002. Development of reflectance spectral libraries for characterization of soil properties. *Soil Science Society of America Journal* 66(3), 988-998.

SHI 2017. Enriching soil, enhancing life: an action plan for soil health. In: *Soil Health Institute* Honeycutt, C. W. (Ed.), p. 48. Available at: https://soilhealthinstitute.org/wp-content/uploads/2017/05/Action-Plan-FINAL-for-flipbook-3.pdf

Sinsabaugh, R. L., Lauber, C. L., Weintraub, M. N., Ahmed, B., Allison, S. D., Crenshaw, C., Contosta, A. R., Cusack, D., Frey, S., Gallo, M. E., Gartner, T. B., Hobbie, S. E., Holland, K., Keeler, B. L., Powers, J. S., Stursova, M., Takacs-Vesbach, C., Waldrop, M. P., Wallenstein, M. D., Zak, D. R. and Zeglin, L. H. 2008. Stoichiometry of soil enzyme activity at global scale. *Ecology Letters* 11(11), 1252-1264.

Six, J., Paustian, K., Elliott, E. T. and Combrink, C. 2000. Soil structure and organic matter I. Distribution of aggregate-size classes and aggregate-associated carbon. *Soil Science Society of America Journal* 64(2), 681-689.

Soil Survey Staff. 2014. Kellogg Soil Survey Laboratory Methods Manual. Soil Survey Investigations Report No. 42, Version 5.0. R. Burt and Soil Survey Staff (ed.). U.S. Department of Agriculture, Natural Resources Conservation Service.

Stenberg, B., Rossel, R. A. V., Mouazen, A. M. and Wetterlind, J. 2010. Visible and near infrared spectroscopy in soil science. *Advances in Agronomy* 107, 163-215.

Udelhoven, T., Emmerling, C. and Jarmer, T. 2003. Quantitative analysis of soil chemical properties with diffuse reflectance spectrometry and partial least-square regression: a feasibility study. *Plant and Soil* 251(2), 319-329.

Vågen, T., Shepherd, K., Walsh, M., Winowiecki, L., Desta, L. and Tondoh, J. 2010. *AfSIS Technical Specifications: Soil Health Surveillance*. World Agroforestry Centre, Nairobi, Kenya.

Vågen, T.-G., Winowiecki, L. A., Tondoh, J. E., Desta, L. T. and Gumbricht, T. 2016. Mapping of soil properties and land degradation risk in Africa using MODIS reflectance. *Geoderma* 263, 216-225.

Van Bavel, C. H. M. 1950. Mean weight-diameter of soil aggregates as a statistical index of aggregation. *Soil Science Society of America Journal* 14(C), 20-23.

Viscarra Rossel, R. A., Behrens, T., Ben-Dor, E., Brown, D. J., Demattê, J. A. M., Shepherd, K. D., Shi, Z., Stenberg, B., Stevens, A., Adamchuk, V., Aïchi, H., Barthès, B. G., Bartholomeus, H. M., Bayer, A. D., Bernoux, M., Böttcher, K., Brodský, L., Du, C. W., Chappell, A., Fouad, Y., Genot, V., Gomez, C., Grunwald, S., Gubler, A., Guerrero, C., Hedley, C. B., Knadel, M., Morrás, H. J. M., Nocita, M., Ramirez-Lopez, L., Roudier, P., Campos, E. M. R., Sanborn, P., Sellitto, V. M., Sudduth, K. A., Rawlins, B. G., Walter, C., Winowiecki, L. A., Hong, S. Y. and Ji, W. 2016. A global spectral library to characterize the world's soil. *Earth-Science Reviews* 155, 198-230.

Viscarra Rossel, R. A. and McBratney, A. B. 1998. Laboratory evaluation of a proximal sensing technique for simultaneous measurement of soil clay and water content. *Geoderma* 85(1), 19-39.

Viscarra Rossel, R. A., Walvoort, D. J. J., McBratney, A. B., Janik, L. J. and Skjemstad, J. O. 2006. Visible, near infrared, mid infrared or combined diffuse reflectance spectroscopy for simultaneous assessment of various soil properties. *Geoderma* 131(1-2), 59-75.

Vos, C., Don, A., Prietz, R., Heidkamp, A. and Freibauer, A. 2016. Field-based soil-texture estimates could replace laboratory analysis. *Geoderma* 267, 215-219.

Wade, J., Culman, S. W., Hurisso, T. T., Miller, R. O., Baker, L. and Horwath, W. R. 2017. Sources of variability that compromise mineralizable carbon as a soil health indicator. *Soil Science Society of America Journal* 82(1), 243–252.

Wetterlind, J., Stenberg, B. and Rossel, R. V. 2013. Soil analysis using visible and Near infrared spectroscopy. In: *Plant Mineral Nutrients* Maathuis, F. J. M. (Ed.). Humana Press, New York, pp. 95–107.

Wijewardane, N. K., Ge, Y., Wills, S. and Libohova, Z. 2018. Predicting physical and chemical properties of US soils with a mid-infrared reflectance spectral library. *Soil Science Society of America Journal* 82(3), 722–731.

Wijewardane, N. K., Ge, Y., Wills, S. and Loecke, T. 2016. Prediction of soil carbon in the conterminous United States: visible and near infrared reflectance spectroscopy analysis of the rapid carbon assessment project. *Soil Science Society of America Journal* 80(4), 973–982.

Workman Jr., J. and Weyer, L. 2012. *Practical Guide and Spectral Atlas for Interpretive Near-Infrared Spectroscopy*. CRC Press, Boca Raton.

Zornoza, R., Guerrero, C., Mataix-Solera, J., Scow, K. M., Arcenegui, V. and Mataix-Beneyto, J. 2008. Near infrared spectroscopy for determination of various physical, chemical and biochemical properties in mediterranean soils. *Soil Biology and Biochemistry* 40(7), 1923–1930.

Chapter 10

Spectral mapping of soil organic carbon

Bas van Wesemael, Université catholique de Louvain, Belgium

1 Introduction

Croplands have lost the majority of their soil organic carbon since the onset of agriculture (Sanderman *et al.*, 2017). The main reasons for this loss are the reduced return of plant residues to the soil compared to forest and grassland systems as well as the increase in microbial activity resulting from the disturbance by tillage, harrowing, fertilization and harvesting (Chenu *et al.*, 2019). Soil organic matter and its main component soil organic carbon (SOC) are essential for preserving a healthy soil and maintaining soil fertility. It is proposed that soil health is dependent on the maintenance of four major functions: carbon transformations; nutrient cycles; soil structure maintenance and the regulation of pests and diseases (Kibblewhite *et al.*, 2008). A meta-analysis by Oldfield *et al.* (2019) showed that SOC content was related to yield potential of wheat and maize and that an increase in SOC concentrations up to region-specific targets can reduce reliance on nitrogen fertilizer and help close the yield gap. Moreover, Minasny *et al.* (2017) demonstrated the potential of atmospheric CO_2 sequestration through increase in SOC stock of a range of agricultural soils along a global climate gradient. Unfortunately, high-resolution maps that indicate areas where SOC levels are critically low and where improved management options, such as cover crops, agroforestry no-till, could be most effective are scarce. Recent SOC maps at regional to global scale consist of rather coarse grid cells that are too coarse to represent heterogeneity at the field scale and are based on legacy data that is generally not up-to-date as

http://dx.doi.org/10.19103/AS.2020.0079.18

collecting and analyzing new soil cores is expensive (Rogge *et al.*, 2018). Apart from indicating zones with critically low SOC contents, SOC maps are particularly valuable for assessing the SOC variability in support of sample designs that are both representative for the region and minimize the number of sample points required (Conant and Paustian, 2002; de Gruijter *et al.*, 2016).

Xiao *et al.* (2019) provided an extensive overview of the role of remote sensing in measuring fluxes and stocks in the terrestrial C cycle. For quantifying SOC stocks, they describe two main techniques: (i) digital soil mapping (DSM) relying on empirical relationships between measured soil properties and spatially distributed co-variates (McBratney *et al.*, 2003) and (ii) imaging spectrometry of bare topsoils based on chemometric techniques. DSM techniques use legacy soil data and relate them to spatially distributed layers describing the so-called SCORPAN (soil, climate, organisms, relief, parent material, age and site) co-variates. Among these SCORPAN co-variates, organisms play a dominant role in explaining the variability in SOC at the regional to landscape scale (Lamichhane, Kumar and Wilson, 2019). Satellite remote sensing products are particularly effective in providing co-variates characterizing the role of organisms such as land use maps and indicators for C input from the vegetation (e.g. NDVI; Poggio *et al.*, 2013).

This chapter focuses on the direct remote sensing of SOC (i.e. optical remote sensing using hyperspectral and high-resolution multispectral sensors), developed since the early 2000s. For croplands in seedbed condition, the uppermost 20–30 cm of the soil is homogenized by annual tillage, and therefore, the reflectance of the soil surface represents the soil properties throughout the plow layer. The spectral characteristics of soils in the visible, near infrared and shortwave infrared domain (i.e. VNIR-SWIR: 400–2500 nm) have proven successful in predicting soil properties since the 1990s (e.g. Ben-Dor and Banin, 1995). Both physical and chemical properties are associated with the continuum spectral signature and specific absorption bands (Stenberg *et al.*, 2010). Soil organic carbon is a so-called chromophore related to soil color and displays absorption features in the visible spectrum (400–700 nm), while weak overtones and combinations of these vibrations are due to stretching and bending of covalent bonds in the NIR and SWIR (700–2500 nm) regions (Angelopoulou *et al.*, 2019). Soil is a complex system with a large variability in chemical and physical composition. The correlation between the soil chromophores and the chemical or physical properties of the soil is not straightforward. The chemometric approach is an empirical method with a strong spectroscopic basis and for which selected bands in the model have specific assignments (Chabrillat *et al.*, 2019). Data mining techniques are used to define these empirical relations between the reflectance at several wavelengths and the soil property. They consist of multivariate analysis of the spectra against the chemical/physical data through principal component analysis regression (Chang *et al.*, 2001), partial least squares regression (PLSR) (Zhao *et al.*, 2015), artificial neural networks (ANN) and random forest among other techniques. Since these techniques are empirical and derived from multivariate statistics, they apply to the population of samples for which they were developed. The accuracy of the models is assessed through (cross) validation using samples for which soil properties were analyzed by means of traditional laboratory techniques. The figures of merit that are most frequently reported are the R^2 of the predicted versus observed SOC contents, the root mean square error (RMSE), the ratio of performance to deviation (RPD, i.e. the standard deviation divided by the RMSE) and the ratio of performance to interquartile range (RPIQ, i.e. similar to the RPD, but using the interquartile range instead of the standard deviation (Bellon-Maurel *et al.*, 2010)). The quality of SOC maps, based on the SOC prediction using the spectra acquired in each grid-cell, is evaluated based on the model uncertainty. After all, the regression models have an inherent distribution of their results and for each pixel this distribution results in an uncertainty that can be mapped. Few examples of uncertainty in spectral mapping of soil properties exist (e.g. Gomez, Drost and Roger, 2015).

A data set containing soil properties analyzed in the laboratory together with their reflectance spectra of dry, ground samples acquired in stable laboratory conditions is referred to as a soil spectral library and is used to derive these multivariate prediction models. While such libraries were in the past often restricted to a limited study area, large-scale spectral libraries are becoming available covering entire countries such as Australia (Baldock *et al.*, 2013), USA (Wijewardane *et al.*, 2016) or Brazil (Demattê *et al.*, 2019), continents such as Europe (land use and coverage area frame survey: LUCAS; Tóth *et al.*, 2013) or Africa (Shepherd and Walsh, 2007) and even a global spectral library with voluntary contributions from many countries (Viscarra Rossel *et al.*, 2016). As the relation between chromophores and soil properties is not always straightforward, applying empirical models developed from a spectral library to predict SOC content for a sample collected anywhere within a large area with heterogeneous soils does not always give satisfactory results. For large regions with a range of soil types, the variation in one or more soil properties (e.g. iron hydroxides or sand content) can influence a prediction model for another soil property such as SOC. In order to deal with the heterogeneity of the soils, Nocita *et al.* (2014) used a limited number of spectra from the LUCAS library based on their similarity with the spectrum of the sample to be predicted and developed a so-called local PLSR model.

Recently, miniature spectrometers and multi-camera systems that can be mounted on unmanned aerial systems (UAS) became available. These systems are generally limited to the VNIR domain (400–1000 nm). Nevertheless, Aldana-Jague *et al.* (2016) were able to map the SOC content in a long-term trial in

Rothamsted (UK) using multispectral imagery (wavelengths 480–550–670–780–880–1000 nm). They obtained a R^2 of 0.95 and an RMSE of 0.31% during cross validation. Furthermore, satellite imagery is now freely available at relatively high spectral (nine bands) and spatial (20 m) resolution from the Sentinel-2 multispectral instrument (MSI) with a revisit time of five days. Hyperspectral satellites such as PRISMA (Loizzo *et al.*, 2018) have been launched or are planned for the near future EnMAP (Guanter *et al.*, 2015) or Shalom (Feingersh and Ben Dor, 2015). These recent developments mean that (hyperspectral) remote sensing imagery is now much more readily available as the flight campaigns are either flexible and low-cost such as for the UAS-borne systems or the images are freely available at regular intervals such as for the new generation of satellites. The potential for either routine mapping of a large number of fields (by UAS systems) or the mapping of large areas (by satellites) means that there is a need for automated procedures for the calibration of the spectral signal and dealing with disturbance of the surface conditions (vegetation, residues, roughness, moisture; Diek *et al.*, 2019)

This chapter first reviews recent pilot studies covering limited areas often with exposed bare soils. Then we focus on the challenges for large-scale application of spectral mapping when the soil and parent material are heterogeneous and surface conditions are unknown. In order to deal with these constraints we discuss (i) calibration of spectral models based on large spectral libraries, (ii) surface conditions that disturb the soil signal and (iii) time series of images in order to delimit cropland fields and increase the extent of bare soil that can be mapped. Finally, a case study deals with a SOC prediction map derived from the spectra of a Sentinel-2 image and calibrated using the LUCAS spectral library.

2 Pilot studies of spectral SOC mapping

Pilot studies have demonstrated that the spectral signal of the soil surface as captured by the sensor can be calibrated using the soil properties of the corresponding pixels. For the calibration, composite samples are generally collected from the plow layer (0–15 cm) covering an area from 6 by 6 m (Gholizadeh *et al.*, 2018), 1 m^2 (Guo *et al.*, 2019) or 3 m radius (Castaldi *et al.*, 2018) in order to take the uncertainty of geo-referencing into account. These samples are then analyzed in the laboratory using standard chemical or physical methodology. Such pilot studies assume that the surface conditions are homogeneous and that the reflectance can be related to one or more soil properties. Chabrillat *et al.* (2019) discussed the potential and limitations of imaging spectroscopy for mapping and monitoring soil properties, while Angelopoulou *et al.* (2019) reviewed pilot studies from remote sensing platforms ranging from UAS to airborne and satellite (Table 1). A tractor-driven

Table 1 Summary of remote sensing platforms for SOC monitoring in terms of their benefits and drawbacks

Platform	Benefits	Drawbacks
Satellites	• Obtain topsoil information from large areas • Provide information for inaccessible areas • Provide auxiliary data • Consistent temporal resolution for creation of time series • Short revisit time • Provide free data	• Atmospheric absorptions interfering with the spectral measurements • Low signal-to-noise ratio due to a short integration time over the target area • Mixed pixels contain more than bare soil surface (e.g. vegetation) • Need for geometric and atmospheric corrections
Airborne	• Provide information for inaccessible areas • Few imagery instruments but becoming more available in the range of (1000–2500 nm) • High spatial resolution	• Need for certain meteorological conditions for remote sensing applications • Limitation of measurements only in a thin layer of topsoil • Legal constraints for the flights • High operational complexity • High cost
LASs	• Flight plan can be scheduled according to weather condition • High spatial resolution	• Limited flight duration • Limited pay load • Need for atmospheric, geometric corrections • Legal constraints for the flights

Source: Angelopoulou *et al.*, 2019.

on-line system of proximal sensing, not mentioned in this review, uses VNIR-SWIR sensors having a sapphire glass window and internal light source mounted at the bottom of a subsoiler. They are thus dragged through the topsoil and collect an on-the-go signal of the soil in the plow layer (Nawar and Mouazen, 2019). Airborne sensors were introduced in the 1990s (Ben-Dor and Banin, 1995) and collect data cubes for pixels of c. 2*2 m containing 200–300 spectral bands in the VNIR-SWIR of the soil surface. While proximal sensing techniques are used for within-field variability, airborne campaigns cover flight strips of up to 10–70 km including a range of soil types (e.g. Stevens *et al.* (2010); Guo *et al.* (2019)).

The Swiss-Belgian Airborne Prism Experiment (APEX) sensor has been flown in several campaigns across Europe (Schaepman *et al.*, 2015). Shi *et al.* (2020) calibrated a PLSR model to predict SOC using the APEX spectra, acquired on 2 September 2018, and the SOC contents as analyzed in the corresponding pixels in a 7 km-wide and 66 km-long SW–NE strip in the Belgian loam belt (R^2= 0.56; RMSE = 0.30 %; n = 43 for an independent validation; Fig. 1). They predicted the SOC contents only for the pixels with an NDVI between 0.10 and

Figure 1 Zoom of the SOC map for the flight strip of the APEX campaign in central Belgium, September 2018. UTM coordinates in the 31N zone are given. Modified after Shi *et al.* (2020).

0.22. These pixels correspond mainly to bare cropland fields that have been plowed and seeded with winter cereals (Fig. 2). The remaining area is either under another land use (grassland, forest or urban) covered by summer crops such as sugar beet, maize and potato, or outside the NDVI range, probably as a result of emerging winter cereals. The zooms show both within-field patterns and differences between adjacent fields that provide a detailed insight in the SOC distribution of the plow layer. The spots with higher SOC content within the fields are traces of historic charcoal kilns dating from before the early nineteenth century when large parts of the areas were under production forest and charcoal was used in iron smelters (Hardy *et al.*, 2017). The clear difference in SOC contents between fields in the southwestern corner of the image are probably due to differences in historic management of the fields. In the same fields a tongue of higher SOC contents appears that is probably related to a third-order valley where the soil is relatively moist. Castaldi *et al.* (2018) provided a more in-depth discussion of similar patterns in SOC contents for fields in central Belgium using a digital terrain model and a semivariogram analysis.

The pilot study of Guo *et al.* (2019) covers a freshly plowed field of c 1.5 by 1 km in Iowa (USA). They acquired spectra from 50 soil samples (0–15 cm) on a

Figure 2 Example of a cropland field in the APEX flight strip for which the SOC content has been predicted.

regular grid both in the laboratory and from an airborne Headwall sensor (400-1700 nm; SpecTIR LLC). They used the PLSR regression to first directly and later indirectly (i.e. by multiplying the predicted soil organic matter (SOM) content with the predicted bulk density) predict the SOC stock. Not surprisingly, the cross validation of the laboratory models yielded the best results. Overall, the R^2 of the prediction of the SOM contents was higher (0.75 in the laboratory and 0.54 airborne) than either the bulk density or the directly predicted SOC stocks (R^2 = 0.66 in the laboratory and 0.42 airborne). The indirect prediction of the SOC stocks gave similar performances. The authors attributed the satisfactory prediction of SOC stocks, which is rare in the literature, to the chromophores of SOM and the strong correlation between SOM and bulk density.

Gholizadeh *et al.* (2018) compared the performance of spectral prediction models acquired in the laboratory, from an airborne campaign using the CASI (370–1040 nm)-SASI (960–2440 nm) sensor (Itres Ltd., Calgary, Canada) and from a Sentinel-2 image. They studied four agricultural sites of c. 2 by

2 km throughout the Czech Republic. Before the airborne campaign and the Sentinel-2 overflight the fields were plowed and harrowed and devoid of vegetation or residues. The prediction accuracy based on lab spectroscopy, airborne and Sentinel-2 spectra was adequate for SOC (for Sentinel 2: RMSE for the independent validation set 0.08–0.23% and an RPD of 1.60–1.92). The best correlation for SOC was obtained with the Sentinel-2 bands B4 (665 nm, width 39nm), B5 (704 nm, width 20 nm), B11 (1610 nm, width 141 nm) and B12 (2186 nm, width 238 nm). The SOC maps produced by Sentinel 2 gave a more accurate prediction of the spatial patterns in zones with high SOC contents compared to the maps produced from the airborne sensor.

Based on an extensive literature review, Angelopoulou *et al.* (2019) summarized the main benefits and drawbacks of remote sensing platforms (Table 1). Up to now, airborne sensors combine the best spectra with high spatial resolution, but the complexity of organizing a flight campaign and its high costs are the main reasons that these images are only available for a few areas. Because of the constraints for planning the flight campaigns in advance, it becomes already more difficult to ensure that the atmospheric conditions are optimal during a period when croplands have been harvested and re-seeded. During this period the surface spectra are most likely to reflect the soil properties such as SOC concentration and the signal is not influenced by (partial) vegetation or residue cover, variation in roughness or soil moisture content.

3 Challenges for SOC mapping over large extents

3.1 Calibration using soil spectral libraries

Until now, calibration of remote sensing hyperspectral imagery relied on developing spectral transfer functions between the signal of the sensor and the soil property analyzed on samples of the corresponding pixel (see previous section and, for example, Gomez, Lagacherie and Coulouma, 2008; Guo *et al.*, 2019). However, as the remote sensing platforms cover larger areas (new generation of hyperspectral satellites) or become more flexible for the acquisition of fields with no prior information (such as UASs), dedicated sampling campaigns and analysis of the soil properties become the main constraints in cost and effort for producing the SOC maps. Hence, attempts have been made to transfer a calibration model based on laboratory spectra contained in one of the large soil spectral libraries (e.g. LUCAS) to the spectra acquired by the airborne sensor. As is the case for wet chemistry analyses, the challenge of the transfer of spectral models developed using a particular spectrometer to another was already recognized by Shenk and Westerhaus (1991). For transfer between laboratory spectra, a series of soil standards that are scanned by both the so-called master and slave instruments are used (Kopačková and Ben-Dor,

2016). The transfer of spectral models derived from spectral libraries to the signal of remote sensing instruments is more complex. After all, for remote sensing instruments, the reflectance is not directly measured, and instead it is calculated after reduction of the optical characteristics of the atmosphere. Parts of the spectra cannot be used from remote sensing instruments as the signal is absorbed by the water vapor in the atmosphere. The spectral resolution and the signal-to-noise ratio (the latter in particular in the SWIR region) are, in general, poorer for the remote sensing instruments. Even when the soil surface is dry, smooth and free from residues, its signal can still be different from the signal of a dry, sieved ground sample measured under controlled conditions in the laboratory. Below are three examples that illustrate different approaches for the transfer of models built using spectral libraries to the remote sensing instruments, airborne and UAS-borne sensors in this case.

Nouri *et al.* (2017) used several transfer functions whereby they measured both the spectra from a number of samples (so-called standards) in the laboratory and the spectra at the corresponding pixel acquired from the airborne sensor. Good predictions of clay content were obtained using different techniques, such as model updating Rep file, transfer by orthogonal projection and piecewise direct standardization, with only 15 standards for an area of c. 24 km^2. However, the transfer functions are specific to the flight campaign and have to be established for each future campaign.

Castaldi *et al.* (2018) observed that airborne spectra acquired over tarmac or white paved surface differed between flight campaigns (APEX in the Belgian loam belt in 2013 and 2015), even though simultaneously spectra acquired with an ASD spectrometer remained stable. Hence, they proposed a different approach, the so-called bottom up approach, in order to integrate the LUCAS spectral library for the calibration of the airborne APEX spectra. First, they used PLSR models to predict the SOC content in 54 calibration samples based on the spectra of the LUCAS library, and then they used these predicted SOC contents to develop a calibration model using the APEX spectra of the corresponding pixels. This approach yielded SOC maps for the flight strips in Luxembourg and the Belgian loam belt with a satisfactory level of accuracy (RMSE = 1.5–4.9 g kg^{-1}; RPD = 1.4–1.7).

Although the spectral range of most UAS-borne spectrometers or multispectral cameras is limited to the VNIR region (400–1000 nm), there are certain advantages compared to the airborne sensors (Table 1). Before take-off, the spectrometer can be calibrated using a white reference (spectralon) meaning that the reflectance is measured and not calculated as is the case for airborne instruments. Moreover, a soil standard can be scanned before each flight. Kopačková and Ben-Dor (2016) suggest using perfect silica sand from an Australian beach as soil standard. Crucil *et al.* (2019) used the soil standard both in outdoor scans with a STS-VIS (350–800 nm; Ocean Optics Inc., Dunedin,

FL, USA) and in the laboratory using an ASD Fieldspec 3 FR spectro-radiometer (Analytical Spectral Devices Inc.). They demonstrated that using a local spectral library for a long-term trial in Rothamsted (UK) the soil standard allowed applying the laboratory model to predict the SOC content of the long-term trial in outdoor conditions using the small STS spectrometer. Further tests with the STS spectrometer flown by a UAS are ongoing.

The examples in the previous paragraphs dealt with aligning the spectra acquired by remote sensing instruments to the ones of a spectral library. The heterogeneity of the soil and parent material determines the selection of optimal pattern and number of samples for the calibration of spectral models. Optimal sampling schemes aim to cover the variation in spectral characteristics of the study area (i.e. the feature space) rather than covering its spatial extent (i.e. the geographical space; Ramirez-Lopez *et al.* (2014)). Castaldi, Chabrillat and van Wesemael (2019) demonstrated the use of Sentinel-2 imagery in order to characterize the spectral feature space of a target area. They advocate using algorithms that simultaneously calculate the number of samples and their spatial distribution such as the Puchwein algorithm. Thus, an efficiency of sample selection, expressed as ratio between accuracy and sampling density can be calculated.

When large spectral libraries, such as LUCAS topsoil database, exist, the heterogeneity of the soils and parent materials prevent a straightforward interpretation of the spectra in terms of soil properties. The application of these spectral libraries for predicting SOC in a specific cropland site requires a selection of the spectra that best reflect those of the target area. Ward *et al.* (2019) applied both a k-means clustering and a local PLSR approach in order to build spectral models that do not rely on external co-variates. Ward *et al.* (2019) removed the water vapor bands from the LUCAS spectra and re-sampled them to the spectral resolution of the forthcoming EnMAP satellite in order to evaluate the potential for using the LUCAS spectral library to build calibration models for EnMAP. Overall, the local PLSR re-sampled to the EnMAP characteristics gave good results (R^2=0.66, RMSEP = 5.78 g kg^{-1} and RPIQ = 1.93).

3.2 Dealing with surface conditions that disturb the soil signal

For most of the pilot studies it was assumed that the pixels on the SOC map represented smooth, bare and dry soils. For these pilot studies, cropland soils were targeted when they had just been harrowed and seeded (Chabrillat *et al.*, 2019). However, the availability of and the area covered by the images provided by satellite systems such as MSI/Sentinel-2 and the upcoming hyperspectral satellites has increased dramatically. Vaudour *et al.* (2019), working in the Versailles plain (France) demonstrated the impact of acquisition date of Sentinel-2 imagery and related weather and surface conditions on the prediction of topsoil SOC in croplands. Not surprisingly, they derived the best SOC models for spring images when the soil moisture and soil roughness were lowest. NDVI values did not influence the prediction performance as long as they were below a threshold of 0.35. This example demonstrates that there is a need for automated procedures that ensure that the pixels under consideration represent dry, smooth and bare cropland soils (Diek *et al.*, 2019). After all, only soils in these conditions are similar to the dry, ground samples scanned to populate the spectral libraries. At this stage, it will most likely not be possible (or only to a limited extent) to correct for roughness, moisture, green and dry vegetation or residues and therefore we suggest to mask all pixels that do not represent similar conditions to the ones in the spectral library (Chabrillat *et al.*, 2019). Here we will only discuss the methods to evaluate soil surface conditions using spectral indices.

The spectral signal of the earth surface is anisotropic. Roughness produces shadow areas where solar beams do not reach the surface, and hence the reflectance is much lower (Chabrillat *et al.*, 2019). It is particularly difficult to correct for the shading effects in furrows created by tillage, as their orientation toward the sun and the sensor determines the degree of shading and reduction of the reflectance. Although techniques exist for creating detailed digital terrain models representing the furrows, as far as we know, no attempts have been made to use these DTMs for each individual field in order to correct the tillage-induced reflection. For temperate croplands, summer crops are sown in April and winter crops are sown in September (see, for example, Diek *et al.*, (2016)). Depending on the weather conditions these are the months with the largest extent of bare soils. Nowadays, farmers harrow and seed their fields within hours after plowing, and therefore most bare soils are smooth during the harvest and seeding periods.

Soil moisture has a profound effect on the shape of a spectrum, as it not only reduces the overall albedo, but also enhances the water absorption peaks around 1400 nm and 1920 nm. Minasny *et al.* (2011) developed the external parameter orthogonalization approach (EPO) in order to split a spectrum of a moist soil in a spectrum that can be attributed to its inherent soil properties and one that reflects the effect of its moisture content. Unfortunately, the EPO approach cannot be used for assessing the variability in soil moisture content of remote sensing image. First of all, the atmospheric moisture produces so much noise at the wavelengths of the moisture absorption peaks that they have to be deleted prior to further treatment of the spectra. Then, the EPO approach requires spectra, acquired with the same instrument, from both dry soils and soils with varying soil moisture content. When using remote sensing instruments the only spectra that can be acquired are the ones of the soil surface with an unknown moisture content. Therefore, Haubrock *et al.* (2008) suggested characterizing the soil moisture content through the normalized soil moisture index (NSMI) using the shoulders of the water absorption peak in the SWIR (eq. 1).

$$\mathrm{NSMI} = \frac{\mathrm{Ri} - \mathrm{Rj}}{\mathrm{Ri} + \mathrm{Rj}} \tag{1}$$

Where R_i and R_j are the reflectance values at 1800 nm and 2119 nm.

A linear relation between NSMI and gravimetric soil moisture content (R^2 = 0.7; RMSE 0.03 g g^{-1}) was obtained by Haubrock *et al.* (2008) and Nocita *et al.* (2013). Nowadays, soil surface moisture can be cautiously estimated based on the NSMI. Validation of NSMI maps derived from remote sensing instruments remains difficult, as the instruments only register the surface moisture content, while other techniques (e.g. capacitive soil moisture probes, ground penetrating radar, sampling and gravimetric analysis) consider a volume of soil that extends to at least 5–10 cm. During a sunny, cloud free day required for a useful satellite image the top 1–2 cm of a bare cropland can quickly dry out creating a steep soil moisture gradient. Hence, correcting for the variation in soil moisture when developing prediction models for organic or mineral soil properties applied to entire remote sensing images is still not possible (Chabrillat *et al.*, 2019). Nevertheless, several authors used laboratory and field instruments measuring the spectra from soils with a range of moisture contents to demonstrate that SOC prediction improves when separate models are developed by NSMI class (Nocita *et al.*, 2013; Rodionov *et al.*, 2014; Fabre *et al.*, 2015; Bablet *et al.*, 2018; Yue *et al.*, 2019).

SOC prediction models are very sensitive to fractional cover of the soil surface by green vegetation, dry vegetation or residues and soil surface crusts (Bartholomeus *et al.*, 2011; Rodionov *et al.*, 2014; Chabrillat *et al.*, 2019). In particular, green, so-called photosynthetically active, vegetation, has a large influence on the performance of SOC models. Bartholomeus *et al.* (2011) reported a relative error of 58% in SOC prediction for a cover of 5–10%, while Rodionov *et al.* (2014) found that SOC was over estimated by 30 g kg^{-1} for only 5% surface cover by green leaves. While Bartholomeus *et al.* (2011) proposed a spectral unmixing approach to correct SOC predictions for vegetation cover up to 25%, Rodionov *et al.* (2014) calculated correction values based on field spectra obtained in long-term trial with a range of vegetation cover. The NDVI effectively separates the signal of green vegetation from the soil signal using the sharp increase in reflectance at the red edge (at 700 nm Fig. 3). To characterize straw cover, they use the cellulose absorption index (CAI) based on the cellulose absorption at wavelengths around 2000 and 2400 nm (eq. 2; Fig. 3).

$$\mathrm{CAI} = 0.5\left(R_{2.0} + R_{2.2}\right) - R_{2.1} \tag{2}$$

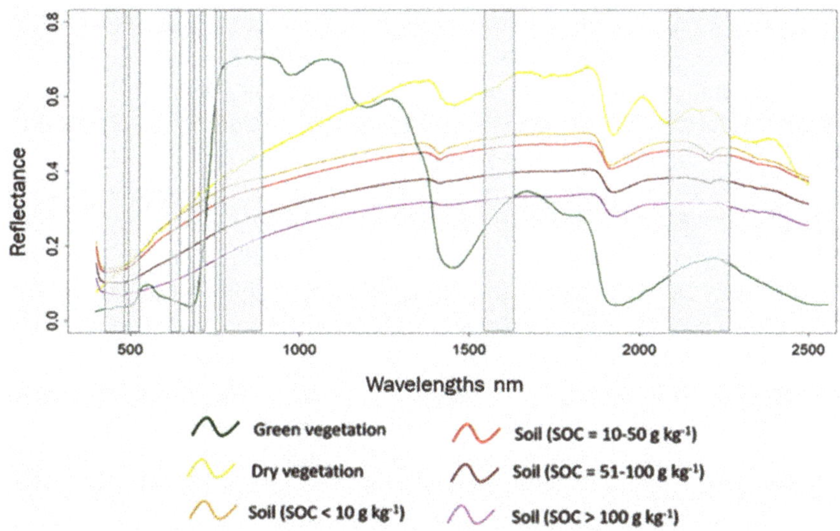

Figure 3 Laboratory soil spectra (400–2500 nm) from LUCAS topsoil database with different organic carbon content compared to dry and green vegetation spectra. The vertical grey bars indicate the Sentinel-2 MSI bands (Castaldi *et al.*, 2019).

Where $R_{2.0}$, $R_{2.2}$, $R_{2.1}$ are reflectance factors in bands at 2000–2050 nm, 2089–2130 nm and 2190–2240 nm, respectively (Rodionov *et al.*, 2016). As a bare soil does not have an absorption peak at 2100 nm, its CAI is usually negative in contrast to the positive CAI of a straw-covered soil. Unfortunately, satellite sensors, such as Landsat or Sentinel-2, do not very well cover the SWIR wavelengths (Fig. 3). The normalized burn ratio 2 (NBR2) index based on the two SWIR bands uses wavelengths (e.g. for Sentinel-2: B11 (center 1610.4 nm, bandwidth 141 nm) and B12 (center 2185.7 nm, bandwidth 238 nm) are somewhat similar to the ones used for the NSMI and the CAI index (Fig. 3).

Ben-Dor *et al.* (2004) have shown that soil crusts and their implication for infiltration rates can be determined from hyperspectral imagery of loess agricultural soils. These crusts develop gradually on cropland soils as a result of raindrop impact throughout the growing season. Thus, well-developed crusts will occur on fields that have been harvested, but they are destroyed by tillage and the preparation of the seedbed. Soil crusts in agricultural soils mostly co-exist with stubble and residue cover, and therefore these conditions can be detected using the CAI (eq. 2).

4 Synthetic bare soil images

One of the main disadvantages of spectral mapping of soil properties is that the maps rely on the spectral signal of bare soils. In particular, in temperate environments, such conditions prevail in croplands during a limited period when the soils are harvested, plowed and seeded. Therefore, single images typically contain only a limited number of such bare soil pixels.

Diek *et al.* (2016) acquired three airborne hyperspectral images (APEX) during the spring and autumn sowing and harvest periods in 2013, 2014 and 2015 with the aim of creating a mosaic and thus extending the areal coverage of bare soils. After selecting the bare soil pixels using a mask based on NDVI and CAI thresholds, they applied the empirical line method in order to correct the 2013 and 2015 spectra in such a way that they coincided with the 2014 spectrum for a number of pixels meeting the bare soil condition in each image. The resulting image contains synthetic spectra for the bare soil pixels that cover twice the area compared to the one covered by a single image. However, the synthetic spectra were probably too much influenced by variation in soil moisture between the years to yield good SOC prediction models ($R^2 = 0.39$ for the three year composite).

Rogge *et al.* (2018) developed a multi-temporal approach, the soil composite mapping processor (SCMAP), based on 15 years of LandSat imagery for the definition of croplands. Croplands change from exposed soil (after plowing) to photosynthetically active vegetation cover at least once per year. For multi-temporal composites, they defined minimum and maximum NDVI thresholds to separate cropland from other land uses. They tested the approach in Germany and extend it to other countries. Surface reflectance composites can be used to evaluate management practices such as the frequency of temporary grassland in the rotation or the management of crop residues, while averages of reflectance of exposed soil pixels relate to broad soil groups.

Demattê *et al.* (2018) proposed the geospatial soil sensing system (GEOS3) in order to bring together the fragmented information of bare soils in Brazil into a single synthetic image. The procedure is based on the following steps: (i) collection of Landsat 5 legacy data, (ii) selection of images from the dry seasons, (iii) development of a set of rules based on the NDVI and NBR2 to select bare soils, (iv) calculation of a median spectrum based from the time series of images for each exposed soil pixel, (v) aggregation to a synthetic soil image and (vi) validation of the quality of this image through correlation between laboratory and median spectra of the exposed soil pixels. While only 5% of the area was classified as bare soil in a single image, this area increased to 68% of the synthetic image.

These techniques to create synthetic images from time series of satellite images clearly result in more continuous reflectance maps. These delineate the croplands, that is, the only areas with a strong annual variation in NDVI. Furthermore, the median reflectance of exposed soils in the time series already shows a correlation to the main soil types. Unfortunately, these median spectra are influenced by the variation in soil moisture between the individual images and cannot, yet, be used for SOC prediction models.

5 Case study

The case study is an example of high-resolution mapping of the SOC content in the topsoil of croplands (0–20 cm) covering a large area (100*100 km) at low costs (Castaldi, Chabrillat, Don *et al.*, 2019). A Sentinel-2B image was acquired on 30 August, 2017, covering the area around the town of Demmin in northern Germany, part of the Tereno NE long-term monitoring site (Zacharias *et al.*, 2011). This image was selected because it was largely cloud-free and the time window corresponded to a period in which the soils are likely to be in seedbed condition after sowing of winter cereals. The site is dominated by cropland (80.3%) and part of the young morainic soil-landscape of northern Germany characterized by geomorphological features, including (sandy-) loamy morainic parent material of the late Pleistocene, kettle holes and slightly undulating relief. The calibration data set consisted of samples in croplands covered by the Sentinel-2 image extracted from the LUCAS spectral library. The laboratory spectra were re-sampled according to the spectral bands of the MSI/Sentinel-2B sensor. They used a combination of two bands of the re-sampled spectra to fit an exponential regression model to the SOC content. These models were later applied to real MSI/Sentinel-2 spectra to predict the SOC content in the bare pixels of the satellite image. Bare pixels meet all of the following criteria: (i) zero clouds probability, (ii) NDVI lower than 0.35, (iii) differences between band 3 (B3) and B2 as well as between B4 and B3 larger than 0 in order to improve the soil mask and (iv) different NBR2 index thresholds were used to exclude spectra from pixels with high moisture content and covered by straw or dry vegetation. The validation data set consisted of a maximum of 253 samples collected in the area as part of the Tereno long-term monitoring and the German agricultural soil inventory (Vos *et al.*, 2019).

Overall, 12% of the pixels were covered by clouds, and of the cloud free area 26% of the pixels was not covered by green vegetation. The best ratio of performance to deviation (RPD) was obtained using the average between bands B6 (740 nm) and B5 (704 nm; RPD: 4.4; RMSE 6.8 g kg^{-1}) and between B4 (665 nm) and B5 (RPD: 2.9; RMSE 10.3 g kg^{-1}) adopting a very strict NBR2 threshold (0.05). The strict NBR2 threshold reduced the area that could be mapped to 10% of the pixels. Increasing the NBR2 to 0.1 resulted in a larger area that could be mapped (15%) at the expense of the accuracy (RMSE 15.7 g kg^{-1}). This application uses a spectral model derived from the LUCAS spectral library that is then applied to the MSI/Sentinel 2B sensor. In contrast to the pilot studies, there is no need for sampling and analysis of SOC content. In theory, the approach could be applied to other Sentinel-2 images within the area covered by the LUCAS spectral library. However, the soils should preferably be quite homogeneous in terms of parent material and pedogenic oxides in order to avoid the need for clustering or local regressions (Ward *et al.*, 2019). Moreover, as the NBR2 index is less specific than the CAI or NSMI indices, its capacity to distinguish between the effect of residues and soil moisture still has to be investigated. In any case Vaudour *et al.* (2019) have demonstrated that the performance of SOC predictions based on Sentinel-2 are poor under suboptimal conditions, that is, in winter, for freshly plowed soils.

6 Summary and future trends

Spectral mapping of soil properties and in particular SOC has provided high-resolution maps of bare croplands for pilot studies. Although strictly speaking only the signal of the soil surface is registered, most authors have worked on fields that have been harrowed and seeded. Thus, conditions that disturb the soil signal such as roughness, variation in moisture content, residues or vegetation hardly influence the performance of the prediction models for well-controlled pilot studies. As more imagery becomes available with the launch of hyperspectral and high resolution multispectral satellites and the miniaturization of spectrometers that can be mounted on UASs, there is a need for automated procedures for calibration and for detection of conditions disturbing the soil signal. Spectral models are already calibrated using spectral libraries (e.g. LUCAS) and clustering or local PLSR approaches have been developed to deal with the heterogeneity of the soil matrix that in the past often led to inaccurate prediction of the soil property. Transfer of these models developed under laboratory conditions to the signal of bare soil pixel acquired by a remote sensor is still in a research stage. Spectral indices that allow distinguishing a bare, dry soil from a wet soil or a soil covered with green vegetation or residue exist and are mainly based on the NDVI and a number of indices using SWIR bands.

The Sentinel 2 program is a unique opportunity for the development of SOC prediction models. Pilot studies have demonstrated that adequate SOC models can be derived for agricultural areas when the soils have been plowed and harrowed. The Czech case study demonstrated that due to its high spatial resolution and stable signal, the SOC maps produced with a spectral model based on Sentinel-2 were even better than the ones using spectra from an airborne platform. The short revisit time (5 days) increases the potential for the acquisition of (a series of) images covering a large extent of exposed soils. Although an adequate SOC prediction model for an entire Sentinel-2 image in Northern Germany was already produced, the possibility to use composite images consisting of pixels acquired at different dates and thus extending the spatial coverage still needs to be confirmed. In order to facilitate the processing of Sentinel 2 (and future hyperspectral) satellite imagery at the regional scale, current research focuses on the transfer of spectral models calibrated on large-scale spectral libraries to the signal of remote sensing instruments. This is vital as for routine applications a large-scale spectral library would reduce (or even eliminate) the cost for sampling and analysis and would provide a standardization of the calibration, as the data in the spectral library are all acquired using a uniform protocol. Furthermore, the focus is on the development of automated procedures to distinguish bare cropland soil pixels, as these are the ones to which the spectral SOC models can be applied. The NDVI can be easily calculated from the spectra of multispectral satellites. However, the indices that characterize soil moisture or residue cover use SWIR bands that are not registered by the current generation of satellites. Last but not least, the conditions for acquiring imagery that best reflect the composition of the bare soil have not yet been objectively defined, although the analysis of time series has provided cropland masks and synthetic images of multiple dates and tending towards a continuous cropland cover.

7 Where to look for further information

7.1 Further reading

A comprehensive review of the latest research on remote sensing for SOC estimation is given by Angelopoulou *et al.* (2019). For full bibliographic details we refer to the next section (References).

An overview of the principles behind imaging spectroscopy for soil mapping and monitoring is given by Chabrillat *et al.* (2019).

Another overview of remote sensing applied to soils is given by Dematte et al. (2015) in Thenkabail, P. S. (ed.). *Remote sensing handbook–three volume set: land resources monitoring, modeling, and mapping with remote sensing*. CRC Press, Boca Raton, pp. 661–732. ISBN-10: 1482217953.

7.2 Key conferences

The European Association of Remote Sensing Laboratories (EARSel) has a special interest group on imaging spectroscopy (SIG IS; http://atcor.dlr.de/SIG-IS.htm). This group organizes bi-annual conferences on hyperspectral remote sensing for mapping and monitoring soil properties.

7.3 Useful websites

Sentinel-2 imagery is freely available and can be accessed through the Copernicus open access hub (https://scihub.copernicus.eu/).

The Sentinel-2 Toolbox can be used for the visualization and analysis of products. More information can be found at: https://sentinel.esa.int/web/sentinel/toolboxes/sentinel-2.

Information on freely available Landsat images can be found at https://landsat.gsfc.nasa.gov/data/where-to-get-data/.

The German Aerospace Center (DLR) has developed a tool for mapping exposed soil surfaces based on Landsat imagery referred to as the soil composite processor (ScMaP): https://geoservice.dlr.de/data-assets/c5ng9ynnyx29.html.

The European soil spectral library developed within the LUCAS soil framework is provided by the EU Joint Research Center. Please consult the website from the European Soil Data Center (ESDAC): https://esdac.jrc.ec.europa.eu/.

The Hyperspectral SOil Mapper (HYSOMA) software interface is developed by the German Research Center for Geosciences (GFZ) and can be downloaded from http://www-app2.gfz-potsdam.de/hysoma.

Antoine Stevens and Leonardo Ramirez Lopez (2014) wrote a useful R package for pre-treating spectra and selecting calibration algorithms: Prospectr at https://cran.r-project.org/web/packages/prospectr/vignettes/prospectr-intro.pdf.

8 References

Aldana-Jague, E., Sommer, M., Saby, N. P. A., Cornelis, J.-T., van Wesemael, B. and van Oost, K. (2016). High resolution characterization of the soil organic carbon depth profile in a soil landscape affected by erosion. *Soil and Tillage Research* 156, 185-193. https://doi.org/10.1016/j.still.2015.05.014.

Angelopoulou, T., Tziolas, N., Balafoutis, A., Zalidis, G. and Bochtis, D. (2019). Remote sensing techniques for soil organic carbon estimation: a review. *Remote Sensing* 11(6). https://doi.org/10.3390/rs11060676.

Bablet, A., Vu, P. V. H., Jacquemoud, S., Viallefont-Robinet, F., Fabre, S., Briottet, X., Sadeghi, M., Whiting, M. L., Baret, F. and Tian, J. (2018). MARMIT: a multilayer radiative transfer model of soil reflectance to estimate surface soil moisture content in the solar domain (400–2500 nm). *Remote Sensing of Environment* 217, 1–17. https://doi.org/10.1016/j.rse.2018.07.031.

Baldock, J. A., Hawke, B., Sanderman, J. and MacDonald, L. M. (2013). Predicting contents of carbon and its component fractions in Australian soils from diffuse reflectance mid-infrared spectra. *Soil Research* 51(7–8), 577–595. https://doi.org/10.1071/SR13077.

Bartholomeus, H., Kooistra, L., Stevens, A., van Leeuwen, M., van Wesemael, B., Ben-Dor, E. and Tychon, B. (2011). Soil organic carbon mapping of partially vegetated agricultural fields with imaging spectroscopy. *International Journal of Applied Earth Observation and Geoinformation* 13(1), 81–88. https://doi.org/10.1016/j.jag.2010.06.009.

Bellon-Maurel, V., Fernandez-Ahumada, E., Palagos, B., Roger, J.-M. and McBratney, A. (2010). Critical review of chemometric indicators commonly used for assessing the quality of the prediction of soil attributes by NIR spectroscopy. *TrAC Trends in Analytical Chemistry* 29(9), 1073–1081. https://doi.org/10.1016/j.trac.2010.05.006.

Ben-Dor, E. and Banin, A. (1995). Near infrared analysis (NIRA) as a method to simultaneously evaluate spectral featureless constituents in soils. *Soil Science* 159(4), 259–270. https://doi.org/10.1097/00010694-199504000-00005.

Ben-Dor, E., Goldshleger, N., Braun, O., Kindel, B., Goetz, A. F. H., Bonfil, D., Margalit, N., Binaymini, Y., Karnieli, A. and Agassi, M. (2004). Monitoring infiltration rates in semiarid soils using airborne hyperspectral technology. *International Journal of Remote Sensing* 25(13), 2607–2624. https://doi.org/10.1080/01431160310001642322.

Castaldi, F., Chabrillat, S., Don, A. and van Wesemael, B. (2019). Soil organic carbon mapping using LUCAS topsoil database and Sentinel-2 data: an approach to reduce soil moisture and crop residue effects. *Remote Sensing* 11(18). https://doi.org/10.3390/rs11182121.

Castaldi, F., Chabrillat, S., Jones, A., Vreys, K., Bomans, B. and van Wesemael, B. (2018). Soil organic carbon estimation in croplands by hyperspectral remote APEX data using the LUCAS topsoil database. *Remote Sensing* 10(2). https://doi.org/10.3390/rs10020153.

Castaldi, F., Chabrillat, S. and van Wesemael, B. (2019). Sampling strategies for soil property mapping using multispectral Sentinel-2 and hyperspectral EnMAP satellite data. *Remote Sensing* 11(3). https://doi.org/10.3390/rs11030309.

Chabrillat, S., Ben-Dor, E., Cierniewski, J., Gomez, C., Schmid, T. and van Wesemael, B. (2019). Imaging spectroscopy for soil mapping and monitoring. *Surveys in Geophysics* 40(3), 361–399. https://doi.org/10.1007/s10712-019-09524-0.

Chang, C.-W., Laird, D. A., Mausbach, M. J. and Hurburgh, C. R., Jr. (2001). Near-infrared reflectance spectroscopy – principal components regression analyses of soil properties. *Soil Science Society of America Journal* 65(2), 480–490.

Chenu, C., Angers, D. A., Barré, P., Derrien, D., Arrouays, D. and Balesdent, J. (2019). Increasing organic stocks in agricultural soils: knowledge gaps and potential innovations. *Soil and Tillage Research* 188, 41–52. https://doi.org/10.1016/j.still.2018.04.011.

Conant, R. T. and Paustian, K. (2002). Spatial variability of soil organic carbon in grasslands: implications for detecting change at different scales. *Environmental Pollution* 116 (Suppl. 1), S127–S135. https://doi.org/10.1016/S0269-7491(01)00265-2.

Crucil, G., Castaldi, F., Aldana-Jague, E., van Wesemael, B., Macdonald, A. and Oos, K. V. (2019). Assessing the performance of UAS-Compatible multispectral and hyperspectral sensors for soil organic carbon prediction. *Sustainability* 11(7). https://doi.org/10.3390/su11071889.

de Gruijter, J. J., McBratney, A. B., Minasny, B., Wheeler, I., Malone, B. P. and Stockmann, U. (2016). Farm-scale soil carbon auditing. *Geoderma* 265, 120–130. https://doi.org/10.1016/j.geoderma.2015.11.010.

Demattê, J. A. M., Dotto, A. C., Paiva, A. F. S., Sato, M. V., Dalmolin, R. S. D., de Araújo, MdS. B., da Silva, E. B., Nanni, M. R., ten Caten, A., Noronha, N. C., Lacerda, M. P. C., de Araújo Filho, J. C., Rizzo, R., Bellinaso, H., Francelino, M. R., Schaefer, C. E. G. R., Vicente, L. E., dos Santos, U. J., de Sá Barretto Sampaio, E. V., Menezes, R. S. C., de Souza, J. J. L. L., Abrahão, W. A. P., Coelho, R. M., Grego, C. R., Lani, J. L., Fernandes, A. R., Gonçalves, D. A. M., Silva, S. H. G., de Menezes, M. D., Curi, N., Couto, E. G., dos Anjos, L. H. C., Ceddia, M. B., Pinheiro, ÉF. M., Grunwald, S., Vasques, G. M.,

Marques Júnior, J., da Silva, A. J., Barreto, M. CdV., Nóbrega, G. N., da Silva, M. Z., de Souza, S. F., Valladares, G. S., Viana, J. H. M., da Silva Terra, F., Horák-Terra, I., Fiorio, P. R., da Silva, R. C., Frade Júnior, E. F., Lima, R. H. C., Alba, J. M. F., de Souza Junior, V. S., Brefin, M. D. L. M. S., Ruivo, M. D. L. P., Ferreira, T. O., Brait, M. A., Caetano, N. R., Bringhenti, I., de Sousa Mendes, W., Safanelli, J. L., Guimarães, C. C. B., Poppiel, R. R., e Souza, A. B., Quesada, C. A. and do Couto, H. T. Z. (2019). The Brazilian Soil Spectral Library (BSSL): a general view, application and challenges. *Geoderma* 354. https://doi.org/10.1016/j.geoderma.2019.05.043.

Demattê, J. A. M., Fongaro, C. T., Rizzo, R. and Safanelli, J. L. (2018). Geospatial Soil Sensing System (GEOS3): a powerful data mining procedure to retrieve soil spectral reflectance from satellite images. *Remote Sensing of Environment* 212, 161–175. https://doi.org/10.1016/J.RSE.2018.04.047.

Demattê, J. A. M., Morgan, C. L. S., Chabrillat, S., Rizzo, R., Franceschini, M. H. D., Terra, F. d., Vasques, G. M. and Wetterlind, J. (2015). Spectral sensing from ground to space in soil science: state of the art, applications, potential and perspectives. In: Thenkabail, P. S. (Ed.), *Remote Sensing Handbook–Three Volume Set: Land Resources Monitoring, Modeling, and Mapping with Remote Sensing*. CRC Press, Boca Raton, pp. 661–732. ISBN-10: 1482217953.

Diek, S., Chabrillat, S., Nocita, M., Schaepman, M. E. and de Jong, R. (2019). Minimizing soil moisture variations in multi-temporal airborne imaging spectrometer data for digital soil mapping. *Geoderma* 337, 607–621. https://doi.org/10.1016/j.geoderma.2018.09.052.

Diek, S., Schaepman, M. E. and de Jong, R. (2016). Creating multi-temporal composites of airborne imaging spectroscopy data in support of digital soil mapping. *Remote Sensing* 8(11). https://doi.org/10.3390/rs8110906.

Fabre, S., Briottet, X. and Lesaignoux, A. (2015). Estimation of Soil Moisture Content from the Spectral Reflectance of Bare Soils in the 0.4–2.5 μm Domain. *Sensors* 15(2), 3262–3281. https://doi.org/10.3390/s150203262.

Feingersh, T. and ben Dor, E. (2015). Shalom – a commercial hyperspectral space mission. In: *Optical Payloads for Space Missions* (pp. 247–263). John Wiley & Sons, Ltd. https://doi.org/10.1002/9781118945179.ch11.

Gholizadeh, A., Žižala, D., Saberioon, M. and Borůvka, L. (2018). Soil organic carbon and texture retrieving and mapping using proximal, airborne and Sentinel-2 spectral imaging. *Remote Sensing of Environment* 218, 89–103. https://doi.org/10.1016/j.rse.2018.09.015.

Gomez, C., Drost, A. P. A. and Roger, J. -M. (2015). Analysis of the uncertainties affecting predictions of clay contents from VNIR/SWIR hyperspectral data. *Remote Sensing of Environment* 156, 58–70. https://doi.org/10.1016/j.rse.2014.09.032.

Gomez, C., Lagacherie, P. and Coulouma, G. (2008). Continuum removal versus PLSR method for clay and calcium carbonate content estimation from laboratory and airborne hyperspectral measurements. *Geoderma* 148(2), 141–148. https://doi.org/10.1016/J.GEODERMA.2008.09.016.

Guanter, L., Kaufmann, H., Segl, K., Foerster, S., Rogass, C., Chabrillat, S., Kuester, T., Hollstein, A., Rossner, G., Chlebek, C., Straif, C., Fischer, S., Schrader, S., Storch, T., Heiden, U., Mueller, A., Bachmann, M., Mühle, H., Müller, R., Habermeyer, M., Ohndorf, A., Hill, J., Buddenbaum, H., Hostert, P., van der Linden, S., Leitão, P., Rabe, A., Doerffer, R., Krasemann, H., Xi, H., Mauser, W., Hank, T., Locherer, M., Rast, M., Staenz, K. and

Sang, B. (2015). The EnMAP spaceborne imaging spectroscopy mission for earth observation. *Remote Sensing* 7(7), 8830–8857. https://doi.org/10.3390/rs70708830.

Guo, L., Zhang, H., Shi, T., Chen, Y., Jiang, Q. and Linderman, M. (2019). Prediction of soil organic carbon stock by laboratory spectral data and airborne hyperspectral images. *Geoderma* 337, 32–41. https://doi.org/10.1016/J.GEODERMA.2018.09.003.

Hardy, B., Cornelis, J.-T., Houben, D., Leifeld, J., Lambert, R. and Dufey, J. E. (2017). Evaluation of the long-term effect of biochar on properties of temperate agricultural soil at pre-industrial charcoal kiln sites in Wallonia, Belgium. *European Journal of Soil Science* 68(1), 80–89. https://doi.org/10.1111/ejss.12395.

Haubrock, S.-N., Chabrillat, S., Lemmnitz, C. and Kaufmann, H. (2008). Surface soil moisture quantification models from reflectance data under field conditions. *International Journal of Remote Sensing* 29(1), 3–29. https://doi.org/10.1080/01431160701294695.

Kibblewhite, M. G., Ritz, K. and Swift, M. J. (2008). Soil health in agricultural systems. In: *Philosophical Transactions of the Royal Society B: Biological Sciences* 363(1492), 685–701. Royal Society. https://doi.org/10.1098/rstb.2007.2178.

Kopačková, V. and Ben-Dor, E. (2016). Normalizing reflectance from different spectrometers and protocols with an internal soil standard. *International Journal of Remote Sensing* 37(6), 1276–1290. https://doi.org/10.1080/01431161.2016.1148291.

Lamichhane, S., Kumar, L. and Wilson, B. (2019). Digital soil mapping algorithms and covariates for soil organic carbon mapping and their implications: a review. *Geoderma* 352, 395–413. https://doi.org/10.1016/J.GEODERMA.2019.05.031.

Loizzo, R., Guarini, R., Longo, F., Scopa, T., Formaro, R., Facchinetti, C. and Varacalli, G. (2018). Prisma: the Italian hyperspectral mission. *International Geoscience and Remote Sensing Symposium (IGARSS)*, 2018-July, 175–178. https://doi.org/10.1109/IGARSS.2018.8518512.

McBratney, A. B., Mendonça Santos, M. L. and Minasny, B. (2003). On digital soil mapping. *Geoderma* 117(1–2), 3–52. https://doi.org/10.1016/S0016-7061(03)00223-4.

Minasny, B., Malone, B. P., McBratney, A. B., Angers, D. A., Arrouays, D., Chambers, A., Chaplot, V., Chen, Z., Cheng, K., Das, B. S., Field, D. J., Gimona, A., Hedley, C. B., Hong, S. Y., Mandal, B., Marchant, B. P., Martin, M., McConkey, B. G., Mulder, V. L., O'Rourke, S., Richer-de-Forges, A. C., Odeh, I., Padarian, J., Paustian, K., Pan, G., Poggio, L., Savin, I., Stolbovoy, V., Stockmann, U., Sulaeman, Y., Tsui, C., Vågen, T., van Wesemael, B. and Winowiecki, L. (2017). Soil carbon 4 per mille. *Geoderma* 292, 59–86. https://doi.org/10.1016/j.geoderma.2017.01.002.

Minasny, B., McBratney, A. B., Bellon-Maurel, V., Roger, J.-M., Gobrecht, A., Ferrand, L. and Joalland, S. (2011). Removing the effect of soil moisture from NIR diffuse reflectance spectra for the prediction of soil organic carbon. *Geoderma* 167–168, 118–124. https://doi.org/10.1016/J.GEODERMA.2011.09.008.

Nawar, S. and Mouazen, A. M. (2019). On-line vis-NIR spectroscopy prediction of soil organic carbon using machine learning. *Soil and Tillage Research* 190, 120–127. https://doi.org/10.1016/J.STILL.2019.03.006.

Nocita, M., Stevens, A., Noon, C. and van Wesemael, B. (2013). Prediction of soil organic carbon for different levels of soil moisture using Vis-NIR spectroscopy. *Geoderma* 199, 37–42. https://doi.org/10.1016/j.geoderma.2012.07.020.

Nocita, M., Stevens, A., Toth, G., Panagos, P., van Wesemael, B. and Montanarella, L. (2014). Prediction of soil organic carbon content by diffuse reflectance spectroscopy using a local partial least square regression approach. *Soil Biology and Biochemistry* 68, 337–347. https://doi.org/10.1016/j.soilbio.2013.10.022.

Nouri, M., Gomez, C., Gorretta, N. and Roger, J. M. (2017). Clay content mapping from airborne hyperspectral Vis-NIR data by transferring a laboratory regression model. *Geoderma* 298, 54–66. https://doi.org/10.1016/J.GEODERMA.2017.03.011.

Oldfield, E. E., Bradford, M. A. and Wood, S. A. (2019). Global meta-analysis of the relationship between soil organic matter and crop yields. *SOIL* 5(1), 15–32. https://doi.org/10.5194/soil-5-15-2019.

Poggio, L., Gimona, A. and Brewer, M. J. (2013). Regional scale mapping of soil properties and their uncertainty with a large number of satellite-derived covariates. *Geoderma* 209–210, 1–14. https://doi.org/10.1016/J.GEODERMA.2013.05.029.

Ramirez-Lopez, L., Schmidt, K., Behrens, T., van Wesemael, B., Demattê, J. A. M. and Scholten, T. (2014). Sampling optimal calibration sets in soil infrared spectroscopy. *Geoderma* 226–227, 140–150. https://doi.org/10.1016/J.GEODERMA.2014.02.002.

Rodionov, A., Pätzold, S., Welp, G., Pallares, R. C., Damerow, L. and Amelung, W. (2014). Sensing of soil organic carbon using visible and near-infrared spectroscopy at variable moisture and surface roughness. *Soil Science Society of America Journal* 78(3), 949–957. https://doi.org/10.2136/sssaj2013.07.0264.

Rodionov, A., Pätzold, S., Welp, G., Pude, R. and Amelung, W. (2016). Proximal field Vis-NIR spectroscopy of soil organic carbon: A solution to clear obstacles related to vegetation and straw cover. *Soil and Tillage Research* 163, 89–98. https://doi.org/10.1016/J.STILL.2016.05.008.

Rogge, D., Bauer, A., Zeidler, J., Mueller, A., Esch, T. and Heiden, U. (2018). Building an exposed soil composite processor (SCMaP) for mapping spatial and temporal characteristics of soils with Landsat imagery (1984–2014). *Remote Sensing of Environment* 205, 1–17. https://doi.org/10.1016/j.rse.2017.11.004.

Sanderman, J., Hengl, T. and Fiske, G. J. (2017). Soil carbon debt of 12,000 years of human land use. *Proceedings of the National Academy of Sciences of the United States of America* 114(36), 9575–9580. https://doi.org/10.1073/pnas.1706103114.

Schaepman, M. E., Jehle, M., Hueni, A., D'Odorico, P., Damm, A., Weyermann, J., Schneider, F. D., Laurent, V., Popp, C., Seidel, F. C., Lenhard, K., Gege, P., Küchler, C., Brazile, J., Kohler, P., de Vos, L., Meuleman, K., Meynart, R., Schläpfer, D., Kneubühler, M. and Itten, K. I. (2015). Advanced radiometry measurements and earth science applications with the Airborne Prism Experiment (APEX). *Remote Sensing of Environment* 158, 207–219. https://doi.org/10.1016/J.RSE.2014.11.014.

Shenk, J. S. and Westerhaus, M. O. (1991). Population definition, sample selection and calibration procedures for Near infrared spectroscopy. *Crop Science* 31(2), 469–474.

Shepherd, K. D. and Walsh, M. G. (2007). Infrared spectroscopy–enabling an evidence-based diagnostic surveillance approach to agricultural and environmental management in developing countries. *Journal of Near Infrared Spectroscopy* 15(1), 1–19. https://doi.org/10.1255/jnirs.716.

Shi, P., Castaldi, F., van Wesemael, B. and van Oost, K. (2020). Large-scale, high-resolution mapping of soil aggregate stability in croplands using APEX hyperspectral imagery. *Remote Sensing* 12(4). https://doi.org/10.3390/rs12040666.

Stenberg, B., Viscarra Rossel, R. A., Mouazen, A. M. and Wetterlind, J. (2010). Visible and near infrared spectroscopy in soil science. In: *Advances in Agronomy* 107(C), 163-215. https://doi.org/10.1016/S0065-2113(10)07005-7.

Stevens, A., Udelhoven, T., Denis, A., Tychon, B., Lioy, R., Hoffmann, L. and van Wesemael, B. (2010). Measuring soil organic carbon in croplands at regional scale using airborne imaging spectroscopy. *Geoderma* 158(1-2), 32-45. https://doi.org/10.1016/j.geoderma.2009.11.032.

Tóth, G., Jones, A. and Montanarella, L. (2013). The LUCAS topsoil database and derived information on the regional variability of cropland topsoil properties in the European Union. *Environmental Monitoring and Assessment* 185(9), 7409-7425. https://doi.org/10.1007/s10661-013-3109-3.

Vaudour, E., Gomez, C., Loiseau, T., Baghdadi, N., Loubet, B., Arrouays, D., Ali, L. and Lagacherie, P. (2019). The impact of acquisition date on the prediction performance of topsoil organic carbon from Sentinel-2 for croplands. *Remote Sensing* 11(18). https://doi.org/10.3390/rs11182143.

Viscarra Rossel, R. A., Behrens, T., Ben-Dor, E., Brown, D. J., Demattê, J. A. M., Shepherd, K. D., Shi, Z., Stenberg, B., Stevens, A., Adamchuk, V., Aïchi, H., Barthès, B. G., Bartholomeus, H. M., Bayer, A. D., Bernoux, M., Böttcher, K., Brodský, L., Du, C. W., Chappell, A., Fouad, Y., Genot, V., Gomez, C., Grunwald, S., Gubler, A., Guerrero, C., Hedley, C. B., Knadel, M., Morrás, H. J. M., Nocita, M., Ramirez-Lopez, L., Roudier, P., Campos, E. M. R., Sanborn, P., Sellitto, V. M., Sudduth, K. A., Rawlins, B. G., Walter, C., Winowiecki, L. A., Hong, S. Y. and Ji, W. (2016). A global spectral library to characterize the world's soil. *Earth-Science Reviews* 155, 198-230. https://doi.org/10.1016/j.earscirev.2016.01.012.

Vos, C., Don, A., Hobley, E. U., Prietz, R., Heidkamp, A. and Freibauer, A. (2019). Factors controlling the variation in organic carbon stocks in agricultural soils of Germany. *European Journal of Soil Science* 70(3), 550-564. https://doi.org/10.1111/ejss.12787.

Ward, K. J., Chabrillat, S., Neumann, C. and Foerster, S. (2019). A remote sensing adapted approach for soil organic carbon prediction based on the spectrally clustered LUCAS soil database. *Geoderma* 353, 297-307. https://doi.org/10.1016/j.geoderma.2019.07.010.

Wijewardane, N. K., Ge, Y., Wills, S. and Loecke, T. (2016). Prediction of soil carbon in the conterminous United States: visible and near infrared reflectance spectroscopy analysis of the rapid carbon assessment project. *Soil Science Society of America Journal* 80(4), 973-982. https://doi.org/10.2136/sssaj2016.02.0052.

Xiao, J., Chevallier, F., Gomez, C., Guanter, L., Hicke, J. A., Huete, A. R., Ichii, K., Ni, W., Pang, Y., Rahman, A. F., Sun, G., Yuan, W., Zhang, L. and Zhang, X. (2019). Remote sensing of the terrestrial carbon cycle: a review of advances over 50 years. *Remote Sensing of Environment* 233, 111383. https://doi.org/10.1016/J.RSE.2019.111383, 111383.

Yue, J., Tian, J., Tian, Q., Xu, K. and Xu, N. (2019). Development of soil moisture indices from differences in water absorption between shortwave-infrared bands. *ISPRS Journal of Photogrammetry and Remote Sensing* 154, 216-230. https://doi.org/10.1016/J.ISPRSJPRS.2019.06.012.

Zacharias, S., Bogena, H., Samaniego, L., Mauder, M., Fuß, R., Pütz, T., Frenzel, M., Schwank, M., Baessler, C., Butterbach-Bahl, K., Bens, O., Borg, E., Brauer, A., Dietrich, P., Hajnsek, I., Helle, G., Kiese, R., Kunstmann, H., Klotz, S., Munch, J. C., Papen, H., Priesack, E.,

Schmid, H. P., Steinbrecher, R., Rosenbaum, U., Teutsch, G. and Vereecken, H. (2011). A network of terrestrial environmental observatories in Germany. *Vadose Zone Journal* 10(3), 955–973. https://doi.org/10.2136/vzj2010.0139.

Zhao, X., Xue, J. F., Zhang, X. Q., Kong, F. L., Chen, F., Lal, R. and Zhang, H. L. (2015). Stratification and storage of soil organic carbon and nitrogen as affected by tillage practices in the North China Plain. *PLoS ONE* 10(6), e0128873. https://doi.org/10.1371/journal.pone.0128873.

Part 3

From measurement to management

Chapter 11

Developing soil health indicators for improved soil management on farm

Elizabeth Stockdale, NIAB, UK; Paul Hargreaves, Scotland's Rural College (SRUC), UK; and Anne Bhogal, ADAS Gleadthorpe, UK

1 Introduction

Soil is a dynamic and complex material that 'occurs in nature, on a landscape, within a climate, affected by its physical, biological, social and governmental setting' (Sojka et al., 2003) in which many physical, chemical and biological processes interact. It is therefore important to begin any discussion of the measurement or management of soil health on-farm with the reminder that soils are consequently site-specific in properties and extremely diverse in the extent, direction and range of the on-going processes of transformation, translocation and interaction (addition/loss) with the wider environment. There are well-established approaches which classify and group soils combining field description with laboratory analysis of their inherent properties (Buol et al., 2011) so that this natural variation can be mapped and then used to guide land use decisions at farm and wider scales (Fig. 1). A range of soil properties and processes also interact dynamically with human management; measurement of soil nutrient content has been a cornerstone of soil fertility and consequently fertiliser management for at least 50 years (Stockdale et al., 2013). More recently the focus has expanded and it is now widely recognised that a range

http://dx.doi.org/10.19103/AS.2020.0079.22

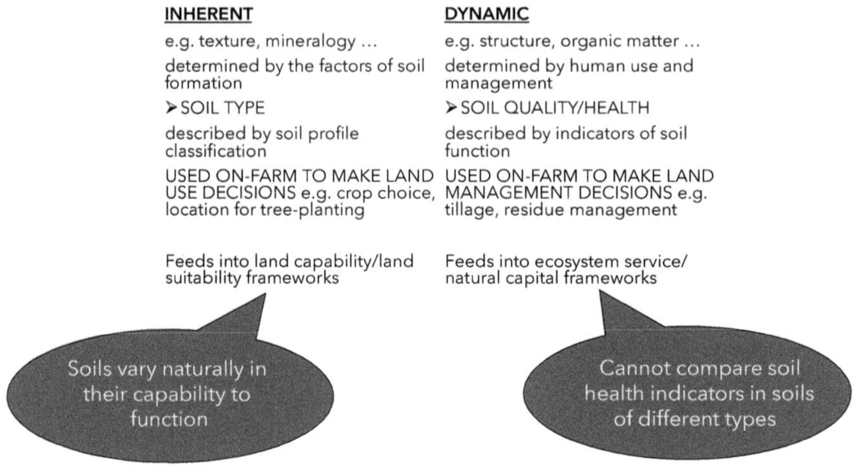

Figure 1 Soil properties can be considered to be either i) inherent (i.e. the result of soil formation) and largely fixed, or ii) dynamic (i.e. the result of interacting human and environmental control) and hence manageable. Different frameworks are therefore used to link inherent and dynamic soil properties to both policy and practice.

of chemical, physical and biological processes are important for sustained productivity and environmental quality (Fig. 1) and the terms soil quality and soil health have come into use.

Rapid developments in the concept of soil quality took place at the end of the 20th century, in parallel with an increasing awareness by land managers, policy-makers and others of the importance of the physical, chemical and biological interactions within soil in delivering many of the unseen environmental goods and services on which society, in part, depends (Barrios, 2007). At landscape scale, soils provide a broad range of ecosystem services simultaneously; consequently, balanced multifunctionality is required of soils, especially in densely occupied landscapes. At the global scale, there is evidence of an increasing focus of policy-makers on both the importance of, and the pressures on, soil resources. Some of the world's soil management challenges are immediate, obvious and serious; other problems are more subtle but equally important in the long term (FAO, 2015). Many of the policy issues are complex and intertwined with other land use and environmental policy objectives, for example:

- Can changes to soil management have a significant impact on atmospheric concentrations of greenhouse gases without jeopardising other functions such as food and fibre production?
- Will the extent and rate of soil degradation threaten food security and the provision of ecosystem services in the coming decades?

For many stakeholders, the sustained ability of soils to carry out these essential and multi-functional roles is the focus of a concept of soil health/quality (Doran and Parkin, 1994).

The general concept of soil health, that soils have a 'quality', which is estimable if not completely measurable, is now widely accepted, perhaps more often by non-specialists than by soil scientists. Recently the term soil health has increasingly replaced the related term soil quality in practical and policy discourses about soils and their management in agricultural systems, for example, the UK government's 25-Year Environment Plan (HM Government, 2018). Some commentators (e.g. Pankhurst et al., 1997) had initially felt that soil health provided an over-arching concept that explicitly considered the soil biological community and hence better reflected 'the living and dynamic nature of soil, and that this differentiates it from soil quality'. The Food and Agriculture Organisation has more recently defined soil health in relation to key soil functions as: '*The capacity of soil to function as a living system, within ecosystem and land use boundaries, to sustain plant and animal productivity, maintain or enhance water and air quality, and promote plant and animal health. Healthy soils maintain a diverse community of soil organisms that help to control plant disease, insect and weed pests, form beneficial symbiotic associations with plant roots; recycle essential plant nutrients; improve soil structure with positive repercussions for soil water and nutrient holding capacity, and ultimately improve crop production*' (FAO, 2008). As this definition shows, the concepts of soil health and soil quality are now virtually indistinguishable; Bünemann et al. (2018) conclude that the distinction between soil quality and soil health has now turned 'from a matter of principle to a matter of preference'. Romig et al. (1996) found that the term soil health had more resonance than the term soil quality with farmers. Hence, despite a preference for the term soil quality in some technical/scientific applications, in this chapter focussed on the development of soil health indicators to improve soil management on-farm, we will use the term soil health as broadly defined by the FAO (2008).

In parallel with the academic and policy discussions about soil function and soil health, farmers' and growers' interest in soils and their management has been increasing and many have taken the initiative to understand the health of their own soils (Stockdale and Watson, 2012). Innovation to create locally adapted crop and soil management by farmers and land managers is common and leads to complex multi-factorial operating systems. This range of innovative approaches to site-specific soil management has often leap-frogged the emergence of detailed underpinning understanding of the physical, chemical and biological processes in the plant-soil system and the interactions with the environmental and management drivers.

A particular concern of the science, policy and practitioner communities is the lack of robust measures of soil health that can be used to describe the current state of soil, to underpin decision-making and support better management. There is no shortage of soil characteristics that can be measured, though some can be measured more easily and cheaply than others. However, the debate is focussed around which of these possible measures can be used as indicators of the state of soil health and/or its direction of change within the local context for example, taking account of climate, land use and soil type. A range of policy-focused initiatives have led to a demand for practical assessment tools at an appropriate scale to inform policy development and the effectiveness of policy implementation. For example, the UK Government's 25-Year Environment Plan (HM Government, 2018) specifically asks for a measured improvement in soil health within a generation. However, soil scientists are only just beginning to work with policy-makers and regulators to deliver the tools that might support effective policy development and delivery (Bünemann et al., 2018). There is also a need to provide clear information for farmers on the impacts of management practices on soil health and its implications for crop productivity and environmental impacts. On-farm initiatives to promote understanding of soils therefore commonly move hand in hand with the development of practical assessment tools at an appropriate scale (Karlen et al., 2001).

In this chapter we will therefore:

1. Describe the indicators and frameworks for soil health that are currently in use;
2. Describe and evaluate the principles underpinning current approaches to monitoring soil quality/health in agricultural systems; and
3. Describe the application of these principles in a particular case study of the development of a soil health toolkit for soils under agricultural management in the UK.

2 Frameworks from policy and practice where soils are considered

2.1 Land use capability and suitability approaches

For many years, land use capability approaches have underpinned land use allocation decisions, in particular whether agricultural land should be released for building development (e.g. www.agric.wa.gov.au/land-use-planning/land-use-planning-overview). Land evaluation using a land capability classification approach was developed for farm planning in the United States of America (USA) by Klingebiel and Montgomery (1961) and has since been adapted and widely used across the globe. Permanent constraints (which may vary in importance in different climates/regions) are defined as limiting factors and used in the assessment of land use capability. Within this approach it is the negative/limiting features of land which are considered when assigning a land capability class. In the UK, climate and gradient are used as defining characteristics. In addition, soil factors that may lie beyond the control of the farmer, including extremes of soil texture, soil depth and stoniness are included. Interactive characteristics, such as erosion risk and wetness (risk of flooding and/or waterlogging), can then also be assessed.

In the assessment of land suitability in relation to particular defined land uses, that is, 'Fitness for Purpose' (FAO, 1976), consideration of both permanent (inherent) and manageable constraints is included. As most suitability assessments have been carried out for various types of crop production, the factors used to define subclasses are commonly those associated with definitions of soil fertility (e.g. droughtiness, soil nutrient deficiencies, waterlogging, ease of cultivation). Definitions of suitability classes are expressed in terms of expected yields and/or the level of inputs required to achieve good yields. Common to all these approaches is the identification of constraints to crop production (and the development of inputs and/or management practices which overcome these constraints), so that crop yield potential defined by genetic and climatic factors can be achieved. Similar, though often less formal, approaches are often used on-farm to allocate land to particular crops/rotations (e.g. land suited for rental for field vegetable production or for tree-planting).

More recently these approaches have been extended to take account of the wider role of soils in supporting ecosystem services (McBratney et al., 2014). Bouma et al. (2017) developed a series of clear examples of how the soil capability approach can be used to quantify the intrinsic capacity of a soil to contribute to ecosystem services and to provide a neutral assessment of what soils can do and how their potential can be reached. A similar concept is embedded within the interactive land potential knowledge system (LandPKS), where the inherent potential of the land to generate ecosystem services sustainably (there called land potential) is assessed through the interactive combination of locally collected data and information from global databases on soils, climate and topography (Herrick et al., 2016).

2.2 Legal frameworks for environmental quality monitoring

The development of environmental law has been an *ad hoc* and largely reactive process in each legal jurisdiction, and private law, statute, criminal and public law have each taken their place. In the UK, the Alkali Acts 1863, the Public Health Act 1875 and the Rivers (Prevention of Pollution) Act 1876 together laid the foundation of the regulatory approach which now dominates environmental law, whereby environmental standards are set and enforced through regulation backed by dominantly criminal procedures (McEldowney and McEldowney 2001). Within this framework, environmental quality standards are set with reference to the potential effect on a particular target; even in environmental law the protected target is often human beings and the standard is set with reference to the effect on them. Targets are also set with reference to the effect of a pollutant on the receiving environment or a particular medium within it (Table 1). Such environmental standards can have direct impacts on farm practice; for example the UK government has introduced rules for farmers that are soil management-focussed but that seek to address water quality targets (Defra, 2018). The term soil quality may therefore have a transferred resonance with a wide range of stakeholders because of the widespread use of the terms air and water quality in relation to environment management and protection.

Environmental quality standards are mainly defined by purity criteria that is, the absence of pollutants which directly impact human or animal health. With water or air, it is possible to define a 'normal' condition in which these materials

Table 1 Examples of UK legal frameworks governing air and water quality, together with the more focused framework governing contaminated land

Air	• Air quality standards are defined by maximum atmospheric concentrations. • Quality objectives have been set for 12 pollutants (sulphur dioxide, nitrogen dioxide, small particulates PM_{10}, lead, ozone, carbon monoxide, benzene, polyaromatic hydrocarbons, cadmium, arsenic, nickel and mercury). • Target concentrations are set by considering the potential effect of the pollutant on human health with some consideration of the effect on the wider environment. • Targets are set for each local area in the UK (Air Quality (England) Regulations 2000; SI 2000/928).
Water	• The Water Framework Directive is the EU's key legislation for water quality and was established for the assessment, management, protection and improvement of water quality. It stipulates that EU member states should aim to achieve good status of water in all bodies of surface and ground water. Good status involves meeting standards for water ecology, chemistry and quantity of waters The Directive is transposed in England and Wales by the Water Environment (Water Framework Directive) (England and Wales) Regulations 2017 (as amended; SI 2007/407). • In practice, water quality is largely defined by target values for a range of specific chemical and biological measures, such as biological oxygen demand, pH, ammonia, Cu and Zn concentrations. • The targets are often set with reference to the intended use of water (drinking, bathing etc.).
Land	• The legal definition of contaminated land is narrow and framed in such a way as to require clean-up action to be taken only when the contamination is posing unacceptable risks to human health or the environment (s 78A(2) Environmental Protection Act 1990). • Pollutant types and soil properties are considered within the risk assessment procedures to determine whether there is a significant pollutant linkage, the nature of the potential harm and the possibility of significant harm arising. • Where a remediation notice is issued, the standard required is often measured by levels of pollutants in soils or percolating waters which indicate that the land is 'suitable for use'.

are relatively pure in a chemical or biological sense. Establishing quality criteria is then largely a matter of defining acceptable maximum concentrations of chemical species or organisms, set on the basis of their impact on human health or ecosystem functioning. However, it is not possible to simply apply these concepts and frameworks to soil. With soil the situation is more complex, not only because of the inherent complexity (including considerable spatial and temporal heterogeneity) of the medium but also because of the multiple uses to which soil is put and the multiple functions required of it, often simultaneously.

2.3 Applying concepts from food quality measurement

A more useful parallel for the definition and application of the soil health concept may well be the complex definitions and criteria (both quantitative and qualitative) used in the assessment of food quality. Food quality is one of the major factors affecting consumers' food purchasing decisions alongside price and convenience. However, as for soil, its definition is not agreed. Food quality is known to result from the complex linkage between a food product's sensory attributes, chemical composition, level of microbiological and toxicological contamination, shelf-life, packaging and other consumer requirements (Molnar, 1995). Freshness, taste and nutritional value are the attributes most often prioritised by consumers (e.g. Wandel and Bugge, 1997). However, it is also well known that consumer expectations are also affected by factors such as the place where a purchase is made (supermarket, local shop, catering outlet), brand and price (Cardello, 1995). Consequently food quality is a 'fuzzy and relative term that is in constant motion' (Fisken, 1990) and clearly involves both scientific and social construction. However 'food quality' is generally understood and the term is widely used to describe the quantitative measurement of a limited number of properties which indicate fitness for a defined purpose (e.g. baking quality of wheat), together with properties which are underpinned by legal standards to ensure food is safe, unadulterated and not subjected to any fraudulent practices intended to deceive the consumer (e.g. maximum mycotoxin levels in cereal-based foodstuffs) and also qualitative assessment of flavours carried out by consumer panels.

Soils are referred to in existing legal (environmental standards and land use planning) and practical land-use suitability decisions. But no clear single structure or framework that could be implemented to underpin the monitoring of soil health emerges from this review. As suggested above, it seems most likely that the frameworks required will need to be flexible and multi-faceted to meet the distinct needs of a range of stakeholders. The overall concept of soil health may well continue to appear vague and fuzzy, but any framework developed for use in policy and/or practice will need to be described and have a sharp definition of scope and clear reference to other frameworks, so that both opportunities for data-sharing between frameworks and the limits of applicability and the limitations to interpretation are well understood.

3 Approaches to monitoring soil quality/health in agricultural systems

3.1 Defining the context

The diverse pattern of soil types, landscapes and farming systems means that any approach to measure and manage soil health on-farm needs to be grounded in an understanding of the threats to soil health and the key functions required of a soil at the particular scale of interest (e.g. farm, drinking water catchment, region). Identification of, and agreement over the prioritisation, of soil functions is a matter of policy. Stakeholders from scientists, land-users (including farmers), regulators and policy-makers to the general public have an interest in some of the functions of soils and many recognise the importance of soil but may have a limited understanding of soil science. Differences in stakeholder requirements and/or perspectives might also arise because of their different scales of interest. Contrasting approaches to soil health and what is needed from its assessment, may not be conflicting, but reflect these differences of scale. In some instances, there may be conflicts to be resolved, or at least choices to be made. For example, the development of tools with a focus on measuring soil health to improve soil management on-farm requires clear links to be established with productivity, but this doesn't remove the need to simultaneously consider the impacts of soil properties/processes on greenhouse gas balance, water quality, flood risk, biodiversity/ecosystems etc.

Bünemann et al. (2018) have recently reviewed major soil quality assessment approaches as part of the EU Horizon-2020 project 'Interactive Soil Quality Assessment in Europe and China for Agricultural Productivity and Environmental Resilience' (iSQAPER - http://www.isqaper-project.eu/). There are approaches that operate at national scales and those that operate at more local scales; some target the policy community and some are focused on land users (usually farmers). Despite the definitions of soil quality used, which all refer to the multifunctional role of soils, it is interesting to note that these systems are all dominantly framed in terms of the use of soils within agricultural production systems. It is likely that in the future, especially to effectively inform policy-making, a more explicit framing at landscape scale will be required. In this context, 'soil quality assessment provides the scientific tools for evaluation of the management of soil resources, also considering the societal demands of the various benefits that soils, if managed well, can provide to humankind' (Bünemann et al., 2018). Soil scientists are ideally qualified to translate the requirements of stakeholder groups into a set of required soil functions and to identify any conflicts or trade-offs within the selection and prioritisation amongst soil functions. Bünemann et al. (2018) showed that while farmers/land managers are expected to be important end-users of soil quality monitoring, to date, they have generally played an insignificant role in the development of soil quality assessment schemes; the initiation and development phases have been dominated by scientists and policy/government agencies. However, more effective approaches to interactive design and decision-making with end users are beginning to be adopted from an early stage. As part of such approaches, soil scientists need to take a proactive role as mediators during the framing of the debate, ensuring that the perspectives of all stakeholders are included and that the range of approaches taken is complementary. Taking a participatory approach to the development of soil health monitoring has also been shown to open up communication about management constraints and opportunities amongst farmers and between farmers and other stakeholders (Moebius-Clune et al., 2016).

In the on-farm approach developed for the integration of soil quality into soil conservation plans (USDA, 2001), the Natural Resources Conservation Service highlights the importance of working with the land manager to put together a site-specific inventory of the local resources (e.g. yield mapping, machinery options, current management practices) and background information (e.g. land use history, soil survey, climate data) and also to identify the main problems and opportunities that arise within this local context. If the measurement of soil health is to lead to improved soil management on-farm, then the farmer must be engaged in determining the emerging objectives/targets for management for example to stop/reverse a particular soil degradation process, to improve poorly-yielding areas or, a more general aim, to adopt practices that improve soil health.

3.2 Identification of appropriate indicators

Only when the key functions required of soils for a site/region have been identified, can appropriate indicators be identified and only then can measurements be selected that are practically based and that give a robust description of soil health in the required context (Fig. 2). Without a clear understanding of the critical functions to which a measurement of soil health must link, any selection of key quantifiable soil properties will largely be a matter of floundering in the dark. *If* the physical, chemical and biological soil properties necessary for a specific set of functions can be defined, it should be possible to base an assessment of soil health on a suitable set of measures and observations. This is a very big *if* indeed. Quantification of soil health by measurement of a range of soil properties can very quickly move towards an almost unimaginable indexing complexity – with an infinite number of different

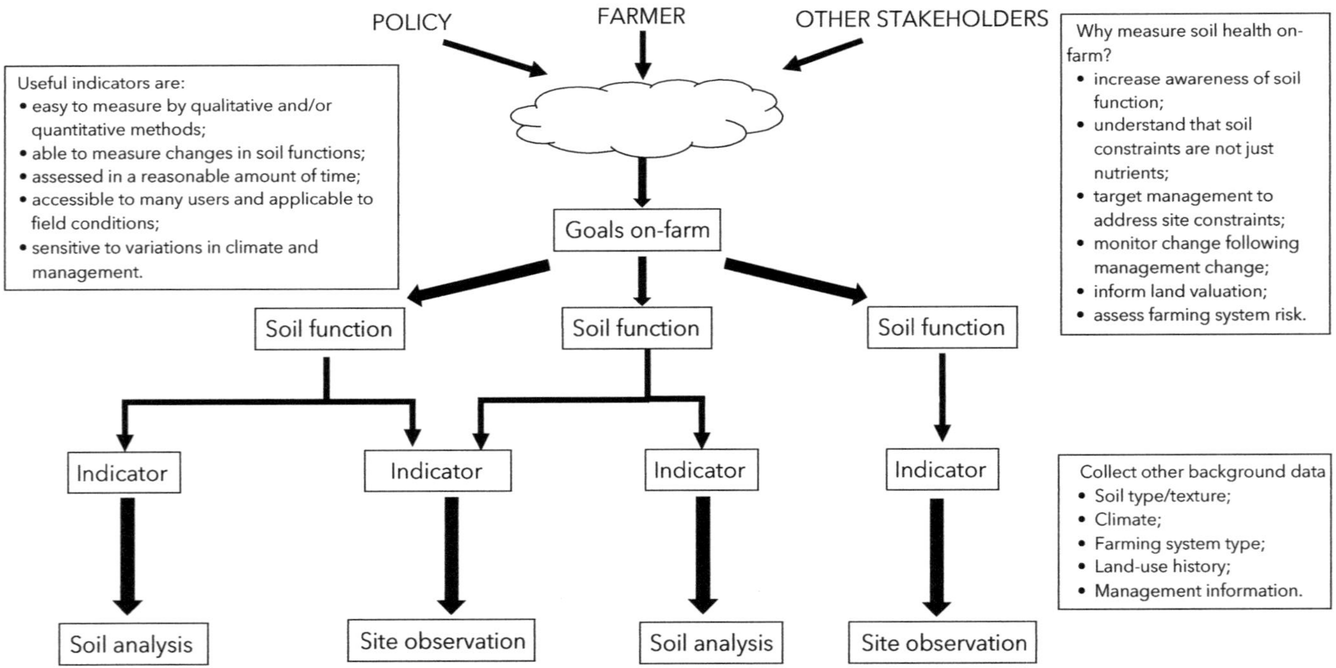

Figure 2 A soil function-focussed approach to developing on-farm soil health monitoring is shown schematically, adapted from USDA (2001). Once the goals on farm for soil health are established, the soil functions supporting those goals are identified. For each soil function, measurable soil properties that influence the capacity of the soil to perform that function are defined. One or more indicators may be needed to describe the change in a specific soil function; not all measurements are useful indicators. For example, a goal of the farmer may be to improve infiltration of rainfall. A soil function relating to this goal would be partitioning rainfall at the soil surface. An indicator of this could be infiltration rate. A methodology for this indicator could be the single ring method or observations of ponding or runoff during a rainfall. Context factors, for example soil texture, should also be measured to allow indicator interpretation.

target indices needed to fit the range of distinct soil types, management and environment scenarios for each given use (Sojka et al., 2003).

The identification of a sensitive set of soil attributes that are (or are combined into) indicators is an important component of soil health assessment on-farm (Fig. 2). Not all measurable soil properties are suitable as soil health indicators, for example, some properties change only slowly if at all (inherent e.g. soil texture, mineralogy); others (often measures of soil processes) are very episodic and extremely dynamic (e.g. CO_2 evolution). Ease of use, the time taken to sample and cost of analysis/sampling are also an important consideration for a monitoring programme, in contrast to a research study. So, for example, a need for undisturbed soil samples to measure some possible soil attributes means they are often not used as indicators (Idowu et al., 2008). Some indicators are always better measured directly in the field, for example, compaction and earthworm numbers. However, field-based indicators are often more susceptible to user bias, and detailed guidelines are needed to facilitate both uniform approaches to sample collection and scoring, where appropriate (Moebius-Clune et al., 2016).

Both a lack of indicators (e.g. for soil biological properties) and the failure of possible measures because of practical difficulties highlight the need for continuing work to develop and evaluate new indicators (e.g. molecular-based measures of soil biological communities, soil-borne disease risk) and new methods (e.g. rapid surface sensing of bulk density/porosity). The development of new techniques may bring previously unattainable indicators into play. It may also be possible to use some measurements together with models or pedo-transfer functions to provide estimates of the unmeasurable but desirable. However, the accuracy and scope of verification for such proxy indicators needs to be known and considered carefully during application.

For on-farm soil health assessment, the focus is usually on identifying the most appropriate minimum data set, that is the smallest set of indicators needed to provide a practical assessment of the soil functions that characterise soil health. Each minimum data set should be tailored to a particular region or soil map unit (soil type), although there are very likely to be common indicators, and includes only those properties relevant to the soil types, farming system and land uses of the areas being evaluated. For example, a minimum data set for the Northeast United States would probably not include such indicators as salt accumulation and electrical conductivity, while a data set for areas with arid and semi-arid soils would include these indicators (USDA, 2001). Compiling a minimum data set helps to identify locally relevant indicators and to evaluate the link between indicators selected and significant soil and plant properties for the region. Where participatory processes are used to develop regionally relevant soil health monitoring procedures then this also often improves communication amongst farmers and between farmers and other stakeholders

(USDA, 2001). In the Cornell Soil Health Assessment, Moebius-Clune et al. (2016) identify a minimum data set but also present add-on options including in-field scoring approaches and more detailed analysis.

Bünemann et al. (2018) reviewed 65 recommended indicator sets for soil quality – despite minor differences, there is generally at least one indicator for each of the chemical, physical and biological properties of the soil. Organic matter (carbon), pH, extractable phosphorus, and various indicators of soil structure/water storage are used. However, Bünemann et al. (2018) note that there is much less attention paid to the need for parallel records of extrinsic factors (climate, management, site data) or outcomes/impacts that is soil functions/ecosystem services, even crop yield.

3.3 Development of an interpretation framework for indicators

For soil health assessment not only are a set of indicators needed, these must be accompanied by a clear interpretation framework if they are to be effectively included in practical soil monitoring schemes. Bünemann et al. (2018) found that currently 'few approaches provide clear interpretation schemes of measured indicator values'.

Any framework of interpretation must clearly show whether the values of each indicator measured or calculated are acceptable that is the ranges over which there is no loss or significant impairment of functioning and contrast these with ranges where one or more soil functions are constrained. For some indicators, very clear links have been established with function. For example, there is consensus that a penetrometer resistance (measured at or near field capacity) greater than 2 MPa will significantly impede root growth. Taylor (1971) showed that 2.5 MPa was an upper limit for cotton taproots to grow through soil, with rooting density reduced at 1.0–2.0 MPa. More recent studies of UK soils have found that at >2.0 MPa root elongation was reduced to <50% of its unimpeded rate (Valentine et al., 2012). For available plant nutrients and pH, data gathered over many years and integrated by agricultural advisory services into nutrient management recommendations will often provide the necessary framework for interpretation (e.g. AHDB, 2017). However, in many circumstances, thresholds of available plant nutrients developed for crop production are now coupled to limits relating to pollution risk for example for extractable P (AHDB, 2017).

For many biological properties (e.g. soil organic matter, soil microbial biomass) an increase in measured values is considered to be beneficial, whereas a decline is considered to be detrimental; however, this level of interpretation is not sufficient if these measures are to be used as indicators of soil health. It is recognised that differences in soil type and land use need to be taken into account for building an understanding at the field/

farm level. Ingram and Fernandes (2001) proposed a conceptual framework for the estimation of thresholds for soil organic matter which was extended by Gonzalez-Quiñones et al. (2011) to apply to soil microbial biomass. This conceptual framework identifies potential, attainable, constraining and critical limits for the biological indicators which are conceptually linked to soil function and also to the site constraints set by extrinsic factors - such as parent material, climate, topography, and hydrology. Andrews et al. (2004) described how scoring functions can be established and then monitored/refined through a combination of expert judgement and literature review to provide a basis for a regional soil quality assessment approach. Moebius-Clune et al. (2016) have developed this approach using the range of measured values for each indicator determined in an integrated set of replicated research trials, grower demonstration trials and from fields of interested growers in the region. The replicated research sites represent the regional range of different vegetable and field crop production systems managed using different practices in various combinations. For example, for the initial development of indicator scoring, samples were collected from the Gates Farm in Geneva, NY, which has three tillage practices (no-till/ridge-till, strip-till and conventional tillage), three cover crops (no cover, rye and vetch) and two rotations (continuous high-value vegetable production, compared with a rotation with season-long soil-building crops) replicated four times; on-farm data then further increased the range of soil types and practices covered by the database (Moebius-Clune et al., 2016). Benchmark sites have also been used in soil quality monitoring programmes, either to provide a focus for national monitoring (a representative range of sites) and/or to provide indicator data against which newly measured sites can be compared for example in New Zealand (https://sindi.landcareresearch.co.nz). In the Netherlands, 10 reference soils have been selected out of 285 sites that had been monitored for over 10 years (Rutgers et al., 2008). These reference soils represent specific combinations of soil type and land use (e.g. arable land on clay soil). Where the database of values for the soil health indicators is large enough, referencing within the dataset itself can be used to give target/typical ranges, with measured values compared with the mean value, given ranges or percentiles of all sites under a given land-use and soil type. Moebius-Clune et al. (2016) describe this approach clearly where scoring curves for each indicator (often separately by soil texture group) were determined from the cumulative normal distribution function of these data and the scoring classes were then allocated based on the associated ventiles (see example using penetrometer data in Fig. 3). There is no single agreed approach for the setting of thresholds and/or targets and different approaches can lead to different interpretations of the data; therefore, it is important that the process is carried out transparently integrating expert and user review (Andrews et al., 2004) and also documented clearly.

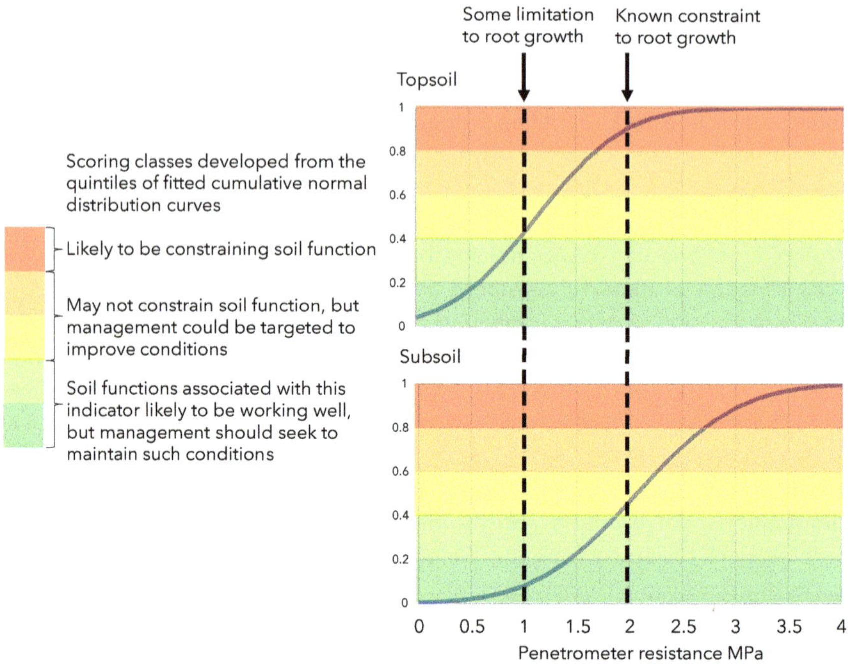

Figure 3 Derivation of scoring classes for soil indicators, here the example of soil hardness (measured by penetrometer resistance) which can limit rooting. A critical value of 2 MPa has been determined from empirical research with some limitation to root growth between 1 MPa and 2 MPa (e.g. Taylor 1971). Where the database of soil indicator data is sufficiently large, scoring classes can be set using the quintiles (or other percentiles) of the cumulative normal distribution. The data presented as cumulative normal distribution curves here with quintiles shown on the y-axis are redrawn from those presented in Moebius-Clune et al. (2016). They found no difference in the cumulative normal distribution curves between soils of different textures. The fitting of scoring classes (shown in the coloured legend) for the topsoil using the database gives classes that are in rough correspondence to the empirical thresholds. However, there is a significant mismatch in the subsoil.

For many indicators, there can be no single scale/critical value; however, these are set. Site factors set the context for the interpretation of indicators (Bünemann et al., 2018) and often influence the attainable values of soil properties to such a large degree that target values need to be set by reference to those constraints (as described by Moebius-Clune et al., 2016). Together with inherent soil properties, such as texture, site factors largely define the ranges within which management is able to control a number of soil properties, for example, the formation and stability of soil structure, soil consistency/workability during tillage, infiltration capacity and depth and penetration of water, and the capacity of soil to supply and retain nutrients. So, for example, it is possible to give some general information on a range of soil characteristics based simply on knowledge of its texture: silty loam will tend to slake and cap at the surface, whereas sandy loam will hold less water and tend to become acidic more quickly and its capacity to hold nutrients will be relatively low. Regardless of management attempts to overcome differences, not all soil constraints will be able to be addressed, some soils simply have much better inherent properties than others; soil health will therefore seem much higher in these soils (USDA, 2001).

Whilst in Europe, developments in soil quality/health monitoring to date have largely been policy-driven, in the United States the focus has been on the empowerment of land users to measure soil quality/health for themselves and then to implement changes in practice. The Natural Resources Conservation Service provide information about in-field tests that can be carried out by land-users themselves and further details about a range of further tests and their interpretation (NCRS 2014). The Soil Management Assessment Framework (SMAF) developed at the Soil Quality Institute (Andrews et al., 2004; http://www.soilquality.org) allows users to select a set of indicators depending on the intended soil function(s). In contrast, the Cornell Soil Health Assessment (Idowu et al., 2008; Moebius-Clune et al., 2016) is much more standardised and targeted directly at land users, offering various soil health testing packages for farmers, landscape managers and others, and supplying them with management advice together with the results of the soil health tests.

3.4 Presenting the results

Presentation of the results is also a key part of soil health measurement. Where monitoring programmes have been developed to inform management action, colour coding for individual indicators or by soil functions (often using a traffic light system e.g. Fig. 3) has been developed to provide links to the need for management action or a more detailed investigation (e.g. Moebius-Clune et al., 2016). In the past a number of schemes have been developed that seek to integrate across indicators by combining the weighted values that have been measured into a single over-arching indicator of soil health (as reviewed by Bünemann et al., 2018). However, such an approach obscures information on any trade-offs and removes any direct link to management actions that are likely to be required. Moebius-Clune et al. (2016) calculate a weighted average soil health score but they also clearly state 'greater attention [should] be paid to the scores of individual indicators and the identification of constraints to proper function of soil processes'. For example, dis-aggregation of a single soil health indicator would be needed so that the user can determine whether liming or sub-soiling is required to improve soil health. In research publications, a range of other graphical presentation methods have also been employed for example, spider diagrams (Stavi et al., 2016).

Advocates of the soil quality concept recognised that in very few situations at any scale will it be possible to make a complete assessment of soil quality, Karlen et al. (2001) suggest that soil quality assessment is therefore a two-stage process 1) identify critical goals and functions that soil resources need to perform at this scale of interest and prioritise them and 2) select appropriate indicator measurements, make measurements, compare different practices on similar soils and/or examine trends on the same soil over time. The aim of such an approach is not to develop absolute but relative measures appropriate to particular defined situations. It is often stressed: 'There never was nor can there be a single value for rating all soils or land uses' (Karlen et al., 2001). Particularly when driven by policy-makers, it is easy for an index developed for soil quality assessment in one situation or for one purpose to be extrapolated; where this occurs, the relevance of the index is lost and its interpretation may be misleading or even wrong (e.g. Sinclair et al., 1996 and its discussion by Karlen et al., 2001 and Sojka et al., 2003). Consequently the caveats attached to any published indicator framework, clearly stating the limits of applicability and the limitations inherent in any interpretation, maybe as important as the framework itself. While farmers (or policy-makers) might want an overall soil health score, the weight of evidence advises against such a calculation.

3.5 Developing guidance to support improved soil management on farms

User experience testing is now embedded in the development processes of most goods and services. However, although farmers/land managers are expected to be important end-users of soil health monitoring, to date, they have generally played an insignificant role in the development of soil quality assessment schemes (Bünemann et al., 2018). Moebius-Clune et al. (2016) describe an effective approach that has integrated interactive design and decision-making with end users from an early stage, together with on-farm evaluation of the indicators. Scientific compromises may be needed to ensure that a set of soil health indicators can be measured on one sampling occasion and as effectively as possible. At the same time the critical importance of the sampling phase cannot be over-stated in the manual for farmers/agronomists. The Cornell Soil Health Assessment manual reminds users to 'ASK YOUR BEST QUESTION!' and to link this to clearly defined sampling goals and the number of necessary samples (Moebius-Clune et al., 2016). The farmer/local agronomist is best-placed to determine their own representative sampling sites (within which soil texture, cropping show limited variation) and where these sites can provide useful data to monitor soil health and inform farm practice in soil management. As for all soil sampling, the area selected should be relatively uniform. Soil observations and samples will continue to be collected on farm for a number of distinct purposes – for example checking the performance of fertilisation/liming strategies, planning new fertiliser applications and determining timeliness for cultivations. Soil health monitoring is intended to fit within those approaches but not necessarily to replace them all; on any farm, there are likely to be further but complementary approaches to soil characterisation for example grid sampling of soil nutrients for precision fertiliser management.

Measuring soil health is only of benefit to the farmer if it is part of an integrated process that removes soil constraints to crop production/quality and supports sustained productivity and environmental quality. Possible management solutions need to be identified and linked to the indicators measured, recognising that as with human health there may be a range of short-term 'treatments' and long-term 'diet and fitness programmes' needed to restore and maintain soil health (Table 2). Hence measuring soil health will mainly be of benefit when part of a sustained monitoring programme.

4 Case study: developing a practical and relevant soil health toolkit for UK agricultural soils

4.1 Moving from principles to practice

In the UK, farmer-levy-funded soil research for many years has focused on the management of soil chemical status to remove constraints to crop and grassland productivity and to ensure optimisation of crop nutrient supply with regard to amount, timing and placement. More recently there has been an increased focus on soil physical condition, links to soil organic matter, and minimising crop constraint resulting from soil structural condition (Hargreaves et al., 2019). With regard to soil biology, the focus of levy-funded research has largely been on the management and minimisation of soil-borne diseases such as common scab, potato cyst nematode, take-all. The Agriculture and Horticulture Development Board (AHDB) and the British Beet Research Organisation (BBRO) have recently funded a 5-year research and knowledge exchange partnership programme (2017–2021) 'Management for Soil Biology and Soil Health' (SBSH) which aims to provide farmers and growers with guidance and better understanding of soil biology and wider soil health measures to improve management for soil health that supports crop and grassland productivity. The project brings together farmers, consultants, and scientists, within a partnership to support development and delivery into practice.

We all want to move beyond talking about soil health in meetings/workshops and the development of general principles, and instead take clear steps to implement a soil health toolkit in practice. Here we will describe the initial stages in the development of a toolkit that can be deployed by farmers/growers and their advisors; however, a similar approach is likely to yield a different but robust framework that can support the needs of policy-makers.

Table 2 Linking indicators, possible problems and management options for soil health. Management solutions may need to be a combination of short-term 'treatments' which fix the immediate problem and long-term 'health promotion' activities that will help to prevent the problem occurring again in the future. The problems listed and management options given are not a comprehensive list. Adapted from a table in USDA (2001)

Indicators Possible problem	Possible reasons for a below optimum soil indicator score	Management options to improve soil indicator score ('treatment')	Management options to avoid the problem in the future ('health promotion')
Indicators showing below optimum scores: Bulk density Penetration resistance Porosity Root growth patterns *Possible problem:* Compacted layers in soil restricting rooting/water movement	Working wet soil Excess traffic Heavy machinery Repeated tillage at same depth Excess animal traffic Poor aggregation Low organic matter	Subsoil or rip just below the compacted layer when soil is not excessively wet or dry	Avoid working wet soil Reduce traffic/tillage operations Use controlled traffic patterns Avoid using heavy machinery Alter tillage depth Add organic residues Add cover crops Use crop rotations Add animal manures
Indicators showing below optimum scores: Aggregate stability Slake test Observations of soil crusts *Possible problem:* Crusting at the soil surface reducing crop establishment	Excess sodium Low organic matter Low crop residues	For sodium problem – apply gypsum and flush with irrigation water	Add organic residue Add animal manure Use cover crops For sodium problem – continue to treat with gypsum regularly, especially for vulnerable crops
Indicators showing below optimum scores: Infiltration rate Hydraulic conductivity *Possible problem:* Waterlogged topsoils	High water table Poor soil structure Compaction Blocked drains (May be associated with high clay content where treatment will have limited impact and problem will need to be managed)	Subsoil to break up compaction or tillage pan Add drainage system	Regularly check and maintain drainage systems Carefully manage grazing intensity Plan crop rotation carefully to reduce the risk of trafficking or working wet soils

Indicators showing below optimum scores: Organic carbon Water content at field capacity Porosity Observed droughtiness in crops *Problem:* Low available water holding capacity	Compaction affecting infiltration or water movement through the profile Low organic matter Excessive drainage Low aggregation Low biological activity (May be associated with high sand content where treatment will have limited impact and problem will need to be managed)	Reduce compaction	Increase organic residues Add animal manure Use cover crops Improve conditions for earthworms/soil life to increase effectiveness of infiltration and soil aggregation Avoid tillage when soil is wet
Indicators showing below optimum scores: Observed rills, gullies Topsoil depth Aggregate stability *Problem:* Erosion	Lack of cover and residue Low organic matter Poor aggregation Tillage pan or compacted layer Tillage practices that move soil down slope Excessive tillage Intensive crop rotation (May be associated with sloping sites where treatment will have limited impact and problem will need to be managed)		Diversify crop rotations Reduce tillage Use animal manure Use cover crops Increase surface residue or roughness Shorten slope length Plant strip crops Use wind breaks
Indicators showing below optimum scores: pH CEC Observed poor crop yields *Problem* Soil acidity	Use of ammonium fertilisers No liming Low buffering capacity; limited duration of liming impacts	Soil test - correct pH levels Add lime for low pH	Alter fertiliser strategy Implement a rotational liming strategy

(*Continued*)

Table 2 *(Continued)*

Indicators Possible problem	Possible reasons for a below optimum soil indicator score	Management options to improve soil indicator score ('treatment')	Management options to avoid the problem in the future ('health promotion')
Indicators showing below optimum scores: Organic carbon Soil pH Soil fertility test CEC Observed poor crop yields *Problem:* Nutrient availability	Nutrient imbalances (deficiencies or excesses) Poor drainage Poor or limited soil microbial activity Incorrect pH Low organic matter	Soil test - correct nutrient and pH levels	Implement a rotational nutrient management plan Increase organic residue Use animal manure Use cover crops & crop rotations
Indicators showing below optimum scores: Earthworms Soil respiration Microbial biomass Pitfall trapping Observed undecomposed residues *Possible problem:* Low biological activity	Excess tillage Low residue crops Too much fallow Excessive tillage Poor aeration		Increase organic residues Increase diversity of cropping Add animal manure Use conservation tillage Add cover crops Improve drainage

4.2 Underpinning approach

Within the SBSH programme, we brought together the scientific knowledge on soil biology and soil health for the UK by undertaking a literature review of articles that were accessible through the Web of Science database or in reports and other "grey" literature known to the team. We tabulated the effects of the different management options identified on the biological, physical and chemical properties of soil. Some of the management options were specific and often described in detail for example reduced tillage, added organic matter, residue retention, fertiliser addition, liming, strip tillage, controlled drainage, controlled traffic, biocides, buffer strips, plastic mulch and adding earthworms; however, others were more general, system-oriented changes such as implementation of conservation agriculture (including cover crops) or organic farming systems. There are still gaps in the knowledge for the less common management options and often for interactions of management options with soil type and climate, as well as for combinations of management options. In addition to drawing out general principles to allow the development of knowledge transfer materials (leaflets/web resources), Griffiths et al. (2018) describe how this information was used in the development of a visual tool to summarise the key physical, chemical and biological interactions affecting soil health in UK agricultural soils. This was based on an underlying descriptive model integrating current understanding of the impacts of site and management factors on soil properties (Stockdale et al., 2019). The visual tool presents the likely magnitude of the effects of representative management options (reduced tillage; no-tillage; cover crops; carbon-rich or nitrogen-rich organic amendments) on a suite of biological, physical, chemical and economic outputs. This tool aims to build farmer/grower understanding to develop improved soil management, and also stimulate a greater awareness amongst growers of (a) the impacts of soil management and crop selection, and (b) the complexity of biological interactions in soil.

At the same time, we used a number of approaches flexibly and actively to work with farmers, growers, advisors and industry partners to shape the direction of the development of a soil health toolkit and wider work on the management of soils in agricultural systems to maintain and improve soil health. Open workshops and on-line questionnaires have been used to add breadth and depth to the viewpoints, concerns and proposed solutions developed within the SBSH team, which is already multi-stakeholder in composition (Fig. 4). Co-creation of knowledge in this way means that it is more likely to be directly implementable and practically embedded (thus benefiting knowledge transfer) but also highlights new research questions and their solutions. Here the soil scientists within the programme seek to use their specialist knowledge, working with stakeholders, to act as informed mediators of the processes

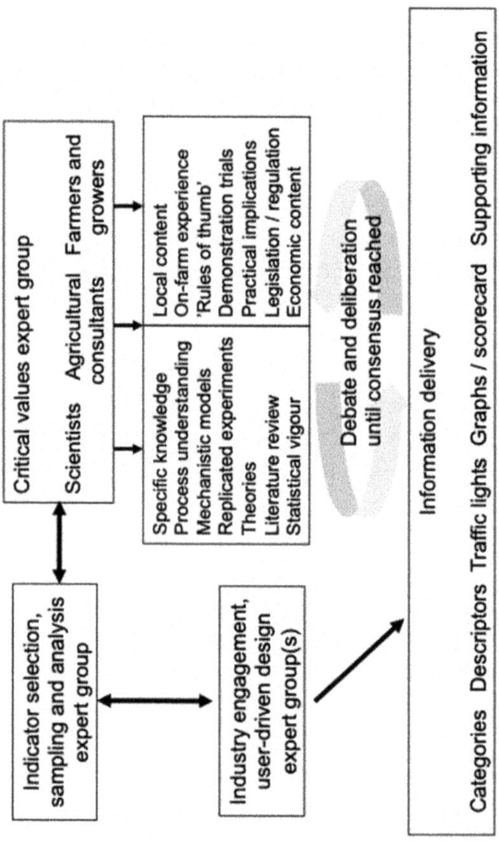

Figure 4 Overall approach taken to develop a soil health scorecard for UK agricultural soils.

to identify conflicts or trade-offs, develop management plans and identify measurable information to form the basis of decision making.

4.3 Selecting indicators

We were able to build on numerous papers in the soil science literature considering and listing possible measurements that may be used as indicators of soil quality/health. These activities have been useful and are a necessary step in deciding on sets of measurements to be used in national or regional soil monitoring schemes. In such schemes there is an inevitable trade-off between what is practicable and affordable and the full range of measurements that research-oriented soil scientists would like. Nonetheless, soil scientists have a vital role in translating the requirements of stakeholders (which may be loosely expressed as soil health for various uses of soil) into definable soil properties or functions. This can then lead to rational decisions on measurements suitable for providing appropriate information on soil functions in different situations. It is important not just to identify the most appropriate indices but also to ensure that these can be quantified with the necessary precision and that farmers and land managers can understand and relate to them (Lal, 2001). Karlen et al. (2001) stress that approaches to soil health (supporting both education and assessment) provide an adaptive management tool supporting the goal of sustainable land management; measurement of soil quality is not a goal in itself.

The aim of the work within the SBSH programme is to develop and test a toolkit for soil health monitoring which allows land users in the UK to take site-specific soil quality issues into account when planning and managing resilient and sustainable crop and grassland production for the long term. Hence in discussion with stakeholders we identified when and how soil observations and samples are collected on farm. Given the overall aim, the stakeholder workshops therefore suggested that sampling for assessment of soil health should take place once per rotation and at the same point in the rotation to maximise comparability between samples so that the farmer/grower can monitor long-term trends. Land users/farmers are in the best place to determine the best point in the rotation to integrate soil health sampling.

To identify a minimum dataset for soil health monitoring which allows land users in the UK to take site-specific soil quality issues into account when planning and managing resilient and sustainable crop and grassland production for the long term, we selected from an initial group of 45 indicators of soil health that had been used previously (Table 3). These prospective indicators were then ranked using a logical sieve approach as pioneered by Ritz et al. (2009) and used by Griffiths et al. (2016). The indicators were rated by members of the project team plus a wide range of stakeholders from industry

Table 3 Soil measures screened for use as part of a scorecard for soil health in agricultural soils in the UK

Physical	Chemical	Biological
Aggregate stability	CEC (cation exchange capacity - capacity of soil to retain cations)	Bait Lamina assay (simple measure of biological activity with depth in top 8cm)
Available water capacity (max. amount of plant available water a soil can provide)	C:N ratio (indicating stability of soil organic matter)	Basal respiration (resting biological activity)
Bulk density	EC (electrical conductivity - measure of available ions)	Earthworms
Depth of soil	Extractable S	Enzyme assays (functional measure of potential activity of C, N P cycling enzymes)
Infiltration rates	Extractable Ca	Functional gene abundance (DNA measure of quantity of the actual enzymes)
Particle density	Heavy metals	Key pathogens
Penetrometer resistance	Hot Water Extractable Carbon (HWEC)	Metabolic quotient (respiration per unit biomass - how stressed the microbes are)
Permeability - possibly only subsoil as the topsoil permeability is so dynamic	Light fraction organic matter (LFOM)	Microarthropods (community structure correlates with soil ecological health)
Porosity/Water filled pore space (WFPS)	Organic matter - LOI (loss on ignition)	Microbial biomass (total amount of microbes in soil)
Rate of erosion	Organic C	Microbial community structure/ diversity of Bacterial, fungal and archaeal communities
Sealing	pH	Mycorrhizal root colonisation
Soil texture	Potentially mineralisable N (PMN)	Nematodes (community structure indices correlate with soil ecological health)
Shear strength (a measure of soil strength)	Routine nutrients (P, K, Mg)	Total fungi and bacteria
Stoniness	Total Nitrogen	
Visual Evaluation of Soil Structure (SRUC method for visual estimation of soil structure; VESS)		
Visual Soil Assessment (New Zealand method for visual estimation of soil structure; VSA)		
Water content at time of sampling		
Water characteristic curve (relation between water content and water potential)		

(including lab. service providers, agronomists and farmers), according to how closely they matched criteria important in three broad categories for indicators of sustainability in agricultural soils:

- First, the measures needed to link with key soil functions relevant to both agricultural production and environmental impact.
- Second, they clearly needed to be relevant to agricultural soils.
- Third, they were rated in terms of practical application for monitoring including: sample throughput; sample storage; necessity of single or multiple visits for sampling; ease of use; understandability of results (i.e. whether the indicator can easily be used to explain issues of soil health); ease of interpretation (i.e. the score can easily be translated into management options); sensitivity; cost; standardisation and availability in the UK.

Soil texture is a fundamental property that is not changed by management and is, therefore, not appropriate as an indicator to monitor changes in soil health. Therefore it is not included as an indicator. However, an underpinning measurement of soil texture is needed, alongside other basic site descriptors, to benchmark the values obtained for the indicators appropriately (Section 3.3). Water content (at time of sampling) was not included in the scorecard as it largely depends on the immediately preceding weather. However, some of the measures (e.g. penetrometer) need to be measured in soil at/close to field capacity; a measure of soil water content at sampling can also provide useful context information when comparing between sites or year-to-year trends

The scorecard aims to capture the interactions between physics, chemistry and biology that underpin soil health and to provide a concise and practical format for the user which also provides useful information to inform management. Hence at least three of the top ranked indicators from each category were chosen for inclusion in the evaluation phase (Table 4). Thus, from the physical category we have taken bulk density, visual assessment of soil structure, (VESS, Ball et al., 2007) and penetrometer resistance. Many of the chemical indicators are in common use in commercial soil analyses, which probably explains a clear preference in the selection process for pH, routine nutrients and soil organic matter (loss-on-ignition). Biological indicators have been the least used to date and so a less clear-cut picture emerged in screening than for the chemical indicators; hence more of these indicators are included for testing and development within the programme (Table 4). We propose to use respiration measured on an incubated sample currently delivered commercially by some laboratories and we will also test the use of a measure of microbial biomass. Although three ecological community measures are being assessed within the programme (nematodes, microarthropods and

Table 4 Soil indicators selected by the SBSH Programme for evaluation

Soil indicator	Sampling/analysis	Evaluation approach within the SBSH programme
pH	Mixed topsoil sample to lab.	Relatively common indicators included; framework for interpretation and usability on farm will be evaluated.
Routine nutrients (extractable P, K, Mg)	Mixed topsoil sample to lab.	
Bulk density	Intact core sample to lab.	
Penetrometer resistance	In-field assessment	
Visual evaluation of soil structure (VESS)	In-field assessment	Less common indicators evaluated; framework for interpretation will be developed; usability on farm will be evaluated.
Loss-on-ignition (soil organic matter)	Mixed topsoil sample to lab.	
Respiration (Solvita CO_2 burst test)	Mixed topsoil sample to lab.	
Earthworms	In-field assessment	
Microbial biomass C	Mixed topsoil sample to lab.	New indicators developed, compared and evaluated; use on farm may be considered
Potentially mineralisable N	Mixed topsoil sample to lab.	
Total N	Mixed topsoil sample to lab.	
Nematodes	Mixed topsoil sample to lab.	
Microarthropods	Intact core sample to lab.	
DNA measures (including pathogens)	Mixed topsoil sample to lab.	

earthworms), it is likely that only one will remain by the end of the Programme. Molecular indicators of overall soil biological community and soil pathogens are also the subject of developmental work within the Programme.

In the scientific literature, each of the selected indicators has not only a standardised laboratory methodology but usually also some guidance for the timing and approach to sample collection (depth, replication, bulking, etc.). In order to develop a full toolkit, we need to move beyond identifying the minimum indicator set and additionally provide a sampling protocol. To the surprise of the project leader, while the principle was readily agreed that in-field assessments and samples for physical, chemical and biological measures (Table 3) should be collected simultaneously to optimise comparability and minimise sampling effort for the farmer, agreement of a protocol with common timing and sampling approach proved far more contentious. The physical and biological measures require sampling while the soil is both warm and moist, hence suggesting (for the UK) a mid-late spring or early-mid-autumn timing, depending on the weather. Farmer workload is high in both seasons and there is no obvious lull into which such sampling would fit. However, the least-worst option for sampling time was agreed to fall in:

- mid-late autumn (after harvest),
- after the topsoil has wetted up in the autumn, and
- at least 1 month after any cultivations/moderate soil disturbance.

This timing may mean that rotational sampling for soil health does not easily fit at all points of the rotation. In some rotations, this may mean sampling in an actively growing cover crop or after drilling of the next main crop. Therefore, feedback from practice on this sample timing and the difficulty/ease of scheduling the sampling in this period is important during the project.

4.4 Setting benchmarks, thresholds and establishing a clear reporting framework

Quantification of soil health by measurement of a range of soil properties linked to the key functions required at any site is complex - with an infinite number of different target indices needed to fit the range of distinct soil types, management and environment scenarios for each given use (Sojka et al., 2003). To make this large umbrella concept of practical use, we have worked with stakeholders and identified some simple grouping variables (describing soil texture groups, climate, cropping system) that are then used for each field/site sampled in UK lowland agricultural systems. More detail on how these grouping variables affect soil characteristics and likely management drivers (organic matter inputs, tillage intensity etc.) are given in Stockdale et al. (2019).

Development of a soil health toolkit needs not only a set of robust and measurable indicators and site descriptor variables but also mutually acceptable and well-informed agreement on the way in which data will be presented and the degree of simplification that will be used. Although there is a substantial literature considering and listing possible measurements that may be used as indicators of soil quality/health, there is much less information on development of appropriate reporting frameworks.

An iterative fitting process was used at a multi-stakeholder workshop in small working groups and then across the workshop as a whole to explore the development of benchmarks and thresholds for each of the selected indicators. First, the full range of possible values for each indicator was established and divided into a small number (5-8) of discrete categories for presentation for example in bar charts. Then the main response shape for each indicator was agreed. Using the categorisation of responses according to Andrews et al. (2004), the shape of the response curve for each indicator was considered. At the end of the workshop the following allocations had been agreed for the indicators:

- more is better (extractable K, loss on ignition, respiration, earthworms),
- optimum range (pH, extractable P, extractable Mg, penetrometer resistance),
- less is better (bulk density, VESS), and
- undesirable range (none of the SBSH programme indicators).

The traffic light approach of presentation adopted in soilquality.org.au was adopted here where the traffic light highlights the level of risk in relation to crop yield or environmental pollution and hence the need to undertake further more detailed investigation associated with an indicator value in that category:

- red (investigate),
- amber (monitor), or
- green (no action needed).

Each category was then allocated to a risk grouping together with consideration of whether risk groupings need to be allocated differently for different climates, farming systems or soil types. In many cases this process was relatively simple as underpinning frameworks for interpretation for agronomic and/or environmental impact are well established (Fig. 5). However, the development of an indicator framework for other measures was more fraught and required substantial iteration and some compromise before a proposal that could be evaluated within the SBSH programme was generated for example Loss-on-ignition, earthworms.

In this project, where the overall aim is to support farmers and growers in planning and managing resilient and sustainable crop and grassland production for the long term, we consider that the codifying of soil health into a simple index is both too aggregated a level to support management decisions and also too complex to apply effectively. Hence in collaboration with stakeholders we have decided that we will present the outcome of the soil health indicators as a scorecard showing a list of measures, measured values and traffic light coding (see above). It was agreed that this approach would probably be more useful, and certainly more transparent, with links made to specific soil functions or properties rather than an abstract measure of soil health/quality.

4.5 Building stakeholder engagement

Knowledge exchange work on-farm regularly highlights that there are a significant minority of farmers already investing in the development of

(a) *Extractable K (ammonium nitrate extract)*

Bar chart classes *mg/L*	Traffic light colour	Description of this class
0-60	Red	Index 0 - risk to production
61-120	Amber	Index 1 - potential risk to production
121-180	Green	Index 2-
181-240	Green	Index 2+
241-400	Green	Index 3
> 400	Green	> Index 3 - no expected benefit of fertiliser K

(b) *Extractable P (Olsen extract)*

Bar chart classes *mg/L*	Traffic light colour	Description of this class
0-9	Red	Index 0 - risk to production
10-15	Amber	Index 1 - potential risk to production
16-25	Green	Index 2
26-45	Green	Index 3 - potential risk to environment, but P required for P-responsive crops including potatoes and some vegetable crops.
46-70	Amber	Index 4 - risk to environment. Consider restriction of rate and/or frequency of manures/biosolid applications
> 71	Red	> Index 4 - risk to environment. Do not add further P in manures/biosolid applications

Figure 5 Using existing nutrient management guidelines to develop the presentation approach (grouping into bar chart classes, associated descriptors and traffic lights) for use in England for indicators of (a) extractable K (more is better) (b) extractable P (optimum range).

techniques to improve soil health by adopting practices which require more of their time and which may also have required significant capital investment. On farm many practices that affect soil health are used for several reasons, including economic or management drivers as well as reflecting a concern for soil function, consequently there is a need to provide breadth of information that can be used to guide uptake and assess cost-effectiveness in the local context. However, in previous farmer consultation work (Stockdale and Watson 2012), we noted that even the most innovative farmers asked for more input, particularly more information and tools that could support them to make more effective decisions for their farming system and evaluate the impact of practices in place.

In the SBSH programme we are seeking to allow farmers to become partners in an adaptive research and knowledge exchange process (Sherwood and Uphoff 2000) by not only testing the soil health scorecard on-farm (user testing and development) but also to allow innovations in soil management developed on-farm to be evaluated robustly in-situ (adaptive co-learning and development) by using the underpinning descriptive model within a clear systematised conceptual framework. Consequently, the descriptive model was discussed in early draft form with industry stakeholders including the agronomists and farmers who are likely to be the main users of the model in the future. The development of a descriptive model was largely supported by the industry and technical workshops. Both made some useful recommendations with regard to how the information is presented (certainty of outcome) and the detail of presentation. Work within the SBSH Programme will continue to develop and improve the presentation of the model as part of a two-tier approach to knowledge exchange, in particular the model will allow farmers to underpin the general principles with additional detail and provide capacity to interrogate the background data/relationships.

4.6 Evaluation in practice

Over the five years of the SBSH programme, the soil health scorecard (and the associated descriptive model) will be evaluated using results from both long-term soil management trials and on-farm monitoring. Seven long-term soil management experiments have been identified which have implemented treatment regimes that directly manipulate the main factors affecting biological function, that is:

- Food source: nutrient and organic matter inputs, cropping choice and sequence,
- Air/water supply – tillage systems, drainage, and
- Chemical environment – pH.

In these experiments, a programme providing detailed monitoring of crop (yield, disease and weed constraints) and rotational soil health is now in place. Soil physical properties are recorded in the field at the time of sampling (VESS, penetrometer resistance) and cores are collected for the measurement of bulk density. A bulk soil sample is also collected to allow extended soil chemical/biological analysis using a commonly available Soil Health package (NRM, 2018) plus direct measures of soil organic carbon, total nitrogen (N), potentially mineralisable N, earthworms, microfauna, nematodes and microbial biomass/respiration. These data are presented in the soil scorecard format for review by stakeholders (Fig. 6). We are also using the long-term trials to aid the development and validation of DNA/eDNA-based soil biological indicators.

To allow control of the factors under study, many research trials have specific and narrowly focused remits, often with limited acknowledgement of rotational impacts and, consequently, the implementation and impact of soil

Soil health parameter	Control (Inorganic fertiliser)	Cattle FYM (23 years)	Slurry (23 years)	Green compost (6 years)	Food-based digestate (9 years)
Soil organic matter (%LOI) *	3.0	4.1	3.6	4.0	3.4
pH *	6.4	7	6.4	7	6.5
Extractable P (mg/l) *	56	73	53	60	65
Extractable K (mg/l) *	80	311	194	187	167
Extractable Mg (mg/l) *	44	87	75	63	46
VESS score **	2	2	2	1	2
Earthworm count (total number in a 20 x 20 x 20 cm block)	11	13	9	11	13

Investigate

Monitor

No action needed

* Attributes that showed a statistically significant difference between treatments ($P<0.05$)
** A lower VESS score indicates better soil structure

Figure 6 Example scorecard for a long-term organic matter additions experimental site at Harper Adams University College. Samples were taken prior to destruction of a 2-year temporary grass-clover ley receiving either recommended rates of inorganic fertilisers ("Control") or annual applications different organic materials. These treatments also received inorganic fertiliser, with N applications reduced to take into account the N supplied by the organic materials (Bhogal et al., 2018).

management at the scale and within the context of farm business have been less well studied. Farmers and growers are already implementing a range of innovative approaches to the management of soil biology and health, often combining a number of the strategies investigated in research projects and integrating new approaches that are adapted for site-specific use (as shown in Stockdale and Watson, 2012). This network of farmer research and innovation can deliver important information to inform both research and practice through participatory approaches to evaluate the impacts on soil biology and health across a broad spectrum of crops, climates, soil types and rotations. Within the SBSH programme farmer research-innovation groups have been established; 8 are operating from autumn 2018. The groups contain a wide range of farms and farming systems across the UK including those with sugar beet, potatoes and/or horticulture in their rotations, as well as reduced/zero-till cereal-oilseed rotations and grass-based systems. Within the farmer groups, a programme providing detailed monitoring of crop (yield, disease and weed constraints) and rotational soil health is now in place to measure the impacts of the broad range of innovations in management of soil health already present on commercial farms. Soil sampling will take place in the autumn post-harvest and after the soil has wetted up using paired field comparisons/split field treatments. In the first year the focus is on evaluation of the sampling requirements and in-field recording of VESS and earthworms with a small number of chemical and biological measurements made on the bulk soil samples collected.

Assessments each season will be used to evaluate the soil health scorecard approach and to update protocols, measures and interpretation as required. In addition, for each treatment (research trial) or field/site sample (on-farm) the descriptive model will be used to generate a descriptive prediction of the relative impact of management (single or multiple factors) compared with the control plots (research trial) or paired/split field (on-farm). These predictions can be evaluated using the measured physical, chemical and biological properties. The data will also be used after the 2019 field season to re-calibrate and update the underpinning descriptive model with a further evaluation phase in 2020–2021.

5 Conclusion and future trends

Since soil health is a term which has resonance with the general public, land managers and policy-makers, it is an ideal peg on which to hang educational initiatives that promote understanding of the role and functioning of soil within the environment. These initiatives may not teach anything new, but any opportunity to increase understanding of the dynamic and living world just beneath our feet should be seized with both hands and quickly; if soil scientists want land users to stop treating the soil like dirt, they must invest time to help others in recognising the importance of soil function and its inherent value (Ball et al., 2018).

Farmers and scientists share a concern with soil degradation, especially where it is associated with yield loss or direct impacts on air and water quality. However how soil health will be addressed is both a technical question – what can be done?, as well as a cultural question – what do the farmers consider worthwhile doing? (Krzywoszynska 2018). An essential explicit, but oft-assumed, step is to define the context of the threats to soil health and the key functions required of a soil at the particular scale of interest (e.g. farm, drinking water catchment, region). The diverse pattern of soil types, landscapes and farming systems means that any approach which seeks to improve the measurement and consequent management of soil health of on-farm soil must take proper account of the diverse pattern of soil types, landscapes and farming systems as well as stakeholder interests. There is no shortage of soil characteristics that can be measured, but the focus must first be on identifying indicators of the state of soil health and/or its direction of change within the local context for example taking account of climate, land use, and soil type. Only then can the focus narrow to the selection of appropriate measures that allow these indicators to be quantified practically, cheaply and robustly so that users can compare different practices on similar soils and/or examine trends on the same soil over time. Despite minor differences in current soil health frameworks across the world, there is generally at least one indicator for each of the chemical, physical and biological properties of the soil. Organic matter (carbon), pH, extractable phosphorus and various indicators of soil structure/water storage are commonly used. The associated framework of interpretation must then clearly show whether the measured values of each indicator are acceptable that is the ranges over which there is no loss or significant impairment of functioning and contrast these with ranges where one or more soil functions are constrained. For some indicators, clear links have been established with function. For many biological indicators (e.g. soil organic matter, soil microbial biomass), although an increase in measured values is considered to be beneficial, more work is needed to move from benchmarks set by reference to typical values and towards thresholds and targets with a clear link to function. New indicators will emerge over the next few years; however, it is essential that the focus is as much on the interpretation framework as the methodological development itself, so that any data generated can quickly be used to inform management decisions. A number of the soil health frameworks in practical use present the soil health indicators in a scorecard showing a list of measures, measured values and traffic light coding as well as directing users towards guidance for improved soil management on farm. Measuring soil health is only of benefit to the farmer if it is part of an integrated process that removes soil constraints to crop production/quality and supports sustained productivity and environmental quality. Measuring soil health will mainly be of benefit on farm when part of a sustained programme of monitoring-led action planning.

The case study given here shows one approach to the co-production of a technically based soil health scorecard to support and inform effective decision-making. This project has developed existing work on selecting soil health indicators further by specifically selecting indicators and tailoring the frameworks for interpretation relevant to UK agriculture (grassland, arable, sugar beet, potato and horticultural crops) under UK climatic and soil conditions. The recommended indicators together provide a soil health scorecard which integrates physical, chemical and biological aspects to give a snapshot overview of soil health akin to a routine car safety check (MOT) or end of school report. To support the management of soil health on farm, indicator results should be benchmarked for comparison within farming systems, soil types and climatic zones and allow the assessment of local trends over time. Benchmarking in this project and elsewhere will quickly (1–5 years) gather a body of data from which the normal operating range of those measures not currently routinely measured (i.e. earthworms, microbial biomass) can be evaluated against soil texture, climate and cropping regime to revise and improve the thresholds and so improve the advice given. It is essential that such on-farm monitoring approaches are directly linked with advice on improved management practices that maintain/improve soil health. In addition to technical development there has also been significant co-learning about the underpinning visions of stakeholders of the agricultural systems of which soils are a key foundation. The approaches used and the resulting outcomes are broadly similar to those developed at Cornell and in Australia; bigger differences would be expected where the approach was applied in very distinctly different soils, climates and farming systems. Nonetheless the process of development is very important to support learning and adoption of soil health monitoring locally and hence should not be short-circuited.

Monitoring soil health is also a concern for the government at national/regional scales and there has been widespread and concentrated effort put into devising practical methods to measure soil health (including soil biodiversity and soil quality) at national levels in the UK and at a European level. Hence there has been some attention paid recently to whether the national soil monitoring and farmer soil health schemes can be brought together effectively. However, soil health monitoring schemes developed to meet these distinct aims often only coincide in name. The difference in scale (landscape/catchment vs field/management zone) and data reporting requirements (national and regional averages vs. field scale comparisons) mean that there will be differences in the detailed implementation of any required monitoring programme for example sampling structure (representative, stratified vs. targeted, problem-directed), not forgetting difficulties in data sharing/ownership.

For many soils and farming systems we can already identify 'best practice' management approaches that minimise deleterious impacts on soil quality, particularly for productivity and in relation to direct impacts on air and water quality. We can also identify situations where maintaining one soil function will be incompatible with another, and a compromise or a decision on priorities is required. The challenge is to get these implemented in the real world where time and money are more immediate controllers of behaviour than any perceived impact on soil quality or function. There is also a need to increase our understanding of the links between land management, soil properties and soil functions, with appropriate consideration of spatial and temporal distribution, to further refine guidance on best (or least/worst) management approaches and policy frameworks supporting them, in order to optimise the delivery of all ecosystem services in the landscape.

6 Where to look for further information

- *Managing Soil Health for Sustainable Agriculture*, Volumes 1 and 2. ed. Don Reicosky. Burleigh Dodds Science Publishing, Cambridge.

These books give an overview of the role of soil as a provider of ecosystem services. In Volume 1, there are reviews of the fundamentals of soil structure and chemistry as well as organic matter, soil microorganisms and fauna. The book also discusses soil dynamics, from water and nutrient cycles to carbon capture and erosion mechanisms. In Volume 2, the latest research on soil monitoring and management is presented. There are reviews on soil classification, sampling and ways of monitoring soil dynamics. The book also surveys key techniques for managing soil, from no-till and conservation tillage techniques to the use of rotations, intercropping and cover crops as well as manure and compost management.

In addition, there are now a wide range of online resources which support increased understanding of soil health and management. The best-developed independently provided examples that are currently available are:

- Soil Health Portal. USDA Natural Resources Conservation Service https://www.nrcs.usda.gov/wps/portal/nrcs/main/national/soils/health/. Last accessed 1 August 2019.
- Soils are Alive, The complete soil health reference for farmers, consultants and researchers. Professor Lyn Abbott. Online book. www.soilhealth.com. Last accessed 1 August 2019.

6.1 For North America

The Soil Health Institute (SHI) is an independent, non-profit organisation charged with coordinating and supporting soil stewardship and advancing soil health in North America. The SHI program is designed to move scientific knowledge and technology from the research laboratory to the farm field by bringing together traditional and non-traditional agricultural industry partners, farmers, ranchers, government agencies, scientists and consumers to focus on one common, clear goal: protecting and enriching our soils. The Soil Health Institute is committed to working with all partners to enhance and disseminate knowledge and technologies directed at key soil processes to increase productivity, resilience and environmental quality; identify research and adoption gaps; coordinate national partnerships to address those gaps; and help drive the transformational changes needed for the betterment of soil health and ultimately society. SHI helps set research priorities, co-ordinates research and promotes action (https://soilhealthinstitute.org/).

Soil Health 360 is a web-news service on soil health provided by the US Soil Health Institute. In order to ensure you have fast access to information about soil health, with access to third-party links by using a computer algorithm to find soil health-related keywords within the content. This can be accessed at: https://soilhealthinstitute.org/soil-health-360/.

7 Acknowledgement

The book chapter has been developed from a presentation to the International Fertiliser Society at its Agronomic Conference in December 2019 and the associated proceeding publication. We acknowledge farmer funding via the levy-boards AHDB and BBRO in the Soil Biology and Soil Health Research and Knowledge Exchange Partnership (AHDB: 91140002; 2017–2021).

We would like to thank all those farmers, agronomists and fellow researchers who have challenged our thinking and contributed to the development of our understanding.

8 References

AHDB, Agricultural and Horticultural Development Board (2017). Nutrient management guide (RB209). Section 1. Principles of nutrient management and fertiliser use. Available at: https://ahdb.org.uk/knowledge-library/rb209-section-1-principles-of-nutrient-management-and-fertiliser-use.

Andrews, S. S., Karlen, D. L. and Cambardella, C. A. (2004). The soil management assessment framework: a quantitative soil quality evaluation method. *Soil Science Society of America Journal* 68(6), 1945–1962.

Ball, B. C., Batey, T. and Munkholm, L. J. (2007). Field assessment of soil structural quality - a development of the Peerlkamp test. *Soil Use and Management* 23(4), 329–337.

Ball, B. C., Hargreaves, P. R. and Watson, C. A. (2018). A framework of connections between soil and people can help improve sustainability of the food system and soil functions. *Ambio* 47(3), 269–283.

Barrios, E. (2007). Soil biota, ecosystem services and land productivity. *Ecological Economics* 64(2), 269–285.

Bhogal, A., Nicholson, F. A., Rollett, A., Taylor, M., Litterick, A., Whittingham, M. J. and Williams, J. R. (2018). Improvements in the quality of agricultural soils following organic material additions depend on both the quantity and quality of the materials applied. *Frontiers in Sustainable Food Systems, Waste Management in Agroecosystems* 19 April 2018. Available at: https://doi.org/10.3389/fsufs.2018.00009.

Bouma, J., van Ittersum, M. K., Stoorvogel, J. J., Batjes, N. H., Droogers, P. and Pulleman, M. M. (2017). Soil capability: exploring the functional potential of soils. In: *Global Soil Security*. Field, D. J., Morgan, C. L. S. and McBratney, A. B. (Eds). Springer, Switzerland, 27–44. Soil quality: a critical review. *Soil Biology and Biochemistry* 120, 105–125.

Buol, S. W., Southard, R. J., Graham, R. C. and McDaniel, P. A. (2011). *Soil Genesis and Classification* (6th edn.). John Wiley & Sons, Ames, IA.

Bünemann, E. K., Bongiorno, G., Bai, Z., Creamer, R. E., De Deyn, G., de Goede, R., Fleskens, L., Geissen, V., Kuyper, T. W., Mäder, P., Pulleman, M., Sukkel, W., van Groenigen, J. W. and Brussaard, L. (2018). Soil quality: a critical review. *Soil Biology and Biochemistry* 120, 105–125.

Cardello, A. V. (1995). Food quality: relativity, context and consumer expectations. *Food Quality and Preference* 6(3), 163–170.

Defra, UK Department of Environment Food and Rural Affairs (2018). Farming rules for water. Available at: https://www.gov.uk/government/publications/farming-rules-for-water-in-england.

Doran, J. W. and Parkin, T. B. (1994). Defining and assessing soil quality. In: *Defining Soil Quality for a Sustainable Environment*. Doran, J. W., Coleman, D. C., Bezdicek, D. F. and Stewart, B. A. (Eds). Soil Science Society of America Special Publication 35. SSSA, Madison, WI, pp. 3–21.

FAO, Food and Agriculture Organisation of the United Nations (1976). A framework for land evaluation. FAO Soils Bulletin No. 32, Rome.

FAO, Food and Agriculture Organisation of the United Nations (2008). An international technical workshop: investing in sustainable crop intensification. The case for improving soil health. 22–24 July 2008. Integrated Crop Management, Vol. 6. FAO, Rome.

FAO, Food and Agriculture Organisation of the United Nations (2015). Status of the World's Soil Resources – Main Report. Food and Agriculture Organization of the United Nations and Intergovernmental Technical Panel on Soils, Rome, Italy.

Fisken, D. (1990). Sensory quality and the consumer: viewpoints and directions. *Journal of Sensory Studies* 5, 203–209.

Gonzalez-Quiñones, V., Stockdale, E. A., Banning, N. C., Hoyle, F. C., Sawada, Y., Wherrett, A. D., Jones, D. L. and Murphy, D. V. (2011). Soil microbial biomass – interpretation and consideration for soil monitoring. *Soil Research* 49(4), 287–304.

Griffiths, B. S., Hargreaves, P. R., Bhogal, A., Elphinstone, J., Stockdale, E. A. and Watson, C. A. (2018). Soil Health – visualising the consequences of management change and developing a scorecard for monitoring. *Soil Improvement: Impact of Management Practices on Soil Function and Quality; Aspects of Applied Biology* 140, 5–8.

Griffiths, B. S., Römbke, J., Schmelz, R. M., Scheffczyk, A., Faber, J. H., Bloem, J., Pérès, G., Cluzeau, D., Chabbi, A., Suhadolc, M., Sousa, J. P., Martins da Silva, P., Carvalho, F., Mendes, S., Morais, P., Francisco, R., Pereira, C., Bonkowski, M., Geisen, S., Bardgett, R. D., de Vries, F. T., Bolger, T., Dirilgen, T., Schmidt, O., Winding, A., Hendriksen, N. B., Johansen, A., Philippot, L., Plassart, P., Bru, D., Thomson, B., Griffiths, R. I., Bailey, M. J., Keith, A., Rutgers, M., Mulder, C., Hannula, S. E., Creamer, R. and Stone, D. (2016). Selecting cost effective and policy-relevant biological indicators for European monitoring of soil biodiversity and ecosystem function. *Ecological Indicators* 69, 213–223.

Hargreaves, P. R., Baker, K. L., Graceson, A., Bonnett, S., Ball, B. C. and Cloy, J. M. (2019). Soil compaction effects on grassland silage yields and soil structure under different levels of compaction over three years. *European Journal of Agronomy* 109, 125916.

Herrick, J. E., Beh, A., Barrios, E., Bouvier, I., Coetzee, M., Dent, D., Elias, E., Hengl, T., Karl, J. W., Liniger, H., Matuszak, J., Neff, J. C., Ndungu, L. W., Obersteiner, M., Shepherd, K. D., Urama, K. C., van den Bosch, R. and Webb, N. P. (2016). The Land- Potential Knowledge System (LandPKS): mobile apps and collaboration for optimizing climate change investments. *Ecosystem Health and Sustainability* 2(3), e01209.

HM Government (2018). *A Green Future: Our 25 Year Plan to Improve the Environment*. Department of Environment Food and Rural Affairs. Available at: https://assets.publishing.service.gov.uk/government/uploads/system/uploads/attachment_data/file/693158/25-year-environment-plan.pdf.

Idowu, O. J., van Es, H. M., Abawi, G. S., Wolfe, D. W., Ball, J. I., Gugino, B. K., Moebius, B. N., Schindelbeck, R. R. and Bilgili, A. V. (2008). Farmer-oriented assessment of soil quality using field, laboratory, and VNIR spectroscopy methods. *Plant and Soil* 307(1–2), 243–253.

Ingram, J. S. I. and Fernandes, E. C. M. (2001). Managing carbon sequestration in soils: concepts and terminology. *Agriculture, Ecosystems and Environment* 87(1), 111–117.

Karlen, D. L., Andrews, S. S. and Doran, J. W. (2001). Soil quality: current concepts and applications. *Advances in Agronomy* 74, 1–40.

Klingebiel, A. A. and Montgomery, P. H. (1961). Land capability classification. *USDA Agriculture Handbook*, vol. 21, US Government Printing Office, Washington, DC, 21 pp.

Krzywoszynska, A. D. (2018). Making knowledge and meaning in communities of practice: what role may science play? The case of sustainable soil management in England. *Soil Use and Management* 35(1), 160–168.

Lal, R. (2001). Managing world soils for food security and environmental quality. *Advances in Agronomy* 74, 155–192.

McBratney, A. B., Field, D. J. and Koch, A. (2014). The dimensions of soil security. *Geoderma* 213, 203–213.

McEldowney, J. F. and McEldowney, S. (2001). *Environmental Law and Regulation*. Blackstone Press, London.

Moebius-Clune, B. N., Moebius-Clune, D. J., Gugino, B. K., Idowu, O. J., Schindelbeck, R. R., Ristow, A. J., van Es, H. M., Thies, J. E., Shayler, H. A., McBride, M. B., Kurtz, K. S. M., Wolfe, D. W. and Abawi, G. S. (2016). *Comprehensive Assessment of Soil Health - The Cornell Framework*, Edition 3.2. Cornell University, Geneva, NY.

Molnar, P. J. (1995). A model for overall description of food quality. *Food Quality and Preference* 6(3), 185–190.

NCRS, Natural Resources Conservation Service (2014). Soil Health Assessment. Available at: https://www.nrcs.usda.gov/wps/portal/nrcs/main/soils/health/assessment/.

NRM Laboratories (2018). Sound Science for Soil Health. Available at: http://www.nrm.uk.com/services.php?service=soil-health.

Pankhurst, C. E., Doube, B. M. and Gupta, V. V. S. R. (1997). Biological indicators of soil health: synthesis. In: *Biological Indicators of Soil Health*. Pankhurst, C. E., Doube, B. M. and Gupta, V. V. S. R. (Eds). CAB International, Wallingford, pp. 419–435.

Ritz, K., Black, H. I. J., Campbell, C. D., Harris, J. A. and Wood, C. (2009). Selecting biological indicators for monitoring soils: a framework for balancing scientific and technical opinion to assist policy development. *Ecological Indicators* 9(6), 1212–1221.

Romig, D. E., Garlynd, M. J. and Harris, R. F. (1996). Farmer-based assessment of soil quality: a soil health scorecard. In: *Methods for Assessing Soil Quality*. Doran, J. W. and Jones, A. J. (Eds). SSSA, Madison, WI, pp. 39–60.

Rutgers, M., Mulder, C. and Schouten, A. J. (2008). Soil Ecosystem Profiling in the Netherlands with Ten References for Soil Biological Quality. RIVM Report 607604009/2008. National Institute for Public Health and the Environment (RIVM).

Sherwood, S. and Uphoff, N. (2000). Soil health: research, practice and policy for a more regenerative agriculture. *Applied Soil Ecology* 15, 85–97.

Sinclair, H. R., Waltman, W. J., Waltman, S. W., Terpstra, H. P. and Reed-Margetan, D. (1996). *Soil Ratings for Plant Growth (SRPG)*. USDA-NRCS, National Soil Survey Center, Lincoln, NE.

Sojka, R. E., Upchurch, D. R. and Borlaug, N. E. (2003). Quality soil management or soil quality management: performance versus semantics. *Advances in Agronomy* 79, 1–68.

Stavi, I., Bel, G. and Zaady, E. (2016). Soil functions and ecosystem services in conventional, conservation, and integrated agricultural systems: a review. *Agronomy for Sustainable Development* 36(2), 32.

Stockdale, E. A. and Watson, C. A. (2012). Managing soil biota to deliver ecosystem services. *Natural England Commissioned Reports 100*. Available at: publications.naturalengland.org.uk/file/2742553.

Stockdale, E. A., Goulding, K. W. T., George, T. S. and Murphy, D. V. (2013). Soil fertility. In: *Soil Conditions and Plant Growth*. Gregory, P. J. and Nortcliff, S. (Eds). Wiley-Blackwell, Oxford, pp. 49–85.

Stockdale, E. A., Griffiths, B. S., Hargreaves, P. R., Bhogal, A., Crotty, F. V. and Watson, C. A. (2019). Conceptual framework underpinning management of soil health - supporting site-specific delivery of sustainable agro-ecosystems. *Food and Energy Security* 8(2), e00158.

Taylor, H. M. (1971). Effect of soil strength on seedling emergence, root growth and crop yield. In: *Compaction of Agricultural Soils*. Barnes, K. K., Carleton, W. M., Taylor H. M., Throckmorton, R. I. and Vanden Berg, G. E. (Eds). American Society of Agricultural Engineers Monograph, pp. 292–305.

USDA, United States Department of Agriculture (2001). *Guidelines for Soil Quality Assessment in Conservation Planning*. Natural Resources Conservation Service and Soil Quality Institute Publication. Available at: https://www.nrcs.usda.gov/Internet/FSE_DOCUMENTS/nrcs142p2_051259.pdf.

Valentine, T. A., Hallett, P. D., Binnie, K., Young, M. W., Squire, G. R., Hawes, C. and Bengough, A. G. (2012). Soil strength and macropore volume limit root elongation rates in many UK agricultural soils. *Annals of Botany* 110(2), 259–270.

Wandel, M. and Bugge, A. (1997). Environmental concern in consumer evaluation of food quality. *Food Quality and Preference* 8(1), 19–26.

Chapter 12

Developing decision support systems (DSS) for farm soil and crop management

Matt Aitkenhead, The James Hutton Institute, UK

1 Introduction

Increases in both human population and area of degraded land mean that farmers are under pressure to produce more food per unit area and to do so with less environmental impact. New crop varieties and increased fertiliser inputs can improve yields but not everywhere and not forever. The twin requirements of increased agricultural intensification and decreased environmental impact can be achieved only if we have greater precision in inputs and reduction of agricultural losses. Managing this will require more agile land management with greater levels of detail and control over processes.

Hochman et al. (2013) identified precision agriculture (monitoring, responding to and controlling intra-field variation in soil and crop conditions) as one of the emerging technologies by which this could be achieved. As precision agriculture cannot be achieved without an increased level of data-gathering and automation within the farming system, this implies a need for better decision support or even decision-taking by farming technology.

The demand for increased food production means that successful strategies will have to be sought for growing crops on marginal land. This implies the development of novel techniques for matching land with suitable

http://dx.doi.org/10.19103/AS.2020.0079.23

crops and providing management strategies that are highly detailed and site-specific (Schroder et al., 2018). The information requirements to solve this challenge can realistically be tackled by using only decision-support tools based on machine learning, with real-time response and control over the management infrastructure and systems. Attempting to develop rule-based heuristics systems or other 'classical' statistical approaches that captured all possible combinations of landscape, crop and management type would require enormous resources and constant updating.

In order to maintain cropping under climate change conditions that imply changing water and temperature regimes, farmers need to consider many new factors and constraints and require access to data resources and models that enable them to adapt (Eigenbrode et al., 2018). The challenge of processing all this additional information means that new technologies and approaches are required that automate at least some of the comprehension and decision-making. While crop selection may be a binary choice, crop management through the growing season can require hour-by-hour decision-making involving multiple pieces of information. Models can be developed to handle the speed and sophistication of response required but need to be parameterised for specific conditions.

The concept of integrated soil fertility management involves the use of multiple combined approaches to maximise soil fertility and efficiency of inputs (Vanlauwe et al., 2015). Achieving this requires consideration of local soil conditions and cropping requirements, as well as access to specific technologies, seed and fertilisers. Successfully dealing with all this different information, in addition to farmer financial constraints and other factors, implies significant computational effort and incorporation of a lot of necessary information, which a lot of farmers in developing countries do not have access to. This concept of local adaptation to complex landscapes is one where decision support systems (DSSs) have the potential to provide an additional, automated layer of interpretation for land managers.

The concept of DSSs ranges from fully integrated sensor-model-control-equipment systems to tools designed to inform decision-making by the land manager (e.g. Hillier et al., 2011). There is a place for each end of this spectrum within the agricultural sector, and positions in between. Additionally, DSSs can be applied from the plant scale to the national/supranational scale (e.g. Ahmed et al., 2018).

Decision support tools can be an extremely useful tool for identifying likely issues with land management, based on integrated knowledge about local conditions (e.g. Shahbazi and Jafarzadeh, 2010). They can also provide useful information on crop selection and management timing and advice on water use based on rainfall and soil modelling (Balderama, 2010). Quantification effects on soil health or functionality from a range of processes and management options may also be possible, in addition to the effects of individual properties. Examples include soil sealing (Piero et al., 2017).

DSSs are therefore vital for future global food security. In addition, Robinson et al. (2012) identified decision support tools as one of the areas of research leading towards sustainable development and understanding the ecosystem value of soils. When restricted to limited information-gathering, land managers tend to adopt practices that result in large areas of uniform conditions in order to maximise the cost-effectiveness of the information they do have. Removing this restriction opens up the possibility of more site-specific, agile farming approaches that reduce the environmental impact without losing yield. However, achieving this requires investment in technology and incentivisation through government subsidies (Wolfe and Richard, 2017).

In this chapter, different aspects of agricultural DSS design, implementation and operation are covered. These aspects include the fundamentally spatial consideration of DSSs, the need for and use of sensor technology within such systems, modelling and software components, system-operator interfaces, the potential (and current) use of DSSs for directly controlling agricultural management and a discussion of what is needed to make DSSs more successful and widely used in agriculture (and why this has not happened yet). The chapter closes with a discussion of possible and likely trends in this topic for the future.

2 Spatial data and sensor requirements for DSS

The spatial variability of soil properties has a profound impact on crop nutrient status and development, and so site-specific nutrient management is important for cost-effective crop growth and reduced environmental damage from run-off. Betteridge et al. (2008) argued that successful site-specific nutrient management is vital from a farming, ecological and political perspective.

Site-specific conditions can be incorporated into adaptive DSSs, allowing widely observed issues such as balancing nitrogen losses and grain yield to be dealt with at a local level beyond the capabilities of even the most detailed process-based models. These models can also potentially be tweaked to satisfy land managers' specific requirements, such as a different level of trade-off between grain yield and nitrogen loss (Ravier et al., 2018).

Within-field precision irrigation requires high-resolution sensor data from crops, soil and weather. Infrared thermal imaging of plants can be used for monitoring plant water stress (Martinez et al., 2017). Katsigiannis et al. (2016) demonstrated an integrated ground- and unmanned aerial vehicle (UAV)/drone-based approach for simultaneous soil and crop monitoring with two cameras operating in the visible and near infrared (Vis-NIR) and thermal infrared ranges. This approach allowed both spatial and temporal variability within crop water stress to be monitored.

Aitkenhead and Coull (2019a,b) demonstrated an approach based on digital soil mapping and modelling that allows the impacts of management options on soil properties, processes, functions and ecosystem services to be evaluated spatially using a relatively simple model. The model links the several layers of abstraction from soil properties to ecosystem services using linear relationships that can be adjusted or added to with new properties as the user requires so that although it does not capture the complexity of individual relationships, it does allow the impacts of changes to multiple soil properties to be explored. This work will be used to underpin the future development of a decision support web-mapping tool that allows users to define and explore their own management options and their impacts. The aim is to ease the identification of management options under specific conditions for maximising productivity and minimising environmental impact.

Jarvis et al. (2017) found that for users of one particular suite of tools (GeoVisage), farmer's use of weather station and remote sensing data was common in Ontario, Canada, and also that these farmers wanted more and better communication and interpretation of the available data and improved access through mobile devices. This implies the need for smartphone and tablet-based apps with GIS functionality and access to existing spatial soil data. An example of how this spatial data could be provided is through the Scotland's Soils website (https://soils.environment.gov.scot/), which allows maps of soil properties in Scotland to be viewed or downloaded.

2.1 Using sensors in DSS

Long-term data from existing in-situ sensors can be used to provide site-specific information about crop and soil condition. Goldstein et al. (2018) used long-term soil moisture and weather data to develop a model capable of providing a short-term irrigation plan from current weather condition data. They explored several modelling approaches to identify the best one (Boosted Tree Classifier), which achieved 95% against 92.7% for another machine-learning model and 62.8% for a linear regression model. This highlights again the importance of appropriate model and data selection for specific site conditions and the potential improvements in using machine-learning approaches instead of regression models.

In addition to making sure that crops receive enough water, reduction of water waste in irrigation is a major potential benefit of automated decision support. Khan et al. (2018) demonstrated a prototype system for using wireless sensors for data collection and decision support, showing how the data streams could be used for manual or automated decision making. They also highlighted that the performance of the system is dependent upon the accuracy and effectiveness of the sensor data. This brings us to a discussion on using the most relevant sensors for the system and ensuring that raw data streams are interpreted in a meaningful manner to present information relevant to the decision-maker.

Spectroscopy, by which the spectral response of materials varies with their properties and can therefore be estimated through non-destructive imaging, can provide a useful alternative to laboratory-based 'wet' chemical analysis (Nocita et al., 2015). Low-cost hand-held spectroscopy has the potential to provide estimates of the properties of soil (organic matter content, pH, magnesium, nitrogen, aluminium and clay content) and crop (crop type, nitrogen, magnesium and manganese content) in the field (e.g. Aitkenhead et al., 2017, 2018). Current efforts by the author and colleagues in this area have demonstrated a working prototype that costs approximately €500. More expensive, but more sophisticated, spectral monitoring systems (e.g. Williams et al., 2017) can be mounted on trolleys or incorporated into mobile systems (e.g. Fig. 1).

Control of pests and pathogens can be informed using DSSs. Recent developments in machine learning are leading towards automated detection and identification of pest/pathogen symptoms, but there are already examples of decision support tools that can allow farmers to identify and understand the impacts of these on the crop. Mondino and Gonzalez-Andujar (2019) describe an example used for apple orchards that provides a visual identification key composed of a library of photographs and drawings, and that was found to work extremely well when evaluated by an expert group.

An important part of DSS is having the necessary information on crop, soil or environment that is needed by the tool, and having this information readily

Figure 1 Tractor-mounted sensor/computer system with arm-mounted backboard providing a consistent image background (Williams et al., 2017).

available to the DSS. Smartphones are an obvious mechanism to facilitate this, as they have both the communication and computing hardware necessary to gather information, process it and pass it on in a visually useful form. This does not have to be done using a large, sophisticated arrangement but can be as straightforward as sensor data being transmitted to a phone for the farmer to view (e.g. Bhatnagar and Poonia, 2018). The use of smartphones for direct estimation of soil properties (Aitkenhead et al., 2014, 2016) could inform rapid decision support in-field (see Fig. 2 for an example of imagery used). Properties we have found amenable to field-based assessment in this way include organic matter, pH, moisture, structure, texture and a small number of elemental concentrations (e.g. Mg, Al, Fe and N). Unfortunately, potassium and phosphorus are proving difficult to assess with any meaningful accuracy using this approach.

3 Models and software for DSS

Environmental and commercial costs and benefits are currently only marginally considered as related within intensive agricultural systems. The closer

Figure 2 Smartphone camera image of the soil profile, showing colour correction card (approx. 10 × 5 cm).

integration of these considerations within agri-environment subsidy schemes will incentivise farmers to become more efficient with inputs, particularly with nitrogen and its associated run-off and implications of environmental damage. Models that calculate whole-farm nitrogen requirement, loss and efficiency based on local conditions are therefore necessary (e.g. Carof and Godinot, 2018). Integration of these models with on-farm sensors and decision support will be an important step towards the cost and environmental impact reduction in modern farming systems.

The model/decision-making algorithm approach used can vary widely depending on several factors. In many cases, the choice of modelling approach will depend simply on the skillset and experience of the researchers and developers implementing it, with many data-mining approaches being relatively interchangeable (e.g. neural networks, support vector machines, random forest and partial least squares). In cases where the model takes multiple numerical inputs and produces one or more numerical outputs, any one of these methods or more is likely to be appropriate. In situations where input variables are discrete or even not always present (e.g. specific features of crop condition), other logical inference approaches may be better. Examples include Dempster-Shafer Inference (Culman et al., 2019).

One of the key challenges to decision support tool development is the integration of data from multiple sources, with different types of data, uncertainty or imprecision in data definitions and often estimated values. Huang et al. (2010) argue for the use of soft computing in tackling this challenge, rather than attempting to hard code exact solutions to every possible problem. Soft computing approaches include fuzzy logic, artificial neural networks, decision trees, genetic algorithms and other methods for generating a response from complex, noisy and often unreliable data.

The ability to estimate crop yield is an important contribution to financial success for farmers as it enables them to plan ahead. Yield prediction involves multiple factors of different kinds (e.g. continuous and categorical), and integrating these factors into a single model is non-trivial. Papageorgiou et al. (2011) demonstrated a fuzzy cognitive map approach to estimate cotton yield. In addition to showing improvement over earlier methods, an important aspect of this work was the ability to visually represent the knowledge implemented within the model. Having models that can be visually represented makes them much easier to develop and use by non-scientists and contributes to stakeholder trust in the model and its decisions.

Soil quality or soil health is a difficult concept to define precisely, as the functions that soil can be expected to carry out vary widely. A soil that is of high quality for carbon sequestration may not be high quality for grain production, and at a finer level of semantic detail, a soil that is of high quality for one crop may be less so for another. As soil quality/health for specific uses is an important consideration in agriculture, this is a challenge that is core to DSS development. One solution is to integrate information on basic soil properties (e.g. pH, texture, organic carbon content) into context-specific estimates of soil quality. Achieving this requires a suitably flexible approach for integrating soil property information such as the weighted additive approach (Triantafyllidis et al., 2018) or matrix multiplication linking soil properties, processes and functions (e.g. Aitkenhead and Coull, 2019b).

Linking sensors with machine learning, models and control systems enables not only automated control but also adaptation to local conditions (e.g. Navarro-Hellin et al., 2016). Liakos et al. (2018) also identified machine learning as a key component in future precision agriculture developments and agriculture management. The application of machine learning to both sensor data interpretation and decision support tools will provide rapid, sophisticated recommendation and direct action for land management.

Puntel et al. (2019) demonstrated a model capable of estimating nitrogen application rates using a combination of static (e.g. soil depth) and dynamic (e.g. soil moisture) variables. This approach integrates the concepts of digital soil mapping and in-field sensing and enhances the performance of statistical models and agronomic recommendations.

Crop rotation selection is based on farmer experience and the constraints of crop suitability and market requirements. Optimising the crop rotation for sustainability within these constraints would require sophisticated knowledge of local conditions and many crop-specific factors. Integrating all of these factors into decision-making is a complex task that may be amenable to modelling and decision support tools (Qureshi et al., 2018).

de la Rosa et al. (2009) used a DSS called MicroLEIS-DSS to evaluate soil quality in the Mediterranean region and demonstrated that this particular DSS was very useful for identifying management options aligned to soil properties in a manner that was appropriate for avoiding and reducing soil degradation.

Low cost of data collection was identified as the main driver of recent rapid changes in agricultural technology (Weersink et al., 2018). However, the same work identified the interpretation and integration of this data as the main challenge to further developments in decision support. Many soil models have been assessed and have varying properties, effectiveness and relevance of application for different land management decision-making (see Chapter 3 of Sang, 2020).

4 DSS user interface design, actuators and systems

Gibbons et al. (2014) demonstrated the importance of communication of trade-offs to farmers in relation to soil nutrient status, greenhouse gas emissions and profitability; in their findings they showed that poor communication of these trade-offs could have impacts on farm profits and aquatic and atmospheric pollution. An important but often overlooked part of the communication of results is an effective interface design.

Networked or smartphone-based decision support tools can also access data and models that would otherwise be inaccessible or meaningless to farmers due to cost, education and language barriers. This is particularly true if the DSS successfully integrates multiple areas of activity (e.g. sensing, site-specific data sourcing, modelling and decision-making) or multiple areas within which management options will have consequences (e.g. yield, environmental impact, economic benefit and nutrition). Clarke et al. (2017) demonstrated the effectiveness of this integrated approach amongst a group of subsistence farmers in Ethiopia.

The potential use of smartphone apps for matching crops and management options to site and soil conditions is vast, given the low cost of smartphone technology. Linking smartphones with server-side processing removes much of the burden of spatial data handling, modelling and processing, and at the same time can resolve important but often-ignored issues with intellectual property and data security. Shirsath et al. (2017) demonstrated a smartphone application with server-side processing that provides crop recommendations for farmers in India based on site-specific soil, water and climate considerations. Applications such as this have the potential to democratize land management decision-making, improve productivity and reduce poverty amongst poorer farmers.

Precision agriculture will only be achievable when tools are available to support the data-rich decision-making requirements it places on farmers. These tools will need to be accessible, freely available and sufficiently flexible to accommodate a wide range of agricultural activities. Leroux et al. (2018) have demonstrated an open-source tool (GeoFIS) for mapping, management option evaluation and decision-making in agriculture. This and other tools still require significant expertise to work with, meaning that farmers cannot use them directly without investment in training and technology. However, some examples such as the KORE system by SoilEssentials can provide useful and intuitive colour-coded maps of farmers' fields showing important information such as historical yield or current/recent crop development (Fig. 3).

4.1 Actuators and systems

Providing a balanced input of nutrients to the soil/crop system to ensure crop yields while minimising losses from the system is the main effort of precision agriculture (Hedley, 2015). A range of technologies are required to accomplish this, including accurate positioning (GPS), appropriate sensing equipment,

Figure 3 KORE interface screenshot from SoilEssentials. With permission from SoilEssentials, 2019.

sufficient computing power and software and equipment capable of distributing nutrients in a controlled manner over small spatial scales.

GPS positioning is now readily available with sub-metre accuracy, and technological advances to farm equipment mean that nutrient inputs can be finely controlled within this scale also. Computer speed is no longer a restriction due to processor advances and the availability of cloud computing, so the main challenge areas are the sensors and software tools capable of advising or directly controlling the nutrient input process.

Another cost consideration is energy consumption, which can affect energy costs and sensor lifespan. Sophisticated power management routines are needed for integrated sensor/decision support/actuator systems (in other words, automated farming systems). Examples of prototype systems of energy-efficient systems are Jawad et al. (2018) who applied a sleep/wake control mechanism into an agricultural sensor network.

The use of networks of wireless sensors to provide real-time information for agricultural resource optimisation is important (Kassim et al., 2014), but only if the data provided is meaningful and can be interpreted appropriately. Given that designing a bespoke decision support tool for each individual farm is not cost-effective, this implies a need for systems that can be adapted to different settings and that can incorporate farmer objectives. A step in this direction is the use of sensors and control systems in controlled-environment agriculture. Shamshiri et al. (2018) highlight the importance of DSSs for production management in this kind of agricultural system and argued that it is moving towards 'smart controlled environment agriculture'.

5 Decision support or decision-making?

Complex management decision-making incorporates crop growth, and environmental impact information requires sophisticated interpretation of multiple factors, even more so than the standard intensive agricultural decision-making as it involves balancing one or more trade-offs between multiple desired outcomes (Carlson et al., 2017).

Intelligent DSSs based on machine-learning methods and technologies are vital for handling the complexities of future precision agriculture. Chlingaryan et al. (2018) highlighted the importance of recent rapid developments in machine learning for sensor data interpretation and decision-making. They also identified the fusion of sensors and knowledge for automated decision-making as likely to be important in the future of precision agriculture.

In some cases, removing the farmer from the loop entirely can be achieved by linking sensors, models and physical mechanisms. Developments in this direction are proceeding rapidly across a range of management actions including irrigation (Mohapatra et al., 2019). A range of short-term planning tools for water management in agriculture in the United States were reviewed by Easton et al. (2017), each of which was based on the integration of weather and catchment hydrology in some manner. They found each to be effective across a range of conditions but highlighted that further development is needed in the quality of system input data and in linking the outputs of process models to the needs of land managers.

Informed decision-making, stakeholder involvement and information integration are vital for the adaptation of agriculture to climate change. Brandt et al. (2017) demonstrated a DSS 'targetCSA' that aims to inform land management in a climate change-relevant manner and identify appropriate locations where these land management options can be successfully implemented. Importantly, it also incorporates stakeholder opinions about specific management options. This transforms such a system from 'telling' the user what is best to 'supporting' their decision-making process. As such, it provides a mechanism to resolve local cultural issues that are otherwise difficult to incorporate into a DSS.

A triangular diagram showing the trade-offs between sophistication, cost and ease of implementation is given in Fig. 4. For the three corners, examples that would fit could be heuristic rule sets (low cost, low sophistication and easy to implement); scientifically published process-based models (low cost, high sophistication, difficult to implement in real systems); and consultant-based mapping applications (high cost, low sophistication, difficult to use for decision-making). For the last corner, examples vary in both the last two 'axes', with more sophisticated, and therefore useful, and less sophisticated, and therefore less useful, examples. The challenge in selecting DSSs is that moving up one axis

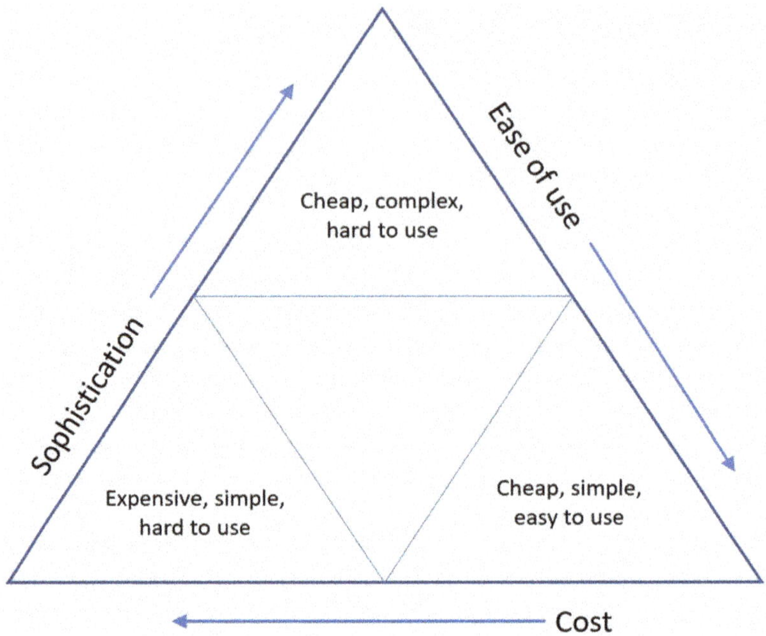

Figure 4 Conceptual visualisation of the three aspects of consideration for DSSs.

tends to move you down the other two and that systems that score highly on two axes tend to score low on the remaining one. No systems yet exist that are affordable and sophisticated enough to deal with all the conditions at a specific site and that provide information from which decisions can easily be made.

6 What reasons are there for low uptake of DSS?

Despite all the technological advances and benefits to agriculture highlighted above, uptake of DSSs remains low. In an investigation of the usage of DSSs in UK agriculture, Rose et al. (2016) found that while all farming advisors used them, only 49% of farmers used some kind of DSS (and of the 395 tools used, 22% were paper-based DSSs). The paper by Rose et al. also contains a downloadable appendix with a list of all of these tools.

Amongst Australia's grain growers, whereas machine guidance/autosteer and yield-monitoring technologies have been readily adopted, sensors that give information about soil characteristics and crop health have not (Bramley and Ouzman, 2019). This is possibly related to the fact that for the most commonly used sensing approach in Australia, electromagnetic sensing of soil properties, awareness and understanding, is low. There are also differences in scale that may make the situation different in the UK (mean area of 54 ha per farm) and Australia (>4000 ha per farm), which imply a bias towards large area rather than small-scale monitoring. Certainly, in Australia technologies such as remote sensing that provide other potentially useful information, alongside yield mapping, are seen as a way to gain farmers' interest and uptake (Bramley and Ouzman, 2019).

Difficulties in the practical implementation of DSSs into agricultural systems (through the factors of profitability, user-friendliness, user's time constraints, credibility, relevance to the specific farm conditions, user expertise) and lack of awareness/trust in them as a viable option have resulted in lower uptake than may have been anticipated by developers of those systems (Kerr, 2004; Rossi et al., 2014). Arguably, the two issues of implementation and trust are connected, as farmers will be reluctant to take risks on buying and then relying on DSSs unless they can be confident that they will be of benefit, and low uptake means that there is a little financial incentive for developers to produce more competitive systems.

The issue currently with low interest in soil and crop sensors may be that farmers do not see the benefit in soil sensing, preferring to determine fertiliser input rates from yield monitoring from a perception that soil does not change while yield does. Highlighting the potential benefits to fertilizer rate calculations through soil sensing might not therefore improve uptake, but identifying other benefits might do so if these benefits relate to farm profitability rather than environmental impact. As farming subsidies shift further towards environmental sustainability, considerations of factors dependent on soil characteristics such as nitrogen run-off and soil erosion will be given more consideration, leading to the uptake of systems that can factor these into the decision-making process. Accurate DSSs imply reduced application rates and less heavy mechanisation, which should implicitly help with these considerations in addition to any explicitly environmentally conscious decision-making.

Schwenke et al. (2019) found that farmers' use of DSSs for N application recommendations in Australia was ad-hoc, with decisions on N application driven by a combination of factors some of which were not incorporated into DSSs (where these were used). A lack of information about the spatial variability of soil N made it difficult for farmers to rely on DSSs, and rules used were generally simple and required less information. This indicates a lack of available, low-cost sensing that would be needed to drive more sophisticated and data-heavy DSSs. Additionally, the integration of data from other sources such as crop type and management practices would be necessary.

Incentivisation is therefore key to the uptake of DSSs in the agricultural industry. An important comment to make here is that agricultural and environmental legislation is often handled separately within governments, meaning that drivers to increase productivity and improve environmental status are neither linked nor usefully aligned. An example is a distinction between

European DGs on agriculture and environment that can cause issues with policymaking and influencing land management decision-making.

Having decision support tools that factor in agricultural and environmental considerations could provide land managers with a way of short-circuiting the issues caused by these policy area silos. It is also possible that this kind of bottom-up decision-making might influence the top-down structures of policy development and regulation within Europe and elsewhere.

In a case study of DSS use on farms in England and Wales, Rose et al. (2018) found that the manner in which they are used is often different from the way they were intended, often due to a lack of faith in the expertise expressed within the tools and its relevance for farm-specific conditions. They also found that that DSS use alters farmer behaviour and interaction with the farm itself. DSS design is therefore a vital consideration in ensuring that both that their usefulness to farmers can be demonstrated and that they can be adapted to the specific conditions within which each farmer operates.

6.1 Why has a perfect system not been achieved?

The primary consideration for uptake and deployment of DSSs will always be cost-effectiveness, which is strongly influenced by both immediate and long-term cost. The ability to purchase module, off-the-shelf components for farm-specific monitoring and decision support has a strong impact on cost in terms of a capital purchase, time and effort. Goap et al. (2018) demonstrated an open-source system for integrating sensors, decision support and actuators for crop irrigation. This kind of system is sufficiently flexible that it can be applied to different crop types, climates and management systems with relative ease.

There will also be no point in installing a DSS if a farmer cannot make use of the recommended actions (e.g. spatially detailed distribution of fertiliser inputs), if they do not have the necessary equipment, or if the equipment and the DSS do not communicate effectively with one another. Various solutions to this issue have been explored, from 'awareness-raising' amongst farmers to promoting the concept amongst fertiliser companies. In the experience of the author, farmers tend to communicate quite effectively amongst themselves through various channels and media, and the sharing of ideas and concepts is often 'too much' rather than 'too little'.

Meanwhile, promoting the concept of DSSs designed to reduce overall application to the companies that produce fertiliser would seem unlikely to succeed. The only thing that appears likely to succeed in driving adoption of DSSs would be the development of a low-cost-effective and easily implemented DSS that could be used 'off the shelf' by a sufficiently broad range of farmers to guarantee a profitable market. So what is needed in order to achieve this?

Developments in cloud computing and mobile phone technology allow farmers access to high-performance computing through application interfaces, enabling sophisticated modelling and data interpretation (Premasudha and Leena, 2017). So why are more farmers not using smartphones in the field to automatically acquire, interpret and make use of soil and crop data to improve their farming efficiency, profitability and environmental sustainability? It appears that all the pieces are in place, and yet nobody has put them together. Are there barriers to further development that have not been acknowledged or even identified?

Integration of soil-crop-atmosphere models with spatial data sets allows spatially discrete modelling and decision support but requires significant computing resources and data preparation/formatting (Thorp et al., 2008). Automating the integration of relevant data sets and models requires either the development of standardised systems (hard and also inflexible to future requirements) or common language protocols enabling translation between multiple systems, formats and stakeholder requirements (also hard, but amenable to modularisation and capable of being adapted to future requirements).

Issues with the uptake of DSSs in the UK were explored by Rose et al. (2016), who found that although a large number of DSSs were available, several factors were responsible for lack of adoption and use. These included usability, relevance and compliance compatibility and highlight the need for DSS development to include stakeholder requirements more robustly and in more depth.

7 What will DSSs of the future look like?

Sensors can accomplish many functions, including crop monitoring, soil moisture sensing, disease detection and weed detection. This last function relies on being able to visually or spectrally discriminate between crops and weed plants. It is more difficult when crops/weeds are smaller and less well-developed, as seedlings of different plants appear more similar to one another. In a review of weed detection systems, Lopez-Granados (2011) evaluated remote sensing and ground-based weed detection systems, both multispectral and hyperspectral, and identified several limitations that still hold to this day. These include education, cost and integration of multiple pieces of equipment, and require concerted research across interdisciplinary research groups to tackle.

Arguably, all of the information and equipment that we need in order to develop a sophisticated, automated DSS that integrates environmental data, farming practices, crop types, weather data and modelling are already in place. But no single programme has ever attempted to put them together. Arguably, what is needed is a framework that allows these components to be integrated in a modular fashion as and when they are available to the farmer so that DSS resources can make use of the data that is available rather than having strict requirements in order to operate at all.

Lindblom et al. (2017) argued that access to, or ownership of technology for precision agriculture was not an issue. Instead they demonstrated that implementation of the technology was lower than would be expected, largely due to issues with design and 'usability'. For this reason, more consideration must be given to what the user needs and how they understand the operation of any DSS. Implementation of scientific principles into a graphical user interface is not enough, even if the data access, modelling and subsequent advice provided are all correct. The manner in which humans interact with such systems must be taken into account and used in their design.

Prioritisation of land use under conflicting demands requires the integration of biophysical, socioeconomic and regulatory information. Traditional approaches to integrate this information for land evaluation generally fail to incorporate participation by stakeholders, meaning that they often do not include local factors of significance (Kerselaers et al., 2015). Successful DSSs not only need to be developed in concert with their eventual users but also their operation and use must incorporate user requirements in a participatory manner (as Kerselaers et al. (2015) demonstrated).

Another requirement of DSSs is that they are sufficiently flexible in dealing with a reasonable range of conditions, including crop types. This requires either sophisticated modelling or detailed accumulation of data on the range of conditions explored. Machet et al. (2017) demonstrated that this can be achieved through a combination of both modelling and parameterisation for nitrogen fertilisation of 40 crops. While demanding, the development of generally applicable tools is therefore not impossible.

Zwickle et al. (2016) highlighted the importance of integrating local knowledge and experience into DSSs and of providing a baseline of scientific evidence to underpin farmer knowledge to clarify benefits and trade-offs within the decision-making process. The Agile approach to software development in which an iterative feedback and implementation loop would seem amenable to achieving this. This approach, while requiring significantly more effort in terms of 'transaction costs' between developer and user, is much better at achieving a product that (a) does what the user wants and (b) does what they actually need. Figure 5 shows a conceptual diagram of what the Agile approach might look like in this case, with the standard paradigm adapted for agricultural DSS development, and also gives a sense of what the resulting generalised DSS might be like and how it might resolve the issues with DSS uptake as described by Rossi et al. (2014) and Kerr (2004).

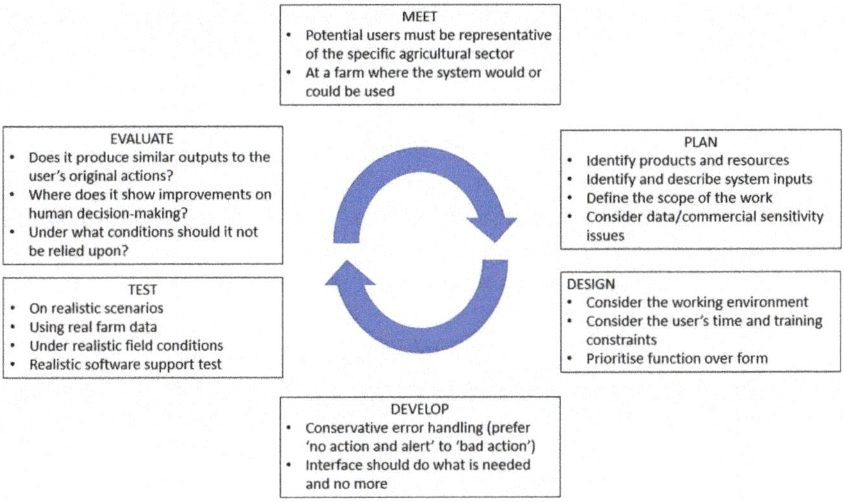

Figure 5 Conceptual diagram of the Agile approach to DSS development.

8 Summary

The use of UAVs for soil and crop monitoring has never truly taken off (pun intended) as UAV operations are notoriously sensitive to weather conditions (wind, daylight and rainfall) (e.g. Martinez et al., 2017), in addition to being subjected to legal restrictions for use (e.g. line of sight operation, height limits, distance from power lines and buildings). For this reason, ground-based sensors or other forms of remote sensing (e.g. satellite) are believed by this author to have more future potential for agricultural monitoring. Automation of mobile platforms within field for crop and soil sensing provides fewer weight and operational restrictions, while making the results of device malfunction less costly.

For DSSs to work in a manner that is cost- and time-effective, there need to be improvements in integration (sensors, biophysical and socioeconomic models, data sets) and flexibility (management options, crop types and environmental conditions). Wolfe and Richard (2017) argue that in addition to technological development, government policies need to change to incentivise more efficient, and, therefore, more detailed and data-hungry land management options.

Using a prototype DSS that integrates spatial data on several factors, Fastelli et al. (2018) were able to demonstrate that spatially explicit ecosystem service provision by agriculture could be determined at the same time as management decision making was informed. This type of management impact dissemination will be vital for future decision- making as agricultural and environmental considerations become more closely integrated.

The use of coupled soil/plant/atmosphere models for simulating specific cropping situations is potentially important in identifying management options that work well under certain conditions. These models, for example, DSSAT, require significant effort to calibrate and run successfully but can be extremely useful in allowing economic evaluations of management options (Tovihoudji et al., 2019).

There are no tools currently available that can provide this kind of service without significant data entry effort, with the main reason for this being that this would require access to (i) a vast resource of high-resolution spatial data on soil properties, (ii) local weather monitoring and (iii) a library of information on crop types and management options sufficiently detailed to incorporate likely farmer options.

While Digital Soil Mapping (DSM) efforts are leading towards global high-resolution spatial data sets for soils and weather monitoring is becoming more dense and detailed, there is still the unsolved challenge of developing a library of crop types and management options that is sufficiently varied and detailed. Because of the rapid pace of agricultural development, the only way such a library could work would be if it could be automatically updated with information on new crop varieties, fertiliser options and tillage systems, amongst other things. To achieve this would require an 'Open Data' approach to sharing information on commercially sensitive agricultural resources that has yet to be attempted.

9 Where to look for further information

9.1 Introductions for non-specialists

- Handbook of Precision Agriculture (Ancha Srinivasan), published in 2008 by CRC Press, provides basic guidelines and recommendations. For more focus on DSS applications and development, another book from 2008, Handbook on Decision Support Systems 1, edited by Burstein and Holsapple and published by Springer may be of use.

9.2 Seminal articles

- Bouma et al. (1999) introduced the important concept of combined modelling/farm management decision-making, preparing the ground for much of the research in this area in the last two decades (DOI 10.2136/sssaj1999.6361763x).
- Cox (2002) identified many information technology requirements and areas of research strongly linked to DSSs (DOI 10.1016/S0168-1699(02)00095-9).

9.3 Key societies

- Euro Working Group on Decision Support Societies (https://ewgdss.wordpress.com/).
- International Society of Precision Agriculture (https://www.ispag.org/).
- International Union of Soil Sciences (https://www.iuss.org/).

9.4 Conferences

- International Conference on Decision Support System Technology (https://icdsst2020.wordpress.com/about/).
- 1st African Conference on Precision Agriculture (https://www.paafrica.org/).

9.5 Key journals

- Sustainability (https://www.mdpi.com/journal/sustainability).
- Agricultural Systems (https://www.journals.elsevier.com/agricultural-systems).
- Computers and Electronics in Agriculture (https://www.sciencedirect.com/journal/computers-and-electronics-in-agriculture).
- Environmental Modelling and Software (https://www.journals.elsevier.com/environmental-modelling-and-software/).
- Precision Agriculture (https://www.springer.com/journal/11119).

9.6 National and international work

- The International Union of Soil Sciences has many members and indeed some working groups focussed on different aspects of decision support for agriculture.
- CGIAR provides support and multiple examples of agricultural DSSs, and runs the INSPIRE Challenge (https://bigdata.cgiar.org/inspire/) which aims to combine data analytics, agri-tech practitioners and researchers to produce innovations in this area.
- The USDA provides multiple examples of decision support systems for farmers in the USA and elsewhere (https://data.nal.usda.gov/nal-terms/decision-support-systems), while CSIRO in Australia also provides some examples through their GrazPlan initiative (https://grazplan.csiro.au/)

10 References

Ahmed, W., Rais, M. N., Bano, R., Tamaddun, K. and Ahmad, S. 2018. Water sharing, governance, and management among the provinces in Pakistan using evidence-based decision support system. In: Kamojjala, S. (Ed.), 18th Annual World

Environmental and Water Resources Congress of the Environmental-and-Water-Resources-Institute (EWRI), Jun 03-07, 2018, Minneapolis, USA, 220-233.

Aitkenhead, M. J., Cameron, C., Gaskin, G., Choisy, B., Coull, M. C. and Black, H. I. J. 2018. Digital RGB photography and visible-range spectroscopy for soil composition analysis. *Geoderma* 313, 265-275.

Aitkenhead, M. J. and Coull, M. C. 2019a. Digital mapping of soil ecosystem services in Scotland using neural networks and relationship modelling. Part 1: Mapping of soil classes. *Soil Use and Management* 35(2), 205-216. DOI 10.1111/sum.12492.

Aitkenhead, M. J. and Coull, M. C. 2019b. Digital mapping of soil ecosystem services in Scotland using neural networks and relationship modelling. Part 2: Mapping of soil ecosystem services. *Soil Use and Management* 35(2), 217-231. 10.1111/sum.12491.

Aitkenhead, M. J., Coull, M. C., Gwatkin, R. and Donnelly, D. 2016. Automated soil physical parameter assessment using smartphone and digital camera imagery. *Journal of Imaging* 2(4), 35. DOI 10.3390/jimaging2040035.

Aitkenhead, M. J., Donnelly, D., Coull, M. and Hastings, E. 2014. Innovations in environmental monitoring using mobile phone technology - a review. *International Journal of Interactive Mobile Technologies* 8(2), 50-58.

Aitkenhead, M. J., Gaskin, G. J., Lafouge, N. and Hawes, C. 2017. PHYLIS: a low-cost portable visible range spectrometer for soil and plants. *Sensors* 17(1), 99. DOI 10.3390/s17010099.

Balderama, O. F. 2010. Development of a decision support system for small reservoir irrigation systems in rainfed and drought prone areas. *Water Science and Technology* 61(11), 2779-2785. DOI 10.2166/wst.2010.193.

Betteridge, K., Schnug, E. and Haneklaus, S. 2008. Will site specific nutrient management live up to expectation? *Landbauforschung Volkenrode* 58(4), 283-293.

Bhatnagar, V. and Poonia, R. C. 2018. A prototype model for decision support system of NPK fertilization. *Journal of Statistics and Management Systems* 21(4), 631-638. DOI 10.1080/09720510.2018.1471266.

Bramley, R. G. V. and Ouzman, J. 2019. Farmer attitudes to the use of sensors and automation in fertilizer decision-making: nitrogen fertilization in the Australian grains sector. *Precision Agriculture* 20(1), 157-175.

Brandt, P., Kvakic, M., Butterbach-Bahl, K. and Rufino, M. C. 2017. How to target climate-smart agriculture? Concept and application of the consensus-driven decision support framework "targetCSA". *Agricultural Systems* 151, 234-245. DOI 10.1016/j.agsy.2015.12.011.

Carlson, B. R., Carpenter-Boggs, L. A., Higgins, S. S., Nelson, R., Stockle, C. O. and Weddell, J. 2017. Development of a web application for estimating carbon footprints of organic farms. *Computers and Electronics in Agriculture* 142, 211-223. DOI 10.1016/j.compag.2017.09.007.

Carof, M. and Godinot, O. 2018. A free online tool to calculate three nitrogen-related indicators for farming systems. *Agricultural Systems* 162, 28-33.

Chlingaryan, A., Sukkarieh, S. and Whelan, B. 2018. Machine learning approaches for crop yield prediction and nitrogen status estimation in precision agriculture: a review. *Computers and Electronics in Agriculture* 151, 61-69.

Clarke, N., Bizimana, J. C., Dile, Y., Worqlul, A., Osorio, J., Herbst, B., Richardson, J. W., Srinivasan, R., Gerik, T. J., Williams, J., Jones, C. A. and Jeong, J. 2017. Evaluation of new farming technologies in Ethiopia using the Integrated Decision Support

System (IDSS). *Agricultural Water Management* 180(B), 267–279. DOI 10.1016/j.agwat.2016.07.023.

Culman, M., de Farias, C. M., Bayona, C. and Cruz, J. D. C. 2019. Using agrometeorological data to assist irrigation management in oil palm crops: a decision support method and results from crop model simulation. *Agricultural Water Management* 213, 1047–1062.

de la Rosa, D., Anaya-Romero, M., Diaz-Pereira, E., Heredia, N. and Shahbazi, F. 2009. Soil-specific agro-ecological strategies for sustainable land use – a case study by using MicroLEIS DSS in Sevilla Province (Spain). *Land Use Policy* 26(4), 1055–1065. DOI 10.1016/j.landusepol.2009.01.004.

Easton, Z. M., Kleinman, P. J. A., Buda, A. R., Goering, D., Emberston, N., Reed, S., Drohan, P. J., Walter, M. T., Guinan, P., Lory, J. A., Sommerlot, A. R. and Sharpley, A. 2017. Short-term forecasting tools for agricultural nutrient management. *Journal of Environmental Quality* 46(6), 1257–1269. DOI 10.2134/jeq2016.09.0377.

Eigenbrode, S. D., Binns, W. P. and Huggins, D. R. 2018. Confronting climate change challenges to dryland cereal production: a call for collaborative, transdisciplinary research, and producer engagement. *Frontiers in Ecology and Evolution* 5, AR UNSP 164. DOI 10.3389/fevo.2017.00164.

Fastelli, L., Rovai, M. and Andreoli, M. 2018. A Spatial Integrated Database for the Enhancement of the Agricultural Custodianship Role (SIDECAR)-some preliminary tests using Tuscany as a case-study region. *Land Use Policy* 78, 791–802.

Gibbons, J. M., Williamson, J. C., Williams, A. P., Withers, P. J. A., Hockley, N., Harris, I. M., Hughes, J. W., Taylor, R. L., Jones, D. L. and Healey, J. R. 2014. Sustainable nutrient management at field, farm and regional level: soil testing, nutrient budgets and the trade-off between lime application and greenhouse gas emissions. *Agriculture, Ecosystems and Environment* 188, 48–56. DI 10.1016/j.agee.2014.02.016.

Goap, A., Sharma, D., Shukla, A. K. and Krishna, C. R. 2018. An IoT based smart irrigation management system using Machine learning and open source technologies. *Computers and Electronics in Agriculture* 155, 41–49.

Goldstein, A., Fink, L., Meitin, A., Bohadana, S., Lutenberg, O. and Ravid, G. 2018. Applying machine learning on sensor data for irrigation recommendations: revealing the agronomist's tacit knowledge. *Precision Agriculture* 19(3), 421–444.

Hedley, C. 2015. The role of precision agriculture for improved nutrient management on farms. *Journal of the Science of Food and Agriculture* 95(1), 12–19. DOI 10.1002/jsfa.6734.

Hillier, J., Walter, C., Malin, D., Garcia-Suarez, T., Mila-i-Canals, L. and Smith, P. 2011. A farm-focused calculator for emissions from crop and livestock production. *Environmental Modelling and Software* 26(9), 1070–1078. DOI 10.1016/j.envsoft.2011.03.014.

Hochman, Z., Carberry, P. S., Robertson, M. J., Gaydon, D. S., Bell, L. W. and McIntosh, P. C. 2013. Prospects for ecological intensification of Australian agriculture. *European Journal of Agronomy* 44, 109–123. DOI 10.1016/j.eja.2011.11.003.

Huang, Y. B., Lan, Y. B., Thomson, S. J., Fang, A., Hoffmann, W. C. and Lacey, R. E. 2010. Development of soft computing and applications in agricultural and biological engineering. *Computers and Electronics in Agriculture* 71(2), 107–127. DOI 10.1016/j.compag.2010.01.001.

Jarvis, D. H., Wachowiak, M. P., Walters, D. F. and Kovacs, J. M. 2017. Adoption of web-based spatial tools by agricultural producers: conversations with seven Northeastern

Ontario farmers using the GeoVisage decision support system. *Agriculture-Basel* 7(8), AR 69. DOI 10.3390/agriculture7080069.

Jawad, H. M., Nordin, R., Gharghan, S. K., Jawad, A. M., Ismail, M. and Abu-AlShaeer, M. J. 2018. Power reduction with sleep/wake on redundant data (SWORD) in a wireless sensor network for energy-efficient precision agriculture. *Sensors* 18(10), AR 3450.

Kassim, M. R. M., Mat, I. and Harun, A. N. 2014. Wireless sensor network in precision agriculture application. International Conference on Computer, Information and Telecommunication Systems (CITS), 07–09 July, 2014, South Korea.

Katsigiannis, P., Galanis, G., Dimitrakos, A., Tsakiridis, N., Kalopesas, C., Alexandridis, T., Chouzouri, A., Patakas, A. and Zalidi, G. 2016. Fusion of spatio-temporal UAV and proximal sensing data for an agricultural decision support system. Proceedings of the 4th International Conference on Remote Sensing and Geoinformation of the Environment (RSCy), 4–8 April 2016, Paphos, Cyprus. DOI 10.1117/12.2244856.

Kerr, D. 2004. Factors influencing the development and adoption of knowledge based decision support systems for small, owner-operated rural business. *Artificial Intelligence Review* 22(2), 127–147.

Kerselaers, E., Rogge, E., Lauwers, L. and Van Huylenbroeck, G. 2015. Decision support for prioritising of land to be preserved for agriculture: can participatory tool development help? *Computers and Electronics in Agriculture* 110, 208–220. DOI 10.1016/j.compag.2014.10.022.

Khan, R., Ali, I., Zakarya, M., Ahmad, M., Imran, M. and Shoaib, M. 2018. Technology-assisted decision support system for efficient water utilization: a real-time testbed for irrigation using wireless sensor networks. *IEEE Access* 6, 25686–25697.

Leroux, C., Jones, H., Pichon, L., Guillaume, S., Lamour, J., Taylor, J., Naud, O., Crestey, T., Lablee, J. L. and Tisseyre, B. 2018. GeoFIS: an open source, decision-support tool for precision agriculture data. *Agriculture-Basel* 8(6), AR 73. DOI 10.3390/agriculture8060073.

Liakos, K. G., Busato, P., Moshou, D., Pearson, S. and Bochtis, D. 2018. Machine learning in agriculture: a review. *Sensors* 18(8), AR 2674.

Lindblom, J., Lundstrom, C., Ljung, M. and Jonsson, A. 2017. Promoting sustainable intensification in precision agriculture: review of decision support systems development and strategies. *Precision Agriculture* 18(3), 309–331. DOI 10.1007/s11119-016-9491-4.

Lopez-Granados, F. 2011. Weed detection for site-specific weed management: mapping and real-time approaches. *Weed Research* 51(1), 1–11. DOI 10.1111/j.1365-3180.2010.00829.x.

Machet, J. M., Dubrulle, P., Damay, N., Duval, R., Julien, J. L. and Recous, S. 2017. A dynamic decision-making tool for calculating the optimal rates of N application for 40 annual crops while minimising the residual level of mineral N at harvest. *Agronomy-Basel* 7(4), AR UNSP 73. DOI 10.3390/agronomy7040073.

Martinez, J., Egea, G., Aguera, J. and Perez-Ruiz, M. 2017. A cost-effective canopy temperature measurement system for precision agriculture: a case study on sugar beet. *Precision Agriculture* 18(1), 95–110. DOI 10.1007/s11119-016-9470-9.

Mohapatra, A. G., Lenka, S. K. and Keswani, B. 2019. Neural network and fuzzy logic based smart DSS model for irrigation notification and control in precision agriculture. *Proceedings of the National Academy of Sciences, India Section A: Physical Sciences* 89(1), 67–76.

Mondino, P. and Gonzalez-Andujar, J. L. 2019. Evaluation of a decision support system for crop protection in apple orchards. *Computers in Industry* 107, 99–103. DOI 10.1016/j.compind.2019.02.005.

Navarro-Hellin, H., Martinez-del-Rincon, J., Domingo-Miguel, R., Soto-Valles, F. and Torres-Sanchez, R. 2016. A decision support system for managing irrigation in agriculture. *Computers and Electronics in Agriculture* 124, 121–131. DOI 10.1016/j.compag.2016.04.003.

Nocita, M., Stevens, A., Wesemael, B. van, Aitkenhead, M., Bachmann, M., Barthes, B., Dor, E. B., Brown, D. J., Clairotte, M., Csorba, A., Dardenne, P., Dematte, J. A. M., Genot, V., Guerrero, C., Knadel, M., Montanarella, L., Noon, C., Ramirez-Lopez, L., Robertson, J., Sakai, H., Soriano-Disla, J. M., Shepherd, K. D., Stenberg, B., Towett, E. K., Vargas, R. and Wetterlind, J. 2015. Soil spectroscopy: an alternative to wet chemistry for soil monitoring. *Advances in Agronomy* 132, 139–159.

Papageorgiou, E. I., Markinos, A. T. and Gemtos, T. A. 2011. Fuzzy cognitive map based approach for predicting yield in cotton crop production as a basis for decision support system in precision agriculture application. *Applied Soft Computing* 11(4), 3643–3657, DOI 10.1016/j.asoc.2011.01.036.

Piero, M., Angelo, B., Antonello, B., Amedeo, D., Carlo, D. M., Michela, I., Giuliano, L., Florindo, M. A., Paolo, P., Simona, V. and Fabio, T. 2017. Soil sealing: quantifying impacts on soil functions by a geospatial decision support system. *Land Degradation and Development* 28(8), 2513–2526. DOI 10.1002/ldr.2802.

Premasudha, B. G. and Leena, H. U. 2017. ICT enabled proposed solutions for soil fertility management in Indian agriculture. *Advances in Intelligent Systems and Computing* 469, 749–757. DOI 10.1007/978-981-10-1678-3_72.

Puntel, L. A., Pagani, A. and Archontoulis, S. V. 2019. Development of a nitrogen recommendation tool for corn considering static and dynamic variables. *European Journal of Agronomy* 105, 189–199.

Qureshi, M. R. N., Singh, R. K. and Hasan, M. A. 2018. Decision support model to select crop pattern for sustainable agricultural practices using fuzzy MCDM. *Environment, Development and Sustainability* 20(2), 641–659.

Ravier, C., Jeuffroy, M. H., Gate, P., Cohan, J. P. and Meynard, J. M. 2018. Combining user involvement with innovative design to develop a radical new method for managing N fertilization. *Nutrient Cycling in Agroecosystems* 110(1), 117–134.

Robinson, D. A., Hockley, N., Dominati, E., Lebron, I., Scow, K. M., Reynolds, B., Emmett, B. A., Keith, A. M., de Jonge, L. W., Schjonning, P., Moldrup, P., Jones, S. B. and Tuller, M. 2012. Natural capital, ecosystem services, and soil change: why soil science must embrace an ecosystems approach. *Vadose Zone Journal* 11(1). DOI 10.2136/vzj2011.0051.

Rose, D. C., Morris, C., Lobley, M., Winter, M., Sutherland, W. J. and Dicks, L. V. 2018. Exploring the spatialities of technological and user re-scripting: the case of decision support tools in UK agriculture. *Geoforum* 89, 11–18. DOI 10.1016/j.geoforum.2017.12.006.

Rose, D. C., Sutherland, W. J., Parker, C., Lobley, M., Winter, M., Morris, C., Twining, S., Ffoulkes, C., Amano, T. and Dicks, L. V. 2016. Decision support tools for agriculture: towards effective design and delivery. *Agricultural Systems* 149, 165–174. DOI 10.1016/j.agsy.2016.09.009.

Rossi, V., Salinari, F., Poni, S., Caffi, T. and Bettati, T. 2014. Addressing the implementation problem in agricultural decision support systems: the example of vite.n et (R). *Computers and Electronics in Agriculture* 100, 88–99. DOI 10.1016/j.compag.2013.10.011.

Sang, N. (Ed.) (2020). Modelling nature-based solutions: integrating computation and participatory scenario modelling for environmental management and planning. Cambridge University Press. DOI 10.1017/9781108553827.

Schroder, P., Beckers, B., Daniels, S., Gnadinger, F., Maestri, E., Marmiroli, N., Mench, M., Millan, R., Obermeier, M. M., Oustriere, N., Persson, T., Poschenrieder, C., Rineau, F., Rutkowska, B., Schmid, T., Szulc, W., Witters, N. and Saebo, A. 2018. Intensify production, transform biomass to energy and novel goods and protect soils in Europe - A vision how to mobilize marginal lands. *Science of the Total Environment* 616–617, 1101–1123.

Schwenke, G., Beange, L., Cameron, J., Bell, M. and Harden, S. 2019. What soil information do crop advisors use to develop nitrogen fertilizer recommendations for grain growers in New South Wales, Australia? *Soil Use and Management* 35(1), 85–93. DOI 10.1111/sum.12469.

Shahbazi, F. and Jafarzadeh, A. A. 2010. Integrated assessment of rural lands for sustainable development using MicroLEIS DSS in West Azerbaijan, Iran. *Geoderma* 157(3–4), 175–184. DOI 10.1016/j.geoderma.2010.04.010.

Shamshiri, R. R., Kalantari, F., C. Ting, K., R. Thorp, K., A. Hameed, I., Weltzien, C., Ahmad, D. and Mojgan Shad, Z. 2018. Advances in greenhouse automation and controlled environment agriculture: a transition to plant factories and urban agriculture. *International Journal of Agricultural and Biological Engineering* 11(1), 1–22. DOI 10.25165/j.ijabe.20181101.3210.

Shirsath, R., Khadke, N., More, D., Patil, P. and Patil, H. 2017. Agriculture decision support system using data mining. Proceedings of the 2017 International Conference on Intelligent Computing and Control (I2C2), June 23–24, 2017, Coimbatore, India.

Thorp, K. R., DeJonge, K. C., Kaleita, A. L., Batchelor, W. D. and Paz, J. O. 2008. Methodology for the use of DSSAT models for precision agriculture decision support. *Computers and Electronics in Agriculture* 64(2), 276–285. DOI 10.1016/j.compag.2008.05.022.

Tovihoudji, P. G., Akponikpe, P. B. I., Agbossou, E. K. and Bielders, C. L. 2019. Using the DSSAT model to support decision making Regarding fertilizer microdosing for maize production in the sub-humid region of Benin. *Frontiers in Environmental Science* 7, AR 13. DOI 10.3389/fenvs.2019.00013.

Triantafyllidis, V., Kosma, A. K. C. and Patakas, A. 2018. An assessment of the soil quality index in a Mediterranean agro ecosystem. *Emirates Journal of Food and Agriculture* 30(12), 1042–1050. DOI 10.9755/ejfa.2018.v30.i12.1886.

Vanlauwe, B., Descheemaeker, K., Giller, K. E., Huising, J., Merckx, R., Nziguheba, G., Wendt, J. and Zingore, S. 2015. Integrated soil fertility management in sub-Saharan Africa: unravelling local adaptation. *SOIL* 1(1), 491–508. DOI 10.5194/soil-1-491-2015.

Weersink, A., Fraser, E., Pannell, D., Duncan, E. and Rotz, S. 2018. Opportunities and challenges for big data in agricultural and environmental analysis. *Annual Review of Resource Economics* 10(1), 19–37.

Williams, D., Britten, A., McCallum, S., Jones, H., Aitkenhead, M., Karley, A., Loades, K., Prashar, A. and Graham, J. 2017. A method for automatic segmentation and splitting of hyperspectral images of raspberry plants collected in field conditions. *Plant Methods* 13, Article No. 74.

Wolfe, M. L. and Richard, T. L. 2017. 21st century engineering for on-farm food-energy-water systems. *Current Opinion in Chemical Engineering* 18, 69–76, DOI 10.1016/j.coche.2017.10.005.

Zwickle, S., Wilson, R., Bessette, D., Herms, C. and Doohan, D. 2016. Facilitating ecological weed management decisions by assessing risk-benefit tradeoffs. *Agroecology and Sustainable Food Systems* 40(7), 635–659. DOI 10.1080/21683565.2016.1166170.

Index

Lightning Source UK Ltd.
Milton Keynes UK
UKHW020620210722
406172UK00001B/11

9 781786 764263